Reflective Cracking in Pavements

Reflective Cracking in Pavements

Design and performance of overlay systems

Proceedings of the
Third International RILEM Conference,
organized by Centre for Research and
Contract Standardization in Civil and
Traffic Engineering (CROW),
Delft University of Technology and
Belgian Road Research Centre

Maastricht, The Netherlands
2–4 October 1996

Edited by

L. Francken
Belgian Road Research Centre, Brussels, Belgium

E. Beuving
*Centre for Research and Contract Standardization in Civil
and Traffic Engineering, Ede, The Netherlands*

and

A.A.A. Molenaar
Delft University of Technology, The Netherlands

CRC Press
Taylor & Francis Group
Boca Raton London New York

CRC Press is an imprint of the
Taylor & Francis Group, an **informa** business
A CHAPMAN & HALL BOOK

First edition 1996 published by E & FN Spon

Published 2020 by CRC Press
Taylor & Francis Group
6000 Broken Sound Parkway NW, Suite 300
Boca Raton, FL 33487-2742

First issued in paperback 2020

CRC Press is an imprint of Taylor & Francis Group, an Informa business

No claim to original U.S. Government works

ISBN 13: 978-0-367-65953-0 (pbk)
ISBN 13: 978-0-419-22260-6 (hbk)

Visit the Taylor & Francis Web site at
http://www.taylorandfrancis.com

and the CRC Press Web site at
http://www.crcpress.com

A catalogue record for this book is available from the British Library

Publisher's Note

This book has been produced from camera ready copy provided by the individual contributors in order to make the book available for the Conference.

Contents

Part one   Origin and prevention of reflective cracking

Part two   Crack resistance of asphalt overlays

Part three   Test methods for interlayers

Contents

Part four Design methods: theory and practice

Part five   Performance of maintenance techniques

Scientific Committee

(RILEM Technical Committee 157-PRC, Systems Prevent Reflective Cracking in Pavements)

L. Francken, Belgian Road Research Centre, Belgium (Chairman)
L. Courard, Université de Liège, Belgium, (Secretary)
A.O. Abd El Halim, Center for Geosynthetics Research Information & Development, Canada
R. Alvarez Loranca, Geocisa Laboratory, Spain
E. Beuving, Centre for Research and Contract Standardization in Civil and Traffic Engineering, The Netherlands
K. Blazejowski, Institute Badawczy Drog i Mostow, Poland
A.H. de Bondt, Delft University of Technology, The Netherlands
G. Colombier, Laboratoire des Ponts et Chaussées d'Autun, France
H.W. Fritz, EMPA, Switzerland
W. Grzybowska, Cracow University of Technology, Poland
W.H. Herbst, Central Road Testing Laboratory, Austria
T. Levy, Ponts et Chaussées, Luxembourg
J. Litzka, Institute für Strassenbau und Strassenverhaltung, Austria
R.L. Lytton, Texas Transportation Institute, USA
A.A.A. Molenaar, Delft University of Technology, The Netherlands
J.F. Potter, TRL, United Kingdom
L. Quaresma, Laboratorio Nacional de Engenharia Civil, Portugal
J.P. Serfass, SCREG, France
J. Silfwerbrand, Royal Institute of Technology, Sweden
H. Sommer, Forschungsinstitut der Vereinigung der Österreichischen Zement industrie, Austria
A. Vanelstraete, Belgian Road Research Centre, Belgium
F. Verhee, Viafrance, France

Conference Organization

Centre for Research and Contract Standardization in Civil and Traffic Engineering (C.R.O.W)
and
Delft University of Technology (TU Delft)
and
Belgian Road Research Centre (BRRC)

Organizing Committee

E. Beuving, C.R.O.W (The Netherlands)
L. Courard, Université de Liège (Belgium)
L. Francken, Belgian Road Research Centre (Belgium)
J.G.L.M. Matser, C.R.O.W (The Netherlands)
A.A.A. Molenaar Delft University of Technology (The Netherlands)

Preface

Crack reflection through a road structure is one of the main causes of premature pavement deterioration.

This phenomenon, which often appears when a layer of asphaltic material is placed on top of a discontinuous base, can have many different aspects, in accordance with the large number of factors governing the mechanism of crack initiation and propagation through a road structure.

At a time when the preservation of the road network and adapting it to an increasingly aggressive traffic are absolute priorities, all the professional circles concerned - road authorities, manufacturers of materials, designers, contractors, researchers - are expected to show their innovativeness by suggesting appropriate ways to take up this challenge.

In this perspective, the RILEM Technical Committee 97-GCR "Application of Geotextiles to Crack Prevention in Roads" was set up under the chairmanship of Professor J.M. Rigo.

Two RILEM Conferences on Reflective Cracking in Pavements, RC89 and RC93, were organized by this committee and the University of Liege in 1989 and 1993, respectively. These events have demonstrated that many innovative systems are now commonly implemented, and the proceedings of these conferences remain unmatched sources of information on the reflective cracking issue :

- RC89 has pointed out the main factors and mechanisms involved in the initiation and propagation of cracks. A first inventory of systems and approaches available to prevent cracking was presented.

- At the RC93 Conference emphasis was on the theoretical and experimental research required in order to optimize solutions. The foundations were laid for a classification of products and overlay systems with the aim of setting up a procedure for selection and design.

The RILEM Technical Committee 157-PRC "Systems to Prevent Reflective Cracking in Pavements" was formed in 1993 with a broader field of tasks; not only does it deal with geotextiles and related products, but it also covers other innovative materials and technical solutions involved in the prevention of crack reflection such as :

- preventive measures to be taken in new construction projects,
- repair and rehabilitation work to be carried out prior to the application of an interlayer,
- practical problems during laying and construction,
- the long-term performance of overlay systems.

The main objective of the RILEM Technical Committee 157-PRC in organizing this Third Conference is to gather the information needed to finalize a state-of-the-art report on the prevention of reflective cracking in pavements.

As an heir of the two former conferences, and after sufficient time has elapsed to look back and evaluate the first applications, the RC96 Conference of Maastricht will pay particular attention to full-scale performance verifications and practical field experience.

Fifty-eight papers have been selected covering the five major topics of the Conference :

- Origin and prevention of reflective cracking.
- Resistance to cracking of asphalt overlays.
- Test methods for interlayers.
- Theoretical and practical aspects of design methods.
- Performance of maintenance techniques.

The number and quality of the contributions proposed at this Conference proves the on-going interest of professionals in innovating rehabilitation techniques and maintenance procedures aiming at longer service lives and cost-effective solutions.

Louis FRANCKEN, Dr. Sc.
Chairman RILEM TC157-PRC

Origin and prevention of reflective cracking

Technique for limiting the consequences of shrinkage in hydraulic-binder-treated bases

M. LEFORT

LROP (Paris West Regional Transport Research Laboratory), Trappes, France

Abstract
The precracking technique is operational today in France. It is based on the principle of causing shrinkage cracking in order to reduce the interval and obtain fine, straight cracks which extend less and are better interlocked.
 The French highway industry has three types of equipment:
  . CRAFT (French acronym for automatic creation of transverse cracks) produces a discontinuity by creating a groove in which a bitumen emulsion is injected.
  . The Active Joint (*Joint Actif*) technique consists in placing a rigid sinusoidal-shaped rigid plastic insert within the thickness of the layer.
  . With the OLIVIA system, the discontinuity is created by placing a flexible plastic film vertically in the thickness of the layer.
 The three techniques are used between the levelling and compacting operations.
 All three are effective for crack location. The precrack interval adopted is generally 3 metres. The operation does not slow down the course laying output.
 If the reduction in the bituminous covering thickness can be considered in certain cases, the technical data at our disposal do not enable us to recommend a reduction in the base thickness.
Key words: Structural design, crack, materials treated with a hydraulic binder, equipment, precracking, shrinkage.

1 Introduction

Shrinkage cracking in pavement bases treated with a hydraulic binder produces a crack whose interval ranges from 8 to 10 metres under French climates. With siliceous materials and the mix designs used, these cracks may be as much as several millimetres wide during cold periods: their extension is accelerated and their maintenance becomes delicate.

Reflective Cracking in Pavements. Edited by L. Francken, E. Beuving and A.A.A. Molenaar. © 1996 RILEM. Published by E & FN Spon, 2–6 Boundary Row, London, SE1 8HN. ISBN 0 419 22260 X.

Precracking is based on the principle of causing the shrinkage crack and imposing a smaller cracking interval in order to obtain fine cracks which are less troublesome and far less detrimental with regard to the behaviour of the structure and with regard to road usage qualities.

The idea of precracking may be associated with the development of the use of hydraulic-binder-treated materials in pavement bases since the first tests date back to the 1960s when wooden slats were placed at the bottom of the base course at intervals of 5 metres.

Developments in the technique took place later, essentially with a precracking technique using notches made by means of a knife on the surface of the compacted material layer. These notches were obtained either with a knife fixed on a heavy vibrating plate, or by means of a double-wheel vibrating roller, one wheel of which is equipped with a cutting disk.

Good results were obtained with notches about 10 cm deep, produced with an interval of 3 metres because, when the crack rises to the surface of the pavement, it is always under a notch. Each notch, even when it has produced a clear shrinkage crack in the base, does not always result in a crack visible on the pavement surface but, even if such should be the case, the crack is finer and straighter than in the case of natural shrinkage cracking.

This notch-based precracking technique is at the origin of present techniques. It has been abandoned owing to difficulties in transferring the equipment between each operation. Its use is currently limited to the provision of joints on compacted concrete roadways.

2 The CRAFT process (acronym for automatic creation of transverse cracks)

This cracking technique is applied before compacting. It consists in creating a discontinuity in the layer of hydraulic-binder-treated material by making a transverse groove in which a bitumen emulsion is injected. The groove normally closes with compacting. The operation makes use of specific devices installed on a shovel dozer (fig. 1). All operations are automated.

The system includes a manipulator arm which guides the tool mounted at the front of the dozer, which also receives a hydraulic system, a compressor, a generating set and the bitumen emulsion tank. The articulated manipulating arm in the form of a compass allows the positioning of the tool in the layer and its horizontal progression perpendicular to the centreline of the pavement over a length of 5 m. The tooth-shaped tool opens the groove and allows the injection of the binder: the movement of the tooth is facilitated by a vibrating device.

fig. 1 CRAFT precracking machine

The emulsion used is a cationic-type fast-breaking bitumen emulsion. The aqueous phase with a low pH creates a zone of lower resistance which allows the location of the shrinkage crack. The bituminous phase creates the permanent discontinuity and improves the material's behaviour with regard to water and abrasion.

The carrying machine moves along the centre of the roadway for precracking at an interval of 2 or 3 m. It is positioned without difficulty at the level of a precrack to extend it onto the adjacent lane. It is incorporated with the other construction equipment, before the compactors, without slowing down the work.

The technique has been covered by a patent filed by the "Laboratoire Central des Ponts et Chaussées - entreprise Cochery Bourdin Chaussé". The method has been evaluated by the French Technical Advice No. 70 of the Road Technical Advice Commission.

3 Active Joints

This precracking technique, developed by the company SACER, is covered by a patent. It consists in placing a rigid insert of undulated form in the thickness of a layer of material treated with hydraulic binder (fig. 2).

Precracking is carried out every 2 metres and the technique is aimed at designing single-layer hydraulic bases of smaller thickness than conventional structures.

fig.2 Section of a base with an active joint

The different phases of the technique are as follows:
. After spreading and levelling the material (before compacting), a V-shaped groove is created which has the length of the insert (2 m or more) placed perpendicular to the centreline of the lane.
. Then the insert is placed and the groove closed while packing the material carefully in the waves of the insert.
. Then the usual laying of the material is completed.

The Active Joint precracking technique is applied with a specific tool mounted at the end of the boom of a self-propelled hydraulic shovel. To precrack a pavement 7 metres wide, the machine moves within the centreline and, by rotating the boom, can place two inserts aligned and placed within the centreline of each traffic lane. The different operations are automated and the work is remote-controlled by a single operator. It is incorporated within the rest of the construction equipment without slowing down the general construction work.

An application has been filed for a Technical Advice (*avis technique*) for this precracking technique, and is currently being processed.

4 The OLIVIA technique

The principle of this precracking is to create a discontinuity in the layer by placing a simple flexible plastic film vertically in its thickness. It is placed by means of a plough which cuts a transverse groove in relation to the centreline of the lane and through which the film unreels. The groove closes behind the plough.
The compacting which follows is not disturbed by this operation (fig. 3).

fig.3 OLIVIA precracking machine

The placement tool is in the form of a specific system mounted on a shovel dozer type machine allowing movement from one precrack to another. The system is controlled by a single operator.

The width of the film placed, usually 80 mm can be adapted to the thickness of the layer (between 1/3 and 1/5 of the thickness). The precracking interval is generally 3 metres. The machine is incorporated within the rest of the construction equipment between the spreading and levelling equipment and the compacting equipment.

Developed by the company Viafrance, the technique is patented.

5 Assessment

French precracking techniques are the consequence of very broad use of materials treated with hydraulic binders, allowing the best utilization of certain local materials. The three processes operational today are in different stages of development but provide general information on precracking.

.The effectiveness of the precracking techniques is unquestionable in producing cracking and reducing its intervals. This contributes to the production of finer, straighter cracks. Consequently, crack propagation is slowed to such a point that specific maintenance can be obviated in certain cases. If necessary, such maintenance would nevertheless be facilitated because of the straightness of these cracks.

. The precracking technique does not justify a modification in standard practices for the manufacture and laying of materials treated with hydraulic binders. The equipment available today makes it possible to follow usual work outputs. The only point to be checked has to do with the workability time, which must be adapted. Incorporating a machine for precracking in the usual roadbuilding equipment extends the project schedule. It must consequently be ensured that compacting can still be carried out before the end of the workability.

. Precracking can be the means of bringing about a different behaviour in semi-rigid structures compared with that of a similar base subjected to natural shrinkage cracking.

The CESAR-LCPC finite-element calculation programme with a 3-dimensional model of a single-layer base can provide for the structural design of precracked structures. Precracking, if the structure can be assimilated with a series of articulated slabs, would allow a reduction in the base thickness.

However, this presupposes operation in conformity with the design assumptions throughout the planned life of the pavement.

Today we are lacking sufficient technical data to ensure this.

. As precracking leads to better load transfer at the level of finer cracks, it is conceivable that the stresses in the wearing course will be reduced and, consequently, reflective cracking will be slower on the surface.

With a comparable cracking growth objective, precracking would then allow a reduction in the thickness of the surface layer. This reduction can only be limited because a minimum thickness of bituminous products is necessary in order to provide sufficient mechanical and thermal protection of the treated base before the complete hardening of the material.

Precracking thus remains a structural design factor, just like the climatic conditions of the site or the type of aggregates. The financial advantage that may be expected is to be sought in a reduction in maintenance costs rather than a reduction in the initial investment.

Bibliography

. Humbert P. (1989) CESAR-LCPC: un code de calcul par éléments finis. Bulletin de liaison des Laboratoire des Ponts et Chaussées n° 160.

. Colombier G. (1990) RILEM Siège - Les procédés utilisés pour maîtriser la remontée des fissures.

. Marchand J.P. (1990) Trois années de préfissuration CRAFT. RGRA n° 680.

. Avis technique n° 70 (1993) Edité par le SETRA (avis technique chaussées).

. Quero J.F./Verhee F. (1993) OLIVIA nouvelle technique Viafrance de préfissuration des matériaux traités aux liants hydrauliques RGRA n° 713.

. Aubert J.L./Faure B. (1995) Le joint actif: un procédé reconnu. RGRA n° 735.

. Verhee F. (1996) RILEM Maastricht. Plastic film precracking system.

. Marchand J.P. (1996) RILEM. Maastricht. Experimental Study and Modelling of a Precracking Process for Semi-rigid Structures.

Effect of highway geometry and construction equipment on the problem of reflection cracks

A.O. ABD EL HALIM
Department of Civil and Environmental Engineering, Carleton University, Ottawa, Ontario, Canada
R.M. ABDEL NABI and A. ABDEL ALEEM
Zagazig University, Banha Branch, Egypt
S.M. EASA
Department of Civil Engineering, Lakehead University, Thunderbay, Ontario, Canada
R. HAAS
University of Waterloo, Ontario, Canada

Abstract
The effect of highway geometry on the problem of transverse cracks of asphalt pavements has not been considered when the main causes of pavement distress are considered. In this study, results of several field surveys carried out in geographically and climatically diverse areas, Canada and Egypt, are presented. The data suggested the existence of strong correlation between highway geometry and transverse cracking of asphalt pavements. The survey results show that roads experience more cracks on horizontal and vertical curves, slopes, and bridges than on its straight flat sections, all other factors being constant. Also, the findings of the field surveys are supported by analytical results obtained from the application of the finite element method to a curved road subjected to thermal stresses.
Keywords: Construction equipment, field survey, highway geometry, reflection cracks, thermal stresses.

1 Introduction

Surface cracks, such as transverse, longitudinal, and alligator, are among the most observed types of distresses in cold regions. Poor asphalt mix design, heavy traffic loads, low/high temperature, reflection cracks, weak subgrade, and poor quality control during construction are reported to be among the main causes of asphalt distress [1]. Any one or more of these factors may cause the pavement to deteriorate to an unacceptable level of service which may require major rehabilitation. However, recent

Reflective Cracking in Pavements. Edited by L. Francken, E. Beuving and A.A.A. Molenaar. © 1996 RILEM. Published by E & FN Spon, 2–6 Boundary Row, London, SE1 8HN. ISBN 0 419 22260 X.

field observations of the surface cracks and their pattern along highway sections in Canada and Egypt suggest that road geometry may in fact play an equal or even more dominant role in the shape and pattern of cracking. For example, when a straight section of a highway is followed by a curved section, the spacing and pattern of the transverse cracks are distinctively different [2,3]. When the two sections represent a continuous road segment with the same asphalt mix, traffic loads, climatic conditions, and construction procedure, the previously identified causes can not adequately explain the reported differences.

2 Geometry of highway sections

The first step in the study was to define the variables and conditions of the field investigation. The term "highway geometry" was defined as: (1) Straight, horizontal (reference) section, (2) Horizontal curve section, (3) Slope or a section on a grade, and (4) Bridge deck section. A bridge was considered in this study as a different geometry despite the fact that it may be on a straight line. The reason for including a bridge deck section is that it represents a discontinuity of the road geometry in the vertical direction. More specifically, bridge decks constructed of asphalt pavement on top of the more rigid concrete slabs or girders follows represents a "geometric distortion" [4] in terms of relative rigidity. It was desired to assess this effect, particularly in terms of construction (i.e., a soft asphalt layer placed on a stiff deck, compacted with steel or vibratory rollers) and its influence on subsequent cracking. After a road section was initially selected based on road maps, information relevant to the design and construction conditions was collected from the regional agency responsible for that specific section. Test sections that failed to meet the selection criteria were eliminated. After the selection process was completed, preliminary site visits were made to each of the selected sections. The details of the surveys are given in a number of published reports [2,3] and therefore will not be repeated here. The results of the field surveys performed on circular curves, slopes and bridge decks are presented herein.

3 Results of Ottawa field survey

The results of the field survey for Ottawa are summarized in Tables 1 and 2. Table 1 gives the results of measurements made on horizontal curve sections. As can be seen in the table, the spacings between transverse cracks on the curve sections were closer on 6 of the 8 sections when compared to similar measurements on the straight or control sections. The mean spacings were statistically significantly different at the 70% confidence level. The spacing range of transverse cracks on the curve test sections were 2.3 to 19.7 m compared to 2.8 to 38.3 m on the straight sections. Also, the density of cracks on the curve sections ranged from 6 to 42 cracks per 100m, while the density on the reference sections ranged from 3 to 35 cracks per 100m. It should be noted that the only difference between the curved and reference straight sections is their geometry.

Results of the survey of the bridge decks showed that the ones constructed with an asphalt mix surface experience the most surface cracking, especially transverse cracks, when compared with the asphalt surfaces of the approaches leading to the bridge. Typical results showed that more than 80% of the crack spacings measured on a given bridge were less than 5 m apart. On the other hand, more than 80% of the crack spacings measured on the approaches leading to the bridge were more than 5 m apart. Table 2 shows the average measurements taken on the 8 bridges included in this study. While the spacings between the transverse cracks observed on the bridge were between 1.8 to 7.8 m, the spacings between the transverse cracks on the reference sections were 2.4 to 18.8 m. The mean spacings were significantly different at the 95% confidence level. In addition to these measurements, visual inspection of the surfaces on the bridges suggested that they are experiencing more rutting than any of the asphalt surfaces included in the study. However, no measurements were made to include this distress type objectively in the analysis. The results of all measurements performed in this study were averaged for each geometry. It was noted that the average spacing between transverse cracks was 30% and 70% smaller on the curves and bridges, respectively. Clearly, closer transverse crack spacing would mean a higher number of these cracks per given unit length.

Table 1. Results of Ottawa field survey for transverse cracks on curve sections.

Section Number	Curve Section		Straight (Control) Section	
	Spacing (m)	Density (no./100m)	Spacing (m)	Density (no./100m)
1	11.4	9	31	4
2	19.7	6	38.3	3
3	2.3	42	2.8	35
4	3.2	27	8.3	14
5	4.9	19	7.2	15
6	6	18	4.8	21
7	7	14	9.3	11
8	7.2	13	5.5	19

4 Results of Cairo field study

The results of the field survey carried out in Cairo showed similar trends to those obtained from the survey of Ottawa, as shown in Table 3. It should be noted that the pattern of similarity with the Ottawa sections existed despite the fact that there are substantial climatic, geographic and other differences between the two cities. For example, Ottawa is known to experience winter temperatures as low as -30°C. In contrast, temperatures in the winters of Cairo were never below freezing. Furthermore, quality control during the asphalt mix production, pavement construction, and traffic operations are different between the two countries. Also, the type of asphalt mixes, base and subbase materials, and type of subgrade are quite different.

Table 2: Summary of results of Ottawa field survey for all geometries.

Geometry	Overall Mean Crack			
	Spacing (m)		Density (no./100m)	
	Range	Average	Range	Average
Curves	2.3-19.7	7.71	6-42	18.5
Reference	2.8-38.3	13.4	3-35	15.25
Ratio		58%		121%
Slopes	3.5-11.5	5.98	10-38	21.25
Reference	3.7-33.3	9.74	3-31	15.92
Ratio				133%
Bridge Deck	2.8-7.8	3.44	13-83	40.63
Reference	2.4-18.8	11.563	6-39	13.75
Ratio				295%

The effect of highway geometry on transverse cracks appears to be more severe in Cairo compared with the same effect in Ottawa. As shown in Table 3, both spacing and density measured on curves and slopes are higher than the average values of Ottawa survey. However, while this result may be explained by pavement age, the performance of these sections relative to their respective reference sections suggests that age could not be the only factor explaining the severity reported on the Egyptian sections. Furthermore, transverse cracks are known to be affected more by cold temperature. Therefore, one should conclude that highway geometry has a major role in contributing to the phenomenon of transverse cracks of asphalt surfaces. This role is discussed in the following sections.

5 Role of highway geometry

The results of the field studies presented in the previous section showed that the number of transverse cracks per 100m length of a given road section can vary depending on the geometry of the road section. The measurements and comparisons are always made between a specific geometry and a horizontal straight portion of the road. The only difference between each two portions of any test section is clearly their geometries. Yet, the results showed that significant differences exist between their respective surface conditions. Furthermore, it is obvious from the results that straight portions of asphalt roads experience the least amount of damage when compared with other geometries of the road. Also, as stated earlier, the selection of the test sections in each country ensured that traffic loadings, climatic conditions, etc., are constant for all pairs of straight sections, slope, curve, or bridge geometries. Therefore, one may ask what would cause a section of road to show 38 cracks per 100m while a few

Table 3: Summary of results of Cairo field survey

Geometry	Spacing (m)	Density (no./100m)
Curves	5.6	24
Reference	20.2	14
Ratio	28%	171%
Slopes	5	25
Reference	10.2	13
Ratio	47%	192
Bridge Deck	4.2	30
Reference	66.8	7
Ratio	6%	429%

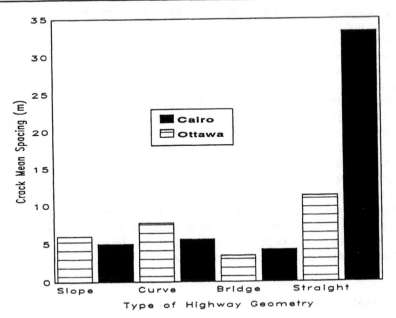

Fig. 1. Agerage reported mean crack spacing from field sections in Ottawa (Canada) and Cairo (Egypt).

meters away the same pavement shows only 10 cracks per 100m? Also, while Ottawa is totally different from Cairo in terms of the main factors affecting pavement cracking, the results of the surveys suggest that there are more similarities in the observed trends and the degree of the reported damage. Figure 1 demonstrates the similarities of reported crack patterns between both cities.

6 Effect of construction equipment

At the present time, there are no special procedures specifying standard construction practices for the unique geometry of the road. The entire operation is left to the contractor to decide which course of action to take. With many constraints on the movement of equipment to and from the site, the main objective of the contractor is to finish the project as soon as possible. Consider the example of laying and compacting asphalt layer on a steeper slope. Clearly, laying the asphalt mat on a steeper slope with the paver can result in significant damage to the mix even before allowing any traffic on it. When the mix is placed while the paver is travelling up the grade, tensile stresses will be present within the asphalt layer. In addition, due to the semi-liquid, semi-solid state of the hot asphalt mix soon after leaving the paver, a significant movement and creep of the asphalt mat can occur downward on the steep slope . The combined effect of both the tensile stresses and the movement of the mix can enhance the initiation of transverse cracks at the time of construction. The problem can be compounded when the steep slope is located on a circular curve as in the case of ramps connected to major highways. This unique geometry provides a challenging task to the operator of the paver. The resulting forces can cause both transverse and longitudinal cracks at the same time.

The use of steel rollers, in vibratory or static modes, on steep slopes or horizontal curves has been shown to cause construction induced cracks on horizontal flat pavements [4,5]. The use of these rollers on slopes, curves, and bridge decks provides the worst working conditions for these rollers. The smaller area of contact between the drum of existing rollers and the asphalt mat will significantly increase all the undesired forces and stresses generated when construction is on a slope or a curve. Bridge decks during compaction of the asphalt layer represent the lowest relative rigidity value compared with any other pavement structure. The stiffness of the asphalt layer during compaction is very small compared to the stiffness of the concrete slab or girders of the bridge. When a stiff roller is used to compact the asphalt mat, surface cracks occur. These cracks increase as the stiffness of the underlying layer increases [4]. Thus, in the case of bridge decks one should not be surprised to observe that the highest number of transverse cracks per unit length was that of bridge decks, as reported in both Canadian and Egyptian field investigations.

7 Analytical verification

The results of the field surveys in Canada and Egypt showed that curved road sections will experience more cracking than their reference straight sections. One of the suggested causes of this observation was that thermal induced stresses may increase on the curves as compared with those induced in the straight sections. The results of using the finite element technique to investigate the effect of the radius of the road section on the induced thermal stresses are shown in Figures 2 and 3. As can be seen from Figure 3, as the radius of the curve increases the calculated thermal stresses drop very rapidly. This supports the results and analysis of the field investigations.

8 Concluding remarks

Based on the research results presented in this paper it is evident that highway geometry has an adverse effect on the phenomenon of transverse cracks of asphalt pavements. Subsequently, the structural design parameters which govern the performance of asphalt layers must be considered in the geometric design of new highways. Also, the results of field surveys carried out in two cities with very different climatic, traffic and economic characteristics showed that the effect of highway geometry is universal. The interaction between pavement geometry and construction procedures and equipment was shown to be one of the main contributors to the observed problem. This paper obviously can not provide all the answers to the problem of transverse cracking. The study upon which it is based was relatively exploratory and further research is needed. For example, field and analytical methods are needed to establish critical geometric design elements such as slope and curve radius, to ensure better pavement performance. Design methods should be developed to deal with these unique geometric characteristics specifically. Reliable construction techniques should be identified for each type of geometry. As pavement engineers begin to recognize the problem of increased cracking on specific geometric sections and its main causes, better quality control during construction, improved maintenance practices and equipment, and significant cost savings can be achieved.

9 Acknowledgments

The authors would like to thank the Egyptian Ministry of Transportation, the Ministry of Transportation of Ontario and the Regional Municipality of Ottawa-Carleton for their assistance and cooperation during the course of the field surveys.

10 References

1. Haas, Ralph, W. Ranold Hudson, and John Zaniewski (1994) *Modern Pavement Management*, Krieger Publ. Inc.
2. Abd El Halim, A. O., H. Hozayen, R. Haas and N. Kassiri (1993) Influence of the Geometry of Highways on the Distress of Asphalt Pavements in Cold Areas, *Proceedings of Paving In Cold Areas*, PICA V, Vol.1, Alberta, Canada, pp. 402-425.

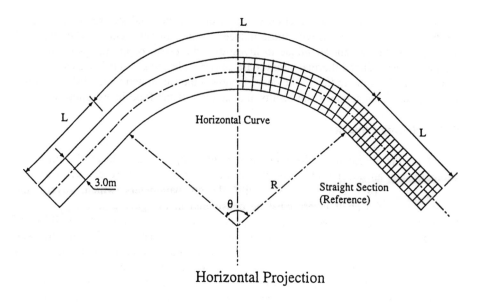

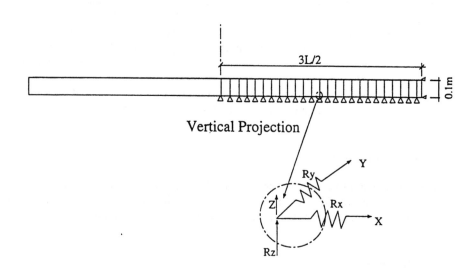

Fig. 2. Modelling of a horizontal curve connected to a straight section.

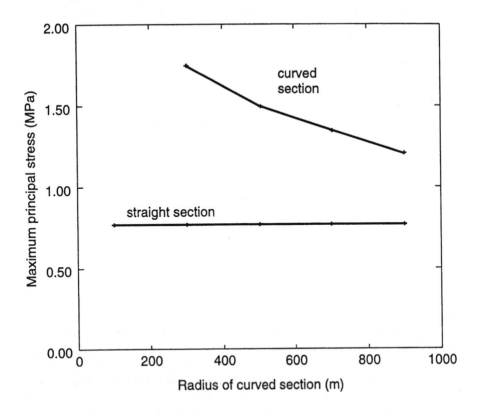

Fig. 3. Effect of curve radius on the maximum principal tensile stress.

3. Abd El Halim, A.O., N. Kassiri and R. Haas (1993) Relating Deterioration of Asphalt Pavements to Highway Geometry, Proceedings, *Al-Azhar Engineering 3rd International Conference*, Cairo, Egypt, pp. 657-572.
4. Abd El Halim, A. O (1986) Experimental and Field Investigation of the Influence of Relative Rigidity on the Problem of Reflection Cracking, *Transportation Research Record 1060*, pp. 88-98.
5. Abd El Halim, A. O, W. Phang and R. Haas (1993) An Unwanted Legacy of Asphalt Pavement Compaction, *Journal of Transportation Engineering*, American Society of Civil Engineers, Vol. 119, No. 6, pp. 914-932.

The development of Emulscement®: a base material with a future

G. ENGBERS
Department of Technology, NBM Zuid B.V., Helmond, Netherlands
J.P. SMALLEGANGE
Department of Technology, NMB-Amstelland Infrastructuur & Milieu B.V., Delft, Netherlands

Abstract
Research has been carried out to establish the possibilities of the cold recycling of tar-contaminated asphalt, in a joint venture by NBM-Amstelland/Vauatol/Smid & Hollander. In the product Emulscement®, the binding agents cement and bitumen emulsion are applied in combination to bind the asphalt granules. The process is based on results of preliminary laboratory research and the experience gained in four practical applications involving the production, processing and subsequent inspection. When recycling old, tar-contaminated asphalt, Emulscement® has two major advantages:
1. *Environment*: it makes the recycling of tar-contaminated asphalt possible and limits the amount of harmful elements which leach into the soil and groundwater.
2. *Construction*: as base material it has a good crack resistance and therefore it prevents crack propagation in the asphalt layer, meaning that thinner overlayers can be used.
The latest results of research, giving more insight into the mechanical properties and leaching behaviour of Emulscement®, are presented.
Keywords: Asphalt granules, cement, bitumen emulsion, crack propagation, rutting, leaching behavior, stiffness module, tensile and compressive strength, design.

Reflective Cracking in Pavements. Edited by L. Francken, E. Beuving and A.A.A. Molenaar. © 1996 RILEM. Published by E & FN Spon, 2–6 Boundary Row, London, SE1 8HN. ISBN 0 419 22260 X.

1 Introduction

Since a long time, a large number of tar products have been used in the construction and maintenance of roads. As well as price, the technical properties of these products also played a role, such as good oil resistance and high water resistance (stripping). In those days, the fact that tar also contained harmful elements was not taken into consideration. The best known group of these elements are the Polycyclic Aromatic Hydrocarbons (PAH). Nowadays, tar products are no longer used. However, when breaking up, renovating or maintaining pavement constructions, road authorities are still confronted with the problem of (coal) tar. Reclaimed asphalt is often contaminated with tar: about 400,000 tons each year, which is too large an amount to be simply transported away to dumps. Controlled dumping of the material is not a viable alternative because it is contrary to the recycling objective for construction waste of 90% by the year 2000. Recycling: processing in a suitable and responsible way is, therefore, definitely the prefered procedure. In this respect, hot recycling into new asphalt mixtures is not possible, because when the old asphalt is heated, harmful vapours are released.

The most realistic solution is, therefore, a cold-recycling process. Experience with milled asphalt mixed with cement (BRAC) has shown that using only cement as a binding agent has several disadvantages: low resistance to cracking, high level of stiffness and not enough immobilization of organic contamination such as PAHs. BRAC remains a viable method for asphalt granules which, for any other reasons than the presence of tar, are not suitable for hot recycling, (due to, for example, the presence of too many stony elements, or wood or plastic). The use of bitumen emulsion alone as a binding agent results in a plastic material with an insufficient load-bearing capacity. The correct combination of cement and bitumen emulsion means that the properties of both binding agents are the most significant elements in Emulscement®. Its application shows how civil engineering can operate in an environmentally responsible way.

The city of Breda ("Stolwerk" material) chose to stabilise the material with cement and a special emulsion (Emulscement®) for applications in base materials. Extensive (preliminary) research has been carried out to gain insight into the particular composition of the foundation material which should be used and its properties.

The research programme has two parts, namely a technical (constructional) part and an environmental part.

The constructional research has been partially carried out in the context of a final year project at the TU Delft (Technical University in Delft) and the environmental research has been carried out by the Intron research institute.

Responsible for the research are NBM Zuid, NBM Infrastructuur & Milieu, Vauatol and Smid en Hollander.

This article summarizes the findings connected with the development of Emulscement® on behalf of the town of Breda and its application in six projects, namely in the towns of Wanroy (in the borough of Sint Anthonis), Bergen op Zoom, Venray, Heerlen and in the provinces of Limburg and Zeeland.

2 The research

As already stated in the introduction, the research has been carried out with the objective of getting more knowledge of the constructional and environmental protection aspects of Emulscement®.

The constructional research concentrated on the various parameters which determine the material's constructional characteristics, with a view to the development and lifespan of pavement constructions which include an Emulscement® base.

Indications that the immobilisation of contaminated elements which can be achieved through stabilisation of asphalt granules with cement and bitumen emulsion is greater than when asphalt granules and cement are used, has meant that research has been carried out using the 'Stolwerk' granules into the leaching behaviour of Emulscement®.

The results of the environmental research are intended to provide further insight into the level of improvement in leaching behaviour which can be achieved using this form of stabilisation. Given the fact that no criteria yet exist for (the determination of) the leaching behaviour of organic materials (Polycyclic Aromatic Hydrocarbons, PAH), the research results have been interpreted using existing procedures which are being developed within the context of the Building Materials standards and the IPO interim policy (IPO = Provincial Coorporation).

Research has been carried out in the laboratory by testing laboratory prepared samples of the Stolwerk material. Compaction of the test samples was carried out using a gyrator.

3 Sampling

The research material originates from a depot for (contaminated) asphalt granules. Approximately 25 tons of material per location was collected from five different locations at the depot and transported to a crushing plant.

The quantity of material (125 tons) is milled in the crusher to a dimension of 0/25 mm.

All the samples to be used to manufacture research test samples are taken from this quantity of 0/25 asphalt granulate.

3.1 Composition and mixture development

The asphalt granules conform to the requirements for asphalt granules type 1, in accordance with the 1995 Standard and contains more than 80% asphalt.

Research into the composition of the asphalt granulate (n=6) has determined that the PAH content is about 150-200 mg/kg dm (dry material). The composition of the test samples is shown in table 1.

For the research of the constructional properties, test samples were made with a diameter of 150 mm and for the environmental research a diameter of 100 mm was selected. The height of the test samples is about 120 mm.

A gyrator was used to compact the test samples. The advantage of this machine is that the desired density of the test samples can be fixed.

Table 1: Composition of test samples

Composition		Portion
. Asphalt granules		85%
- asphalt	92.3%	
- masonry	3.1%	
- concrete	3.0%	
- slags	1.6%	
. Sand		15%
. Cement		3%
. Emulsion		3%
. Moisture content		6%

Table 2 shows the densities which were determined for test samples which were compacted by a gyrator. As a comparison, the table below also includes the densities which have been achieved using a Proctor machine on the same material.

Table 2: Densities of the test samples

Test sample	Wet density kg/m^3	Moisture %	Dry density kg/m^3
Gyrator	2094	6.0	1975
Gyrator	2103	6.0	1984
Proctor	2090	6.4	1964
Proctor	2097	6.4	1971

The results shown in table 2 show that the dry density of the Emulscement® is approximately 1975 kg/m^3. In general, both the constructional and environmental protection properties of Emulscement® appear to a large extent to be dependent on the material's density. It is, therefore, very important to have adequate control over the process of compacting the material. A great deal of attention is paid to this parameter during the preliminary research and company monitoring.

3.2 Constructional properties

3.2.1 Compression strength
The table below shows the 7 day and 28 day compression strength of the test samples which were compacted using the gyrator. Because, as mentioned before,

the compression strength is (to a large extent) depending on the material's density, the table also shows the corresponding densities.

Table 3: Compression strength of Stolwerk Emulscement®

Cement	Emulsion	Dry density	7 day compression strength	28 day compression strength
%	%	kg/m³	MPa	MPa
3	3	1975	1.4	-
3	3	1984	-	2.1

3.2.2 Stiffness module
The stiffness module of Emulscement® has been determined using a dynamic indirect tensile strength test at frequencies from 0.5 to 16 Hz and temperatures from 5-30°C. The results of this research show that at a frequency of 8-16 Hz and a temperature of 20°C, the stiffness module is approximately 5500 MPa. Just like asphalt, the stiffness module of Emulscement® is affected both by temperature and frequency.

3.2.3 Tensile strength
The fact that Emulscement® contains cement and emulsion means, of course, that this material is a bonded base material. Because it is possible, depending on the construction and the working details of the pavement construction, for tensile stress to occur in the base material, it is important to have some knowledge of the material's tensile strength.   The static indirect tensile strength test (split test) revealed an indication for the tensile strenght for Emulscement® of 0.4-0.6 MPa at 0°C.

3.2.4 Cyclical creep
In order to get knowledge of the deformation behaviour of Emulscement®, the dynamic creep test was used to determine the axial deformation of test samples of the material after 7200 load repetitions.   The test was carried out under the following conditions:

. *Stress duration*: 0.2 sec.
. *Period of rest*:   0.8 sec.
. *Load*:   0.25 MPa
. *Temperature*:   40°C

After 7200 stress repetitions under the conditions stated, the axial deformation of the material was measured to be approximately 0.4-0.5 % of the height of the samples. In this way Emulscement® can be classified as a material with a good resistance to deformation (fig. 1.).

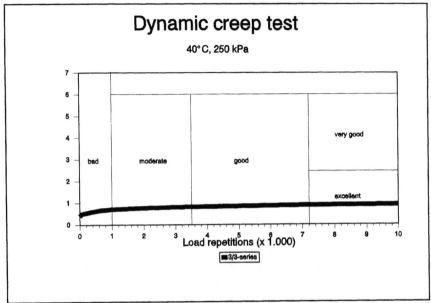

Fig. 1. Results and standard of the dynamic creep test

3.2.5 Fatigue
At the TU-Delft the material Emulscement® has been investigated in order to establish the fatigue behavior. A design line has been found representing the number of axle loads against the strain level at the bottom of the base. Validation of this design line still has to be done.

4 Leaching behaviour

At the moment, both the IPO interim policy and the concept Building Materials standard state that testing the environmental protection quality of organic components in building materials must take place on the basis of composition. This is due to the fact that no agreement has yet been reached concerning the test to be used to determine the leaching behaviour of organic components.
It is precisely because the leaching of contaminated elements out of the material can be considerably limited by the bitumen components contained in Emulscement®, that the leaching behaviour of the material is an interesting parameter.
Testing the results of research into the composition of the Stolwerk material in relation to the regulations which currently apply (IPO interim policy) revealed that non-isolated application of the material is not possible. The PAH content (10 PAH) of the material of approximately 150-200 mg/kg dm exceeds the limit of 75 mg/kg dm which has been set for non-isolated application.
Because testing the environmental protection quality of secondary raw materials should take place first and foremost on the basis of the leaching behaviour

(emission of contaminated components), the leaching behaviour of Emulscement®
from the Stolwerk material was determined.

This leaching research has been carried out with the primary objective of
discovering the extent to which the mobility of contaminated components in the
asphalt granules can be limited through stabilisation using cement and emulsion.
On the other hand, the research results have been used to get a step closer to
finding a suitable method for determining the leaching behaviour of organic
components in secondary materials.

To determine the leaching behaviour of Emulscement®, three diffusion tests have
been carried out, each involving four test samples.

Because no test standards for the leaching of organic components into the soil have
yet been formulated, two indicatory test standards have been calculated on the basis
of the system stated in the Building Materials standard.

Table 4 shows the results of the leaching research. This table shows the PAH
emission after 64 days (E_{64d}) and the calculated (with some assumptions) immission
(I(100 yr).

The table shows that Emulscement® produced from the Stolwerk material has a
leaching value of 1.7 mg/m². This emission results in an immission into the soil
which stays below the immission of unbonded material with a PAH content of 75
mg/kg dm and a layer thickness of 0.3 m. The immission into the soil even stays
below the indicatory limit value which is taken from the standard which assumes a
maximum increase in the target value of 1% over 100 years.

Table 4: Results of the leaching research

Component	E_{64d} (mg/m²)	Standard deviation (mg/m²)	I(100 yr) (mg/m²)
Naphthalene	0.62	0.05	2.0
Phenanthrene	0.37	0.01	1.2
Anthracene	0.19	0.005	0.3
Fluoranthene	0.49	0.03	1.6
10 PAH	1.7	<0.1	5.3

5 Structural design

A design graph was created for Emulscement®. This design graph shows the strain
(ϵ) set against the number of axle loads (N). Unlike the usual design graphs for
asphalt and cement which take account of the number of axle loads to crack
initiation, the design graph which has been created for Emulscement® has mainly
taken account of crack propagation. Although it has been assumed that in the case

of sand cement the base material crumbles after crack initiation (the minimum crack propagation phase), the research into Emulscement® and visual condition surveys during a period of three years have shown that the material displays much more favourable crack propagation behaviour due to its bitumen-emulsion content. Due to its favourable crack propagation behaviour, the minimum asphalt layer on the base material which is normally required to prevent the base from cracking, does not apply to the application of Emulscement® as a base material.

As the constructive behaviour is not yet fully characterized, the design of Emulscement® constructions will be done by the existing design methods. It is common knowledge that the heaviest axle loads create the most damage to cementbound bases, for Emulscement® the design method for cementbound bases is applied. This means that the whole axle load distribution is taken into account and the rule of Miner is used.

6 Emulscement® test cases

On the basis of the available data relating to Emulscement®'s constructional and environmental protection properties, three practical applications were assumed. This involved the selection of a country road, a minor road in a residential area and a heavily used motorway. The basic conditions for these three cases were that a quantity of tar-contaminated asphalt was available and that this material has to be applied in new pavement constructions. In all three cases, a good foundation (sand) was used and the case was not carried out in a water-catchment area. The implementation of the cases involved both environmental protection and road engineering (constructional) aspects.

Table 5 shows the basic conditions which applied to the pavement design in the three cases.

Table 5: Basic conditions for pavement design

	Country road	Minor road	Motorway
Intensity	500 vehicles/day	8,000 vehicles/day	40,000 vehicles/day
Lorry traffic	5%	8%	15%
Axle load distribution/ Axle coefficient			
10 kN	15%/2.0	10%/2.0	10%/3.0
30 kN	20%/2.5	25%/2.5	20%/3.0
50 kN	20%/2.5	20%/2.5	20%/3.0
70 kN	21.5%/3.0	21.5%/3.0	15%/3.0
90 kN	16%/3.0	15%/3.0	15%/3.0
110 kN	6%/3.5	6%/3.5	12%/3.5
130 kN	1%/4.0	2%/4.0	5%/4.0
150 kN	0.5%/4.0	0.5%/4.0	2.5%/4.0
170 kN	-	-	0.5%/4.0
Road width	4.0 m	7.2 m	12.0 m
Lane width	-	3.6 m	3.5 m
Length of road section	1000 m	750 m	5000 m
Development period	20 years	20 years	20 years
No. of working days per year	250	250	250
Traffic growth	1.0%	2.0%	2.5%
Neq-100kN	1.4E+05	4.4E+06	8.8E+07
E-module asphalt	6000 MPa	6000 MPa	6000 MPa
E-module base material	5500 MPa	5500 MPa	5500 MPa
Thickness of base material	200 mm	200 mm	250 mm
E-module sand	100 MPa	100 MPa	100 MPa

6.1 Environmental protection aspects

To ensure that the asphalt granules are reused in an environmentally responsible way, it is essential that the environmental characteristics of the material is determined. On the basis of the current categorisations in the IPO interim policy and the concept Building Materials standard, the material (tar-contaminated asphalt granules) is automatically placed in a special category which means that only isolated application is possible.

On the basis of the material's composition, the Stolwerk material has to be placed in the special category (PAH content > 75 mg/kg dm). Because checks on the environmental protection quality of the material should be carried out primarily on the basis of leaching behaviour, this leaching behaviour was also researched. This research shows that the material does not exceed the indicatory test standards for leaching relating to category 1.

On the basis of these results, it is quite right to propose that the available quantity of tar-contaminated asphalt granules can be placed in category 1 and that non-isolated application is permissable.

With respect to the quantities of tar-contaminated asphalt to be applied, in all three cases more than 1000 tons (the preferred quantity) must be used.

No demands exist with respect to application area since the material is not used in water-catchment areas.

Because the material can be placed in category 1, on the basis of composition and leaching behaviour, a declaration must be submitted to the responsible authority.

6.2 Road engineering aspects (structural design)

On the basis of the base material thickness and pavement properties selected, the required asphalt thickness is determined first of all. After the required asphalt thickness has been determined, the base layer checked for fatigue on the basis of the strains which are expected during the design life. Finally, depending on the conditions of use, the structural design of the pavement constructions is determined.

Based on the design method as described above, and the basic conditions for pavement design with respect to the three cases shown in table 5, the following prototype constructions were devised (fig. 2.):

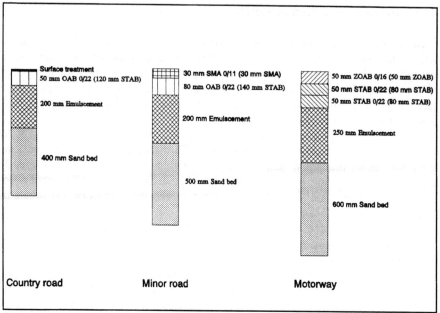

Fig. 2. Prototype constructions

As a comparison, the calculated (recommended) thicknesses[1] in the event of only cement bound granular material being used are shown in brackets.

7 Further developments

Besides the Emulscement® material, which primarily consists of asphalt granules with cement and emulsion, work is now being carried out to develop the Fundec-layer (base material-emulsion-cement) which has as basic ingredients building, demolition waste and metal slag.
This material is also stabilised using cement and emulsion and three projects have already been carried out (in Steenbergen, Etten-Leur and Weert) and a fourth is being planned.

The leaching results which have been achieved are very promising, but more important is the resistance to cracking which also characterises this Fundec layer.

---

[1] See SAG/VNC publication 'Dimensioneren met zandcement', CROW publication 32 'Breekasfaltcement' and the 'Handleiding Wegenbouw-Ontwerp Verhardingen' by the DWW (Department of Civil Engineering).

8 Conclusion

Emulscement® and Fundec are promising, new materials that offer an environmentally friendly solution for the (cold) recycling of tar-contaminated asphalt and slag containing heavy metals and from a construction point of view produce a base material with a high value:

1. *construction properties*: a greater resistance to fatigue, no cracking, thinner asphalt overlayers are possible.
2. *environmental protection properties*: a considerable reduction in the leaching of PAH, energy savings and a reduction in the emissions of harmful vapours.
3. *financial*: savings due to dumping costs being avoided, the reduced extraction of new raw materials and reduced energy costs.

9 Future Developments

It is important to optimise the preliminary research method, especially with respect to the relationship with the realizable on-site density. At the moment, tests are being carried out with the proctor test on 15 cm test samples. In the future, the gyrator will be put into action on such test samples. Future research will, therefore, be directed to get higher densities and a more precise definition of the material's properties, as well as the setting up of a simple method of testing, in order to be able to adequately describe the material. At the moment, mechanical tests are also being carried out such as dynamic and static splitting tensile tests. It may be that this will also lead to improvement of official regulations. In addition, the implementation of higher percentages of bitumen emulsion, for the effective binding of granules with a higher tar content, is being examined closely.

Influence of thermal stresses on construction-induced cracks

A. SHALABY and A.O. ABD EL HALIM
Department of Civil and Environmental Engineering, Carleton University, Ottawa, Ontario, Canada
S. EASA
Department of Civil Engineering, Lakehead University, Thunder Bay, Ontario, Canada

Abstract

Results of a comprehensive research over the past decade indicated that present compaction equipment induce a pattern of hairline cracks in the surface of newly constructed pavements. The presence of these cracks alters the structural behaviour of the asphalt layer under both traffic and thermal stresses. The paper attempts to classify transverse cracks based on geometry and location, and determines the propagation of such cracks under thermal loading. The analysis invokes the finite element method using both two- and three-dimensional models. The results are utilized into two independent fracture mechanics methods to compute the stress intensity factors and hence the amount of crack propagation in each loading cycle. Comparisons between different crack configurations and locations will be helpful in determining which cracks are expected to propagate faster, if at all. The numerical analysis emphasizes the interaction between adjacent cracks and the effects of boundary parameters such as the bond between layers and existence of cracks in various locations in the pavement system. From the three dimensional analysis, horizontal crack growth at increasing crack lengths is accelerated with the existence of an underlying crack but not affected when the underlayer is assumed uncracked.
Keywords: crack interaction, finite element analysis, pavement cracking, stress intensity factor, thermal stresses.

1 Introduction

Numerical analysis of cracked pavement structures is a prerequisite for recent pavement management systems. Decisions to overlay a pavement section, perform a preventive maintenance or crack rout and seal are increasingly in need to be based on mechanistic approaches. Invariably, the formulation of proper models that best suit

Reflective Cracking in Pavements. Edited by L. Francken, E. Beuving and A.A.A. Molenaar. © 1996 RILEM. Published by E & FN Spon, 2–6 Boundary Row, London, SE1 8HN. ISBN 0 419 22260 X.

the given problem is the key element to a successful and economical decision. In the previous two reflection cracking conferences, the analysis of temperature related stresses was presented [1]. Historically, the thermal cracking problem gained attention in the 1960s' [2], and the most recent developments in terms of performance prediction models are given in [3,4]. Finite element analysis of the thermal stresses in pavements were investigated in [5,6]. In this paper, a finite element model for thermal crack propagation is presented. The model features include a more precise treatment of the thermal response and a detailed treatment of crack propagation in pavement layer.

A testing program was conducted to verify the effect of crack location on the remaining life of the asphalt layer. It is important that the use of finite element analysis and fracture mechanics principles to predict the response of the asphalt layer, consider the initiation of surface cracks at the time of construction which are induced by the heavy and stiff rolling equipment [7,8].

2   Model formulation

2.1   Material properties
The thermal properties of pavement and unbound layers are selected according to Kersten models [9]. The models relate the thermal conductivity and specific heat of the material to the frozen and unfrozen water content and the density. For stress analysis, the input to the FEA requires the stiffness-temperature relationship and Poisson's ratio for each different layer, both are assumed in the present analysis based from the literature [5,6]. The coefficient of thermal contraction of the asphalt concrete mixture α_{mix} is determined from similar coefficients for the binder α_b and aggregate α_a and according to their percentage in the mixture as follows,

$$\alpha_{mix} = \frac{\alpha_b \, V_b + \alpha_a \, V_a}{100} \tag{1}$$

where V_b and V_a are the percent of binder and aggregate volumes in the mixture.

2.2   Loading conditions
The change in pavement surface temperature and the corresponding thermal strains in the pavement are the only source of loading being considered. The temperature is applied at the surface in a sinusoidal fashion. Typically each cycle has a duration of one day. The pavement system is subjected to the temperature cycling for 48 hours prior to stress analysis as a soaking period.

2.3   Thermal and stress analysis
The thermal cycling is applied to the pavement surface and multiple solutions for the nodal temperatures on each element are obtained at discrete time intervals. A solver for full transient thermal analysis is used for this purpose. The stress and displacement fields are computed for each time step by applying the thermal strain created by the difference in temperature between the original state and the current time step. A static analysis solver is used for this purpose. Displacements at and around the crack tip are stored for use in the fracture analysis.

32   *A. Shalaby* et al.

2.4 Fracture analysis

The two point displacement formula is used to determine the stress intensity factors at the crack tip [10].

$$K = \frac{2G\left(4u_2 - u_3\right)}{\gamma + 1} \sqrt{\frac{2\pi}{l}} \qquad (2)$$

where $\gamma = 3 - 4\upsilon$ in plane strain, $\gamma = (3 - \upsilon)/(1 + \upsilon)$ in plane stress, l = element length, u_2 and u_3 = crack opening at $l/4$ and l from the crack tip, respectively, υ = Poisson's ratio, and G = shear modulus = $E/2(1 + \upsilon)$.

3 Finite element procedure

The pavement thermal response discussed in the previous section is implemented in a finite element model. The ANSYS program is used throughout the procedure [11].

3.1 Element selection and mesh design

The following factors are found relevant to the sensitivity of the mesh design: 1. Finer mesh elements are required closer to the pavement surface in the thermal analysis. 2. Parabolic elements are required in the stress analysis. 3. Crack tip elements, modified from parabolic elements are necessary around the crack tip to avoid the numerical singularity case (division by zero) around the crack. 4. where possible, the procedure should be automated with capabilities to modify the inputs of material properties, temperature loading and general dimensions but with least user interference as to mesh design, computations of stress intensity factors, element size and types.

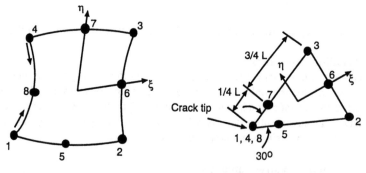

(a) Typical 8-node parabolic isoparametric element, 8 DOF thermal or 16 DOF Structural

(b) Quarter point element (Crack tip) formed by collapsing the side of 1, 4 and 8, and relocating midside nodes 5 and 7.

Fig. 1 Element types used in 2-D finite element modelling.

The selected element type and crack tip element for plane analysis are shown in Fig. 1. Eight-node isoparametric elements are used throughout the meshes, except at the vicinity of the crack where 6-node elements are used and modified to quarter-point elements when in contact with the crack tip. The mesh generation procedure is standardized and a configuration of the crack tip elements is maintained for all solutions, shown in Fig. 2. A program is prepared to interact with the ANSYS program for the data input, the solution procedure. The program activates the post processor of ANSYS and determines the location of the crack tip, and computes the stress intensity factors with virtually no user interference.

4 Crack configurations

In an attempt to classify the severity of different types of cracks and the potential for crack propagation, four distinct crack types are examined as illustrated in Fig. 3. The first type is the vertical crack that initiates from the top of the pavement, extends laterally for a sufficient distance that plane conditions are valid, and propagates downwards through the thickness of the pavement layer or overlay. The crack originates from hair line cracks initiated during the compaction procedure by the mismatch in stiffness and geometry between the soft and flat asphalt mixture and the rigid and round steel or pneumatic rollers [7].

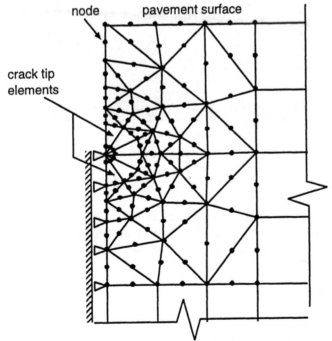

Fig. 2 Detail of the finite element mesh around the crack tip.

The second type is also vertical but originates at the bottom of the pavement layer and propagates upwards through the thickness [12]. This type is generally termed "reflection crack" and originates from traffic induced tensile strains at the bottom of the pavement layer or due to thermal expansion and contraction of the underlayers, specifically when sufficient bond is present between layers and the underlayer is already cracked [12].

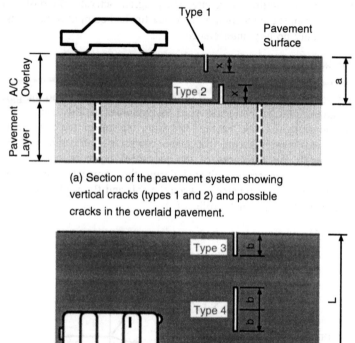

(a) Section of the pavement system showing vertical cracks (types 1 and 2) and possible cracks in the overlaid pavement.

(b) Plan of the pavement surface showing full depth horizontal cracks (types 3 and 4).

Fig. 3 Configuration of cracks used in the analysis.

The third and fourth crack types represent full depth crack that originate at the pavement edge (type 3) or within the layer (type 4) and propagate in the horizontal plane until complete separation occurs. These two types have not been investigated as thermal cracks before.

5   Results and analysis

5.1   Thermal conditions

The thermal response of a single layer system can be checked to analytical formulae. Fig. 4 shows the changes in thermal gradients in the 200 mm close to the pavement surface during a 24-hour thermal cycle. The close agreement between the FEM results and the analytical formula, has led to the extension of the method in multilayer systems with varying thermal properties, where analytical solutions are not available, within practical limits [13].

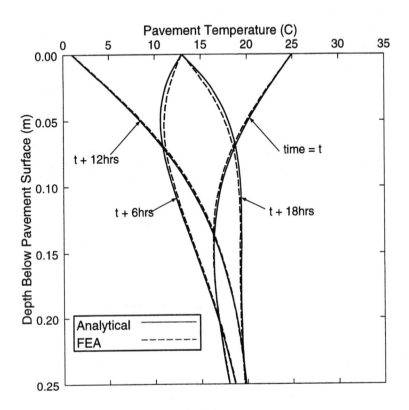

Fig. 4   Change in thermal gradients in the pavement system during a 24-hour period.

5.2   Surface deformations and crack opening
The results shows that the first two types of cracking cause lipping and cupping around the edges of the crack from a slab curling effect, shown in Figures 5 and 6. These deformations affect the travelling speed, and safety and stability of vehicles on the road as previously reported in the literature.

5.3   Stress intensity factors
Fig. 7 illustrates the stress intensity factors obtained from the finite element analysis at different stages of crack propagation due to a temperature drop of 20°C. Fig. 7a compares the stress intensity factors in two crack types, the surface crack propagating downwards (type 1) and the reflected crack propagating upwards. Fig. 7b shows the stress intensity factors for a horizontal crack (type 3), at the surface of the pavement and at the middle of the layer depth. It is shown that the crack will travel faster at the

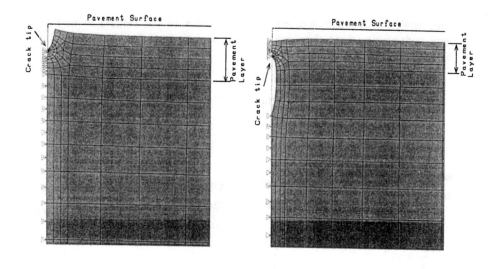

Fig. 5 Deformed pavement system under thermal load: crack type 1.

Fig. 6 Deformed pavement system under thermal load: crack type 2.

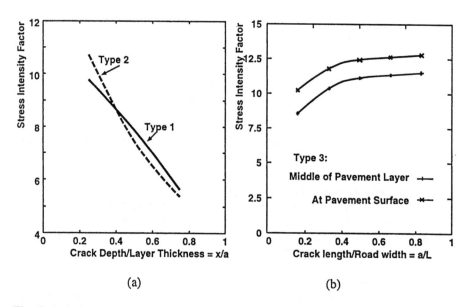

(a)   (b)

Fig. 7 Relationship between the stress intensity factor (KN/m$^{3/2}$) and the crack progression: (a) types 1 and 2 (b) type 3

surface, and the speed decreases with increasing the depth. Also, the stress intensity factors are higher than those of type 1 and 2, which also translates to a more rapid crack growth rate.

6   Laboratory Testing

The effect of crack initiation and propagation is examined by testing field constructed asphalt slabs in cyclic direct tension. The mixture consisted of crushed stone aggregate, 5% asphalt cement content of 85/100 pen, 4.3 % air voids and 16.1 % voids in mineral aggregate. The mixture was poured into large forms at the construction site and compacted with 8 passes of a steel roller to achieve a thickness of 70 mm. The slabs were sawed into test size of 450 mm x 150 mm x 70 mm.

The laboratory test facility comprised a hydraulic actuator and a moving and fixed mounting plates attached to an LVDT and load cell, and a computer-controlled data acquisition system. The loading cycle was triangular, displacement controlled with 0.45 mm opening displacement, 1.0 Hz frequency and no rest periods. The samples are glued from the underside to steel plates which are mounted on the fixed and moving plates of the testing equipment. The only variable tested is the location of crack with respect to load application. The reason is to identify the number of cycles for crack initiation and crack propagation. Each test set contained two identical slabs and tests were performed at room temperature. One set of slabs is notched at the underside, closer to the load application (crack propagation) with the effective thickness at the notch location = 40 mm. A second set is notched on the upper side and maintained the same effective thickness (crack initiation and propagation). Results of the test are shown in Fig. 8 and table 1. The number of cycles consumed by crack initiation is approximately determined from the comparison of slabs notched at the bottom surface to slabs notched at the top surface.

Fig. 8   Test set up and tested samples.

A. Shalaby et al.

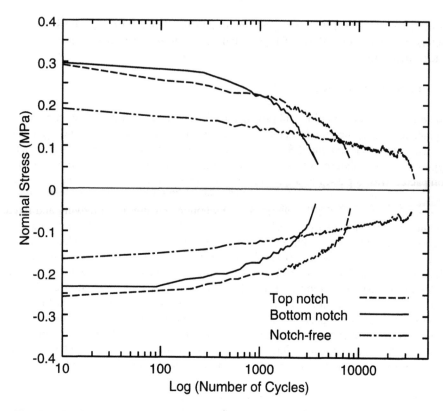

Fig. 9   Number of cycles to failure in displacement controlled cyclic tensile test.

Table 1: Number of cycles and remaining life of asphalt layer

Case	Notch-free	Top-notched	Bottom-notched
No. of Cycles	34 000	7 500	4 000
Remaining life	1.0	0.22	0.12

7   Conclusions

An analytical model has been presented to investigate the effect of thermal cracking. The model benefits from the flexibility of the finite element method and is designed in a way to limit the user interference with the procedure if so required. The model potential lies in a very detailed thermal analysis cycle and a corresponding stress and fracture analyses all performed on the same element mesh. Results of experimental study was presented to demonstrate the important relationship between the location of the load with respect to the initiation and propagation of transverse cracks under cyclic tensile loads. The number of cycles to cause failure decreased significantly when a notch was introduced and again suffered a large decrease when the notch was made on the same side of the applied stress. Further research is required to explain the difference in behaviour between horizonal and vertical cracks and establish the relationships for crack growth models based on the stress intensity factors.

8 References

1. Francken, L. (1989) Laboratory simulation and modelling of overlay systems, in *Reflective cracking in pavements-Assessment and control,* (ed. J.M. Rigo and R. Degeimbre), E & FN Spon, London, pp. 75-99.
2. Haas, R.C.G. and Topper, T.H. (1969) Thermal fracture phenomena in bituminous surfaces, *Highway Research Record, Special Report 101,* National Research Council, Washington, D.C., pp. 136-153.
3. Hiltunen, D. and Roque, R. (1994) A mechanics-based prediction model for thermal cracking of asphalt concrete pavements. *Association of Asphalt Paving Technologists,* Vol. 53.
4. Shalaby, Ahmed and Easa, Said (1995) Modeling pavement thermal cracking: Sensitivity analysis. Proceedings, *Canadian Society for Civil Engineering,* IV, pp. 285-294.
5. Shalaby, A., Abd El Halim, A.O., and Easa, S. (1996) Low-temperature stresses and fracture of asphalt overlays, *75th Annual meeting of the transportation research board,* National Research Council, Washington, D.C.
6. Selvadurai, A.P.S., Au, M.C. and Phang, W.A. (1990) Modelling of low temperature behaviour of cracks in asphalt pavement structures. *Canadian Journal of Civil Engineering,* Vol. 17, pp. 844-858.
7. Abd El Halim, A.O., Phang, W. and Haas, R. (1987) Realizing Structural Design Objectives Through Minimizing Of Construction Induced Cracking. *Sixth International Conference on Structural Design of Asphalt Pavements,* Ann Arbor, Michigan, July 13-16, Vol. I, pp. 965-970.
8. Abd El Halim, A.O., and Razaqpur, A.G. (1993) Minimization of reflection cracking through the use of geogrids and better compaction, in *Reflective Cracking in Pavements,* (ed. J.M. Rigo, R. Degeimbre and L. Francken), E & FN Spon, London. pp. 299-306.
9. Farouki, O.T. (1986) *Thermal Properties of Soils,* Series on rock and soil mechanics, Vol. 11, Trans Tech Publishers, D-3392 Clausthal-Zellerfeld, Germany.
10. Barsoum, R.S. (1976) On the use of isoparametric elements in linear fracture mechanics. *International Journal of Numerical Methods in Engineering,* Vol. 10, pp. 25-37.
11. Swanson Analysis Systems (1993) *Ansys 5.0a computer program: reference manual,* Houston, Pennsylvania.
12. Jaecklin, F.P. (1993) Geotextile use in asphalt Overlays - design and installation techniques for successful applications, in *Reflective Cracking in Pavements,* (ed. J.M. Rigo, R. Degeimbre and L. Francken), E & FN Spon, London. pp. 100-118.
13. Hsu, Tai-Ran (1986) *The Finite Element Method in Thermomechanics.* Allen & Unwin, Boston.

Prevention of cracking in semirigid pavement base by slow setting binder

I. GSCHWENDT
Slovak Technical University Bratislava, Slovak Republic
V. MEDELSKÝ and L.' POLAKOVIČ
Research Inst. VUIS-CESTY, ltd., Bratislava, Slovak Republic

Abstract

This paper describes the results of research programm dealing with mechanical properties of slow setting hydraulic binder made from blast-furnage granulated slags and behaviour of construction (stabilized) mixes with this inorganic binder in pavement structure -as measure for reflective cracking prevention.
Keywords: Bounded materials, Hydraulic binders, Pavement structures, Reflective cracking.

1 Introduction

In analyzing the condition of cracked asphalt pavements special attention should be given to characterization of the cracks. Their differ in shape, configuration, aspect orientation and many other characteristic. We can observe:

- longitudinal cracking,
- transverse cracking,
- alligator and
- block cracking.

There are large number of reasons and mechanismus as origin of the cracks. Longitudinal cracking in the wheelpaths is most likely influenced by traffic [1]. Alligator and block cracking are commonly related to fatigue of the pavement and to the tensile strain at the bottom of the asphalt layer. The basic factors which lead to transverse cracks are thermally induced stresses and strains and repeated traffic loading. The well known example is the development of cracks in cement treated pavement base layer due to hydratation of the binder. Subsequent daily and seasonal

Reflective Cracking in Pavements. Edited by L. Francken, E. Beuving and A.A.A. Molenaar. © 1996 RILEM. Published by E & FN Spon, 2–6 Boundary Row, London, SE1 8HN. ISBN 0 419 22260 X.

temperature variations induce openings and closures of the cracks and propagation the traffic stresses take part. The reflective cracking process in asphalt pavements with cement treated bases is a complicated one.

The rehabilitation of cracked asphalt pavements by simple overlaying is rarely durable solution [1]. In preventing cracks propagation on pavements have been proposed the placement of a stress/strain absorbing membrane interlayer between the surfacing and the overlay and/or modification of the overlay: by the use of modified bitumen, modified asphalt mixes. We have to say it is possible to act on the origin of the cracks by modification of pavement base. The visual condition surveys and performance tests [2] on asphalt pavement structures with hydraulic binder treated bases shows and proved the creation of shrinkage cracks in relation to stiffness of base material. This modification concern the modulus of elasticity, tensile strength and volume changes (thermal susceptibility).

2 Inorganic slow settings binders

Making use of metallurgical slags having hydraulic properties but (and) a very slow process of hydratation was completed in 1992 (with participation of slag producers).

The activity of metallurgical slag has been increased by means of their crushing and grinding with additives having a task of initiating the activity of the hydratation process. The fineness of grinding has been established by means of specific surface (according to Blain) 300-400 $m^2.kg^{-1}$ or by means of a content of 70 to 85 per cent to particles less than 80 μ.

The activity capacity of blast-furnage granulated slag is expressed by means of relation

$$AC = \frac{CaO + MgO + Al_2O_3}{SiO_2} \tag{1}$$

When the minimum amount of CaO is 33 per cent by weight, maximum amount of MgO is 12 per cent and a minimum amount of Al_2O_3 is 5 per cent. Generally

$$1.3 < AC < 1.8 \tag{2}$$

The compressive strength and the flexural strength estimated on flexure-test specimens (beams) of ordinary portland cement and slow setting binder are in table 1. We can compare the two binders and strength changing during one year.

The strength of three slow setting binders with different composition after 28 days and one year are in table 2. The ratio between the compressive strength and tensile strength of binders are in range 3.5 to 5.0 and more after one year.

The recommended strength values of inorganic slowly setting binder are (after 28 days):

compressive strength   min. 4.0 MPa
                                    max. 12.0 MPa
bending tensile strength   min. 2.0 MPa
                                    max. 5.0 MPa

Table 1. Strength of slow setting binder and portland cement, in MPa (standard sand, flexure-test specimens 40 x 40 x 160 mm)

Time (days)	Portland cement		Slow setting binder	
	Compress.	Flexural	Compress.	Flexural
7	14.18	3.53	-	-
28	26.52	7.34	8.63	2.81
60	29.78	7.41	15.31	4.75
90	27.77	7.97	17.15	5.58
180	34.66	8.44	24.27	6.08
360	38.81	9.50	25.88	7.85

Table 2. Strength of slow setting binders (standard test with sand)

Composition of binder	Strength after 28 days		Strength after 360 days		
	Compress (MPa)	Tensile (MPa)	Compress (MPa)	Tensile (MPa)	
	a	b	c	d	$\frac{a:b}{c:d}$
87% granulated (pulverised) slag 3% clinker 10% gypsum	18.5	5.3	35.0	7.2	$\frac{3.5}{4.9}$
80-85% granulated (pulverised) slag 10-15% clinker 4% gypsum	16.0	3.2	57.0	6.1	$\frac{5.0}{9.3}$
55% granulated (pulverised) slag 40% fly-ash 5% lime	17.0	4.4	47.0	9.1	$\frac{3.9}{5.2}$

3 Mechanical properties of mixes

Under laboratory conditions the parameters of new construction mixtures bounded with slow setting binder have been monitored - in certain intervals of time (up to two years). The test specimens, cylinders and small beams from bounded mixes of soils and coarse (granular) materials were used for strength test, for freezing and thawing test and volume (linear) changes.

Inorganic binder has been prepared on the base of very fine crushed and grinded blust-furnage granulated slag with 5 per cent of CaO additive and/or portland cement clinker. Many laboratory test have been made by slow setting binder with materials used for base courses.

On fig. 1 there is the relationship between the time and compressive strength of coarse aggregate (0-32 mm) bounded by slow setting binder and by portland cement. The test results of coarse aggregate 0-16 mm bounded with 6.5 % of slow setting binder are in fig. 2.

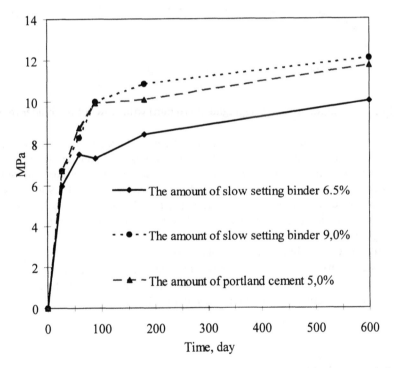

Fig. 1. Coarse aggregate (0-32 mm) bounded by portland cement and slow setting binder, compressive strength.

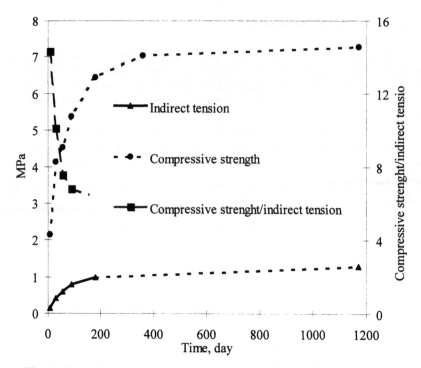

Fig. 2. Strength of coarse aggregate (0-16 mm) with slow setting binder (6.5%)

The resulting statement of these test is, that strength of construction mixtures with this binder after 60 days is adequate to the strength of mixtures bounded with portland cement. The relation between the compressive strength and the bending tensile strength of mixtures with the slow setting binder is within the range of 2.9 to 3.4 while for mixtures bounded with cement this relation reaches values of 5.0 to 6.0. Road base courses bounded with inorganic binder have a lower value of elasticity modulus with a relative higher bending tensile strength.

The time of setting - initial set and final set of binder influence the basic physical properties of construction mixes: the soil stabilization and coarse aggregate. In the center of the interest is the compressive strength of mixes in relation to the time.

Generally we describe the growth of strength by relation:

$$y = b . \log \frac{x}{a} \qquad\qquad\qquad (3)$$

where   y is the strength,
         x - time,
         a and b - material characteristics.
The values of a and b for the stabilized mix with different binder content were:

5% slow setting binder →	a=5.10	b=3.49
7% slow setting binder →	3.20	3.73
9% slow setting binder →	4.00	6.59
5 % portland cement →	0.65	3.62

After many test results evaluation for strength and time we suggest better relation:

$$y = b. \ln (x + 1) \tag{4}$$

4 Test of pavement structures

The performance of pavement structures with base layers using coarse aggregate stabilized with slow setting binder were tested on sections of circular test track. Compositions of pavement structures are on fig. 3. Structure No3 has base layer from soil stabilization with portland cement.

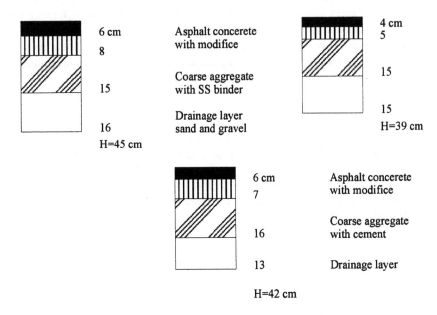

Fig. 3. Pavement structures tested on circular test track VUIS-CESTY, ltd., in Bratislava

The pavement test sections on the circular test track (Research Institution VUIS-CESTY, ltd in Bratislava) have been designed for 700 thousand repetitions of axle loads (10 tons), but the test sections have actually been loaded by 1.4 million drives until the first cracks of the wearing course appeared [2].

The constructed road test sections with application of inorganic slow setting binder in construction mixtures for base and subbase layers of pavements have proved a gradual process of strength increasing what has manifested itself in the extension of service efficiency of pavement structure by 30-45 per cent.

5 Conclusion

The production and the laying up of the stabilized mixtures with the use of the new inorganic binder has been carried out in a similar way as in the case of the portland cement using. The necessity of urgent work, placement and compaction of the construction mixture is not emphasized because there is a slow hydratation process and growth of strength compared with portland cement stabilization.

The developed construction mix for base layers represents the prevention of cracking in new pavements with asphalt surfacing. The application of material allows the pavement design with 40 mm less thickness of asphalt surfacing, total 80 to 120 mm for medium and heavy traffic load.

Circular test track and other full scale tests sections of pavements with structures made with base layers use of inorganic slow setting binder from blast-furnage slag have not show reflection cracking.

The use of new slow setting inorganic binder for bounded base layers means a technical solution as well as an economical contribution for the road administration.

References

1.   Rigo, J.M. (1993) *Main conclusions of the Conference on reflective cracking in Pavements.* Proceedings of the second International RILEM Conference, Liege.
2.   Loveček, Z. and all (1992) Final report of research work *Tuhé a polotuhé konštrukcie dopravných plôch so zvýšenou únosnosťou.* VUIS-CESTY, s.r.o., Bratislava (Slovak Republic).

Préfissuration par film plastique

F. VERHÉE
VIAFRANCE, France

Résumé
Après le développement en 1984 d'un premier système de préfissuration pour matériaux d'assise hydrauliques de plates-formes industrielles, VIAFRANCE a développé en 1992 une nouvelle technique pour chaussées routières.
Le principe de celle-ci est la réduction de la section de la couche traitée. L'amorce de fissuration est créée en insérant un film plastique suffisamment large pour produire cet effet (1/3 à 1/4 de l'épaisseur de la couche) et d'une épaisseur telle qu'elle préserve la fonction transfert de charge.
Un matériel spécial a été conçu et fabriqué. Il peut insérer des bandes de plastiques sur des largeurs de 2 à 5 mètres. Son rendement peut atteindre un kilomètre/jour pour une chaussée de 8 mètres de large. Plus de 600 000 m² de chaussée ont été préfissurés.

Le contenu de cette communication est :
- de présenter la conception de ce matériel
- de donner la méthodologie d'exécution de cette technique
- de fournir le bilan des applications
- de présenter une première évaluation de celles-ci, spécialement le suivi de l'effet de cette préfissuration

Mots clés   : chaussée / comportement / film plastique / retrait hydraulique / matériel / préfissuration.

Abstract
After the development of a first precracking system for hydraulic base materials in 1984 for industrial platforms, VIAFRANCE developed in 1992 a new equipment for road pavements.
The principle of this one is the reduction of the section of the treated course. The crack initiator is created by inserting a film plastic sufficiently wide to produce this effect (1/3 to 1/4 (of the course thickness) and reasonably thin to preserve the load transfer function at this level.
A special equipment has been designed and manufactured . It can insert plastic strips on 2 to 5 meters pavement widths. Its out put can be 1 km of 8 meters with pavement per day. More than 600 0000 m² of precracked pavements have been realised.

Reflective Cracking in Pavements. Edited by L. Francken, E. Beuving and A.A.A. Molenaar. © 1996 RILEM. Published by E & FN Spon, 2–6 Boundary Row, London, SE1 8HN. ISBN 0 419 22260 X.

F. Verhée

The purpose of the paper will be :
- to present the design of this equipment,
- to give the methodology of execution of this technique,
- to supply the assessment of theses applications,
- to present a first evaluation o f the behaviour of theses ones, specially the survey of the effect of this precracking

Key words : behaviour / equipment / hydraulic shrinkage./ pavement / plastic film / precracking.

En 1984, VIAFRANCE a développé et mis au point un système de préfissuration consistant à créer une amorce de fissuration dans la partie supérieure d'une couche traitée au liant hydraulique. Cette amorce était réalisée à l'aide d'un "couteau" constitué par un anneau métallique fixé sur le jante d'un compacteur à main qui introduisait un film plastique. Cette technique a notamment été utilisée pour la réalisation du terre-plein du Port de la Goulette (Tunisie -280 000 m²) [1] et d'un chantier comparatif de diverses techniques antifissures [2].
Cette première expérience a permis de concevoir une évolution de cette technique plus adaptée aux chantiers routiers.

1 Rappel sur la préfissuration

La préfissuration est destinée aux assises traitées aux liants hydrauliques qui, par nature, présentent une fissuration de retrait principalement thermique.

Cette fissuration, uniquement transversale sur les chaussées routières, présente les inconvénients suivants :

- réduction de l'effet de dalle augmentant les contraintes sur le sol support,
- création d'un effet de bord augmentant les contraintes dans les couches de chaussées.

Il en découle :

- la nécessité d'augmenter les épaisseurs de la chaussée,
- l'apparition de fissures en surface et, souvent, leur dégradation sous les effets climatiques et du trafic.

Cette fissuration ne pouvant être évitée, on peut rechercher à limiter ses effets :

- soit en introduisant un matériau d'interposition, limitant la vitesse de remontée des fissures dans la couverture bitumineuse,
- soit en limitant les mouvements horizontaux et transversaux des fissures, par préfissuration,

ce qui :

- limite, comme les interpositions, la vitesse de transmission des fissures,
- mais aussi très certainement, bien que cela n'ait pu encore être démontré de façon formelle, réduit les contraintes et déformations des couches de chaussées.

La préfissuration présente donc plusieurs avantages.

Le principe de la préfissuration est de provoquer des fissures à un pas très inférieur à celui de la fissuration naturelle ; 2 ou 3 mètres est le pas généralement retenu.

Il en découle une ouverture beaucoup plus faible de ces fissures engendrant :

- des mouvements horizontaux limités : ceux-ci sont directement proportionnels au pas de fissuration,
- une réduction des mouvements verticaux, la plus faible ouverture des fissures assurant un meilleur engrèment et donc un meilleur transfert de charge.

2 La technique de prefissuration par film plastique

Elle est basée sur le principe de la réduction de section de la couche traitée : les fissures se créent au droit de cette réduction.

Principe spécifique :
L'amorce de fissuration est créée dans la couche de matériau traité au liant hydraulique par introduction d'une feuille de plastique d'une hauteur assurant l'effet d'amorce : de 1/3 de l'épaisseur à 1/4 voire 1/5. (Dans le cas des chaussées en béton de ciment, on considère que le transfert de charge aux joints est suffisant même si la section est réduite de 1/4).
La réduction d'épaisseur est réalisée sur toute la largeur de la chaussée.
Le film de plastique est choisi suffisamment fin (e = 40 ou 80 µ) pour que l'on puisse espérer la conservation du transfert de charge sur l'épaisseur de la couche de chaussée : la faible épaisseur permet sa déformation par les granulats au compactage

3 Matériel

Le matériel de préfissuration, conçu et réalisé par VIAFRANCE (figures 1 et 2) assure l'introduction verticale d'un film plastique dans la couche de matériau traité aux liants hydrauliques.
Il permet la pose de films de différentes hauteurs (en général 6, 8 ou 10 cm) à des profondeurs variables par simple réglage.
La longueur de pose du ruban plastique est réglable et peut varier de 2 à 4 ou 5 m selon la machine.
Le cycle de poste est entièrement automatisé.

4 Mode de réalisation

La préfissuration est réalisée en principe dans le matériau foisonné et réglé avant le compactage. Pour prendre en compte le foisonnement, le ruban et enfoncé d'environ 8 à 10 cm en dessous de la surface du matériau foisonné. Pour une épaisseur de couche de 25 cm, la largeur du ruban est de 8 cm (25 cm : 3). Le pas de fissuration est celui retenu en général pour la préfissuration, c'est-à-dire 3 mètres. Mais il peut être différent si souhaité.

Figure 1 : Le matériel de préfissuration est constitué par une poutre télescopique supportant l'outil de préfissuration qui réalise la mise en place du film plastique. Ce matériel est porté par un engin automoteur. Les largeurs de travail varient de 2 à 4 ou 5 mètres selon le matériel.

Figure 2 : L'outil de préfissuration proprement dit est formé d'un soc. Il est alimenté par un rouleau de film plastique d'épaisseur 40 à 80 u. Le soc ouvre un sillon et y dépose le film. La mise en oeuvre simple et rapide permet la réalisation de chantiers à hauts rendements (jusqu'à 1 km /jour pour une chaussée de 8 mètres de largeur).

5 Réalisation

Un premier chantier expérimental a été réalisé en octobre 1992. En 1993, cinq chantiers ont été réalisés. Fin 1995, 600 000 m² de chaussée ont été préfissurés avec cette technique. Les matériaux et types de structure sont très variables. A titre d'exemple, le tableau 1 présente les structures des chaussées préfissurées en 1993.
Cette variété permettra d'évaluer la technique dans de nombreux contextes.

Site	Date	Type de Structure	Surface	Trafic
SAPRR A-6	Mars 1993	• Béton Bitumineux sur Béton compacté	7 500 m²	T 3
RD-70 Haute-Saône	Juillet 1993	• Béton Bitumineux sur Grave hydraulique à hautes performances	28 000 m²	T2 210 PL/jour/sens
Piste militaire	Août 1993	• Béton compacté	15 000 m²	
RN-124 Landes	Août 1993	• Béton Bitumineux sur micro Grave hydraulique et Sable hydraulique	56 500 m²	T1 475 PL/jour/sens
RN-57 Vosges	Octobre 1993	• Béton Bitumineux sur Grave hydraulique et Sable hydraulique	70 000 m²	T0 1 600 PL/jour/sens

TABLEAU 1 : Exemples de structures préfissurées : chantiers 1993

Constatations à la réalisation :

• RD-108 : Déviation de St ROGACIEN:

Des sondages et carottages en rive de chaussée ont été réalisés au printemps 1993. Ils ont mis en évidence le bon positionnement du film plastique (Figure 3).

• RD-124 - Déviation de BEEGAR :

Fissuration fine au droit des préfissures avant l'application de la couche de roulement à la suite d'une forte variation de température : ceci montre l'efficacité de la technique : toutes les préfissures sont actives.

6 Comportement dans le temps

Plusieurs chantiers font l'objet d'un suivi. Ces chantiers sont soit des chantiers significatifs par leur importance, les sollicitations dues au trafic, les techniques utilisées, soit des chantiers où des planches expérimentales et / ou comparatives ont été réalisées.
Le tableau 2 présente les constatations faites sur les principaux sites faisant l'objet d'un suivi.

Site	Date de réalisation	Structure	Trafic	Comportement en janvier 1996
RD-108 Charentes-Maritimes Déviation ST ROGATIEN	Octobre 1992	• 4 cm Béton Bitumineux • 4 cm Béton Bitumineux • Géotextile imprégné • 20 cm Grave laitier préfissurée • 22 cm Sable laitier	ClasseT2 150 à 300 PL/jour/sens	- Absence totale de fissures.
RN-124 Landes Déviation BEEGAR	Août 1993	• 6 cm Béton Bitumineux • Géotextile imprégné • 22 cm micro Grave Laitier préfissurée. • 25 cm Sable Laitier	Classe T1 475 PL/jour/sens	• Chaussée courante : absence de fissuration • Planches comparatives - sans géotextile et sans préfissuration : quelques fissures. - avec géotextile et sans préfissuration : quelques fissures - sans géotextile et avec préfisuration : absence de fissures - avec géotextile avec préfissuration : absence de fissures
RD-621 Tarn	Juin 1994	• 3 cm Béton Bitumineux • 15 cm Grave Ciment préfissurée	Classe T2 150 à 300 PL/jour/sens	• Chaussée courante : apparition des fissures en surface. • Section témoin non préfissurée (figure 4) : fissuration également au pas de 3 à 5 m ; mais fissures très erratiques
RN-57 Vosges Déviation EPINAL	Octobre 1993	• 4 cm Béton Bitumineux • 6 cm Béton Bitumineux • Géotextile imprégné • 25 Grave Ciment préfissurée • 20 Sable Ciment D	Classe T0 1 600 PL/jour/sens	• Chaussée courante : absence de fissures. • Planches comparatives. - préfissuration et géotextile, - préfisssuration seule, - géotextile seul, - préfissuration + géotextile, - préfissuration + géotextile avec couche de roulement de 4 cm : (pas de couche de liaison), A ce jour aucune fissuration.

TABLEAU 2 : Caractéristiques et comportement de sites faisant l'objet d'un suivi.

Figure 3 : Sondage réalisé en bord d'une chaussée préfissurée : on distingue le bon positionnement du film.

De ce tableau, on peut retenir :

- Déviation de BEEGAR : les seules planches présentant une fissuration sont celles non préfissurées
- RD 621 : en l'absence de préfissuration, les fissures réapparues sont erratiques. (figure 4); de ce fait elles risquent de provoquer des épaufrures et seront difficiles à entretenir.
- RN 57 : absence de fissures sous seulement 4 cm d'enrobé.

Ces premières constatations sont représentatives de l'effet recherché avec la préfissuration. Le suivi est poursuivi pour évaluer le comportement à long terme.

La RN 57 (Vosges) fera l'objet d'un suivi mécanique très détaillé dans le cadre de la Charte Innovation signée avec la Direction des Routes (Ministère des transports). Ce suivi a pour objectif l'évaluation du comportement mécanique spécifique lié à la préfissuration en vue d'une éventuelle prise en compte dans le dimensionnement des structures.

Il comportera :

- la mesure des températures dans la structure à différents niveaux pour évaluation de la cambrure de l'assise à prendre en compte dans les calculs
- la mesure de battements à chaque préfissure à l'aide d'un inclinomètre
- la définition, à l'aide de ces mesures, de familles de fissures en fonction de leur mode réponse
- à l'aide de ces définitions, le choix de préfissures représentatives pour auscultation fine, spécialement du mode d'articulation, à l'aide d'un inclinomètre après enlèvement de la couche de roulement
- une fois ces données enregistrées, un travail de modélisations à l'aide du modèle aux éléments finis à trois dimensions CESAR 3 D du Laboratoire Central des Ponts et Chaussées.

Figure 4 : RD 621 (TARN) : section témoin non préfissurée : les fissures remontées, de pas de 3 à 5 m, sont très erratiques : elles risquent de générer des épaufrures et seront difficiles à entretenir.

7 Conclusion

La technique de préfissuraton par insertion d'un film de plastique a démontré sa faisabilité et la possibilité de la mettre en oeuvre à grand rendement sans perturber l'avancement du chantier.
Les premières constatations faites sur des chaussées préfissurées avec celle-ci montre son efficacité. Ce suivi doit être poursuivi pour confirmation du comportement à long terme.
Un chantier particulier (RN 57 VOSGES) fait l'objet d'un suivi mécanique détaillé afin de permettre la prise en compte de l'apport éventuel de la préfissuration dans le comportement des chaussées traitées

Bibliographie

1. "Travaux d'extension du port de la GOULETTE à TUNIS" P. DELIGNE Revue Générale des Routes et Aérodromes n° 628 - Mars 1990 - Mars 1986
2. "Techniques antifissures intégrées au renforcement de la RN-190" M. LEFORT, J.P. MARCHAND, M. LEDUFF - Revue Générale des Routes et Aérodromes n°679 - Mars 1990
3. "Les procédés utilisés pour maîtriser la remontée des fissures : bilan actuel" C. COLOMBIER - Second International RILEM - CONFERENCE LIEGE - mars 1993.

Performance of cement bound bases with controlled cracking

M.A. SHAHID and N.H. THOM
Civil Engineering Department, University of Nottingham, Nottingham, United Kingdom

Abstract
This paper presents description and data on the performance of a system of controlled cracking in cement bound bases. This technique is intended to prevent the occurrence of occasional but relatively wide and damaging natural cracks which can easily propagate through bituminous surfacing due to relative vertical movement of the crack edges under trafficking, therefore necessitating thick bituminous surfacing.

Two full scale field trials have been carried out in the UK, in 1994 and 1995, incorporating two different cement bound materials (CBMs). The first trial involved a crushed limestone aggregate whereas a flint gravel sand was used on the second one and the response of each has been studied.

The effectiveness of controlled cracking in cement bound bases has been evaluated through coring and falling weight deflectometer (FWD) surveys. The results of coring surveys have given evidence of the occurrence of frequent cracking with almost negligible width. The FWD results have revealed significantly higher in-situ stiffness and load transfer efficiency at controlled cracks compared to natural cracking, found on adjoining pavement sections.
Keywords: Cement bound bases, controlled cracking, coring survey, cracking pattern, FWD survey, in-situ stiffness, load transfer, site trials.

1 Introduction

Cement bound bases have much higher stiffness and strength than conventional bituminous bases. They perform like a relatively rigid slab and substantially reduce the tensile strain at the bottom of bituminous surfacing.

Reflective Cracking in Pavements. Edited by L. Francken, E. Beuving and A.A. Molenaar. © 1996 RILEM. Published by E & FN Spon, 2–6 Boundary Row, London, SE1 8HN. ISBN 0 419 22260 X.

The perceived problems with cement bound base roads have generally stemmed from the tendency for discrete cracks within the base to propagate through bituminous surfacing layers, giving rise to maintenance concerns. However, this does not happen in every case and in many instances the cracking, though present, does not lead on to major maintenance requirements. In fact, the performance of a cement bound base is often hardly affected by primary cracking, which is the occurrence of transverse cracks due to thermal contraction and shrinkage effects [1]. Traffic causes further deterioration of primary cracks through shear movement of the crack edges and, in some cases, it also induces longitudinal wheel-path cracks. This phenomenon, termed as secondary cracking, is the chief cause of a significant reduction in the performance of a cement bound base and hence, eventually, to pavement distress.

With controlled cracking, a cement bound base can provide a high quality support for bituminous surfacing without incurring the penalty of early life cracking and associated maintenance concerns. Control over the variation of crack size makes it possible to limit the tensile stresses generated within the bituminous layer.

2 Aims of controlled cracking

- To prevent occasional but wide natural cracking.
- To induce frequent cracking due to shrinkage and thermal effects, so that the cracks thus formed will be of minimum width and have maximum load transfer capacity due to superior aggregate interlock between the crack faces.
- To minimize the thickness of bituminous overlay required to counter reflective cracking.
- To avoid the use of expensive modified bituminous mixes.

3 Description of the site trials

3.1 M4 Second Severn Crossing, Bristol
A trial section on the new M4 access route to the Second Severn Crossing consisted of a 150 mm thick CBM3 sub-base involving a crushed limestone with 40 mm maximum aggregate size. The trial section was 120 metres long on a 12 metre wide carriageway. Construction took place in December 1994 when the air temperature was 12 °C.

For comparison purposes, data was also collected from an adjoining section of M4 pavement and another similar pavement, without controlled cracking, on the new M49 access route, also an element of the Second Severn Crossing scheme.

3.2 A11 Wymondham bypass, Norwich
The A11 Wymondham bypass trial section comprised a 150 mm thick CBM3 sub-base involving a crushed flint gravel of 40 mm maximum aggregate size blended with sand. This trial section was 140 metres long on a 12 metre wide carriageway and was constructed in July 1995 when the air temperature was 22 °C.

All of the above mentioned pavement carriageways were subjected to construction traffic during the investigation.

4 Description of the process

A brief description of the controlled cracking process is as follows:

- In fresh CBM, laid by a slipform paver, create 10 mm wide transverse slots at 3 m regular intervals to approximately half the layer depth (Figure 1a). For this purpose, a simple device was used, consisting of a small vibrating plate with a vertical cutting blade attached to its bottom (Figure 1b).
- Introduce bitumen emulsion into the slots before compaction of the surface. The bitumen emulsion creates a weak zone in the CBM layer and helps the shrinkage cracks to be accurately located.
- Close the slots at the time of compaction by rolling, thereby restoring the integrity of the layer as the slot faces are pushed back together. Figure 2 shows the finished surface, on its right, before spraying of emulsion for curing.
- The entire process is completed within two hours time after mixing of the CBM.

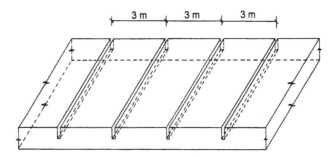

Fig. 1a. Principle of controlled cracking

Fig. 1b. Process of controlled cracking in a cement bound layer.

Fig. 2. A slotted cement bound layer before and after compaction.

5 Investigation of the site trials

5.1 Visual examination

After two months, slot marks were identifiable but in none of them could a crack be visually detected on the M4 Second Severn Crossing. However, coring at the slot locations revealed that 40% had developed cracks throughout the entire thickness of CBM layer. The crack width was less than 0.25 mm. In contrast, there were visible natural cracks on the M49, having a width of more than 1.0 mm, and a crack spacing in the range of 12 to 40 metres.

One month after construction, distinctly visible cracks were found at slot locations on the A11 Wymondham bypass. A coring survey showed that 90% of the slots had developed cracks through the entire layer thickness. The crack width was equal to or less than 0.5 mm. No naturally occurring cracks could be identified between the positions of the induced cracks on either trial section.

5.2 Laboratory testing

Cores were taken from the slot locations (Figure 3) and elsewhere. In addition to the visual examination and density measurement, indirect tensile strength testing was performed on the cores taken from uncracked slots and intact CBM. During this testing, the load was applied diametrically along the film of bitumen emulsion on the cores taken from the uncracked slots. Average results are given in Table 1 and show that the slotted zones are very significantly weaker than intact CBM even where no crack has formed. The difference between the strengths of the two materials is also illustrated.

Fig. 3. Core taken from a slot on M4 Second Severn Crossing trial section.

Table 1. Average splitting strength of cores taken from uncracked slots and intact CBM.

Site trial section	Cores from uncracked slots		Cores from intact CBM layer	
	Saturated density (kg/m^3)	Splitting strength (MPa)	Saturated density (kg/m^3)	Splitting strength (MPa)
M4	2419	0.88	2419	1.90
A11	2359	0.67	2359	1.35

5.3 Falling weight deflectometer survey

Falling weight deflectometer (FWD) testing was performed to investigate the load transfer efficiency of the induced cracks and in-situ stiffness of the CBM layer.

5.3.1 Load transfer investigation

For the load transfer investigation, deflections were measured on either side of each slot at a distance of 500 and 600 mm respectively from the centre of the loading plate (termed d_3 and d_4). The difference between d_3 and d_4 indicates the differential movement at the induced cracks. For comparison purposes, deflections d_3 and d_4 were also measured at the middle of CBM slabs, i.e. between the slots. The magnitudes of all deflections and pressures were normalized to a pressure of 620 kPa, corresponding to a wheel load of 43.8 kN. Figure 4 represents some of the results of differential deflection (d_3-d_4) obtained from the trial section of M4 Second Severn Crossing. Load transfer efficiency was also investigated for natural cracking on a similar limestone CBM pavement on the M49 (Figure 5). Crack spacing was also recorded.

From Figures 4 and 5, it is clear that the differential deflection across natural cracks is almost always greater than across artificially induced cracks. This is mainly attributable to the greater width of the natural cracking.

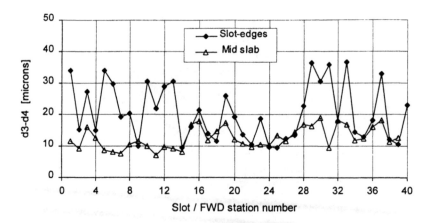

Fig. 4. Comparison between the magnitudes of differential deflection measured across the slots and those from normal CBM slabs between the slots.

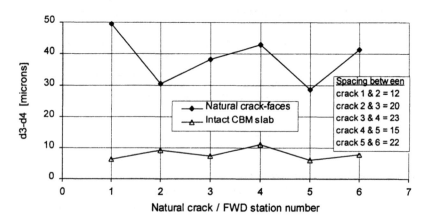

Fig. 5. Comparison of the magnitudes of differential deflection measured across the natural cracks with those from normal CBM slabs between the cracks.

5.3.2 Determination of in-situ stiffness

The in-situ stiffness of the CBM layer was calculated from the FWD data by means of a back-analysis computer programme [2]. Two sets of data were back-analyzed separately, i.e. for FWD pulses applied near the slots or cracks (at 550 mm) and in the middle of slabs, between the slots. Table 2 shows average magnitudes of stiffnesses obtained from the trial section of the M4, the adjacent section of the M4 and from the M49 Second Severn Crossing.

Table 2. Comparison of average in-situ stiffnesses.

Section	Stiffness of CBM layer (MPa)	
	Near cracks / slots	Between cracks / slots
Trial section M4	22888	23587
Adjacent section M4	-------	27531
M49	21865	38513

Table 2 shows a significant difference between the M4 and M49 results, probably due to variations in mix proportions, aggregate grading or CBM thickness. However, of greater note is the fact that the in-situ stiffness near natural cracks on the M49 is only 57% of that away from a crack, whereas near a controlled crack on M4 trial section the stiffness is 83% of that distant from a crack, on the adjacent M4 section. The stiffness between slots is in fact slightly greater, being still only 1.5 m from a controlled crack.

6 Interpretation of results

6.1 Cracking pattern
The coring survey revealed that 90% of the slots of the A11 Wymondham bypass trial section had developed cracks through the entire depth, in comparison with 40% at the M4 Second Severn Crossing. This significant difference in the cracking pattern is due to two major reasons. Firstly, tensile strength of flint gravel sand CBM is nearly half of that for limestone CBM (Table 3). Secondly, the flint gravel sand CBM is more sensitive to thermal variations than limestone CBM because of a considerable difference between the magnitudes of their coefficients of thermal expansion [3] as given below:

$\alpha = 13 \times 10^{-6}$ per °C   for flint gravel concrete
$\alpha = 6 \times 10^{-6}$ per °C   for limestone concrete

Since shrinkage is restrained in a CBM base, it develops tensile stresses σ_t under a temperature drop ΔT in accordance with the following relation [4]:

$$\sigma_t = E\alpha \, \Delta T \tag{1}$$

When these tensile shrinkage stresses approach the tensile strength f_t of a CBM base at any location, they crack the base. The temperature drop ΔT_f required to fracture a CBM base having a modulus of elasticity E is as follows:

$$\Delta T_f = \frac{f_t}{E\alpha} \tag{2}$$

Thus, the magnitude of ΔT_f required to fracture the flint gravel sand CBM is much smaller than that for the limestone CBM, leading to a greater tendency to crack.

Table 3. Comparison between the results of specimens collected from two sites.

Site trial section	Average saturated density (kg/m³)	Average cube compressive strength, f_c (MPa)		Direct tensile strength, f_t (MPa)
		at 7 days	at 28 days	at 28 days
M4	2513	24.2	31.8	2.03
A11	2415	19.3	25.3	1.12

Note: Direct tensile strength was determined from a new direct tensile test for CBMs [1].
        Each value in column 5 represents the average from testing of two specimens.

6.2 Slot spacing

In controlled cracking, a spacing of 3 metres was selected in order to limit shrinkage of the slabs after cracking, and consequently to minimize the crack width. Under a temperature change ΔT, the amount of change in length 'δl' of the slab is directly related to its crack spacing 'L' according to the following relation:

$$\delta l = L \alpha \Delta T \qquad\qquad (3)$$

In fact, the change in length δl is equal to the opening of the cracks and it is obvious that δl increases with spacing L. With 3 metre regular spacing, the width of induced cracks was very small and most of them were observed like hairline cracks.

7 French experience

In France, pre-cracking is included in the recommendations for the construction of underlays incorporating hydraulic binders [5]. The benefit of pre-cracking has been found to save 50 to 60 mm of bituminous overlay (Figure 6) which would otherwise be required to prevent or delay reflective cracking.

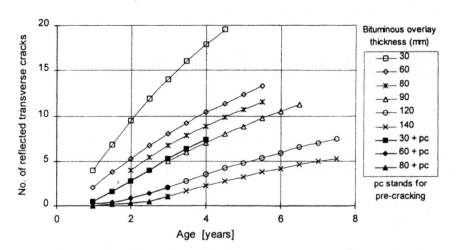

Fig. 6. Number of reflected transverse cracks per 100 m section, with and without pre-cracking, as a function of the thickness of bituminous overlay and age of the underlay [6].

From 1988 to 1992, pre-cracking has been carried out over more than 40 sections in France (nearly 100 km length) and one in Spain. In each case, slotting was performed throughout the entire depth of CBM base. In addition, eight reference sections were also constructed without pre-cracking. Several types and thicknesses of wearing courses were used on different sections. A systematic monitoring programme was applied to all sites in order to establish the results in terms of the number of transverse cracks reflecting to the surface [6], as shown in Figure 6.

8 Effectiveness of controlled cracking

The new controlled cracking system effectively induces frequent cracks of less than 0.5 mm width in a CBM base. FWD results show that the differential deflection of the crack faces is very limited which is due to a good aggregate interlock. In consequence, the induced cracks have superior load transfer characteristics to natural ones. Also, the back-analyzed stiffnesses obtained from FWD pulses applied near induced cracks do not show as great a reduction in stiffness as those near natural cracks.

In the French pre-cracking technique, slotting was performed throughout the entire depth of a CBM layer, which leads to a higher differential deflection at the induced cracks. An average of 50 microns has been reported by Colombier et al. [6], but no mention of the type of CBM and applied pressure has yet been found.

With this new controlled cracking system, an average differential deflection of 20 microns has been found in case of limestone CBM base with an applied pressure of 620 kPa. This value is only a little higher than that measured from the intact material. This benefit is attributed to the fact that slots are created only to half the CBM base thickness which help to:

1. Produce weak zones due to the presence of a film of bitumen emulsion but only in the upper half of the layer thickness.
2. Provide a better aggregate interlock through strong crack faces in the lower half thickness after the cracks are induced.

9 Conclusions

Based on two trial pavements incorporating pre-cracked CBM sub-base, assessed before being overlaid, the following conclusions have been drawn:

• The cracks induced by controlled cracking have superior load transfer characteristics to natural cracks.
• Controlled cracking limits the crack width to less than 0.5 mm, whereas the width of natural cracking is more than 1.0 mm, depending upon the aggregate type.
• The performance data suggests that the improvement is due to better aggregate interlock between the crack faces.
• The stiffness of a CBM base is only slightly affected by controlled cracking, so that it does not influence the structural performance of the CBM base.

- For a controlled cracking system, a spacing of 3 metres is adequate to minimize the crack width. It also enables site work to proceed at the normal speed.
- Controlled cracking brings a new concept to the design of flexible composite pavements, in which the bituminous surfacing has a much reduced role in terms of reflection crack control.

10 References

1. Shahid, M.A. and Thom, N.H. (1994) Pavements with cement bound bases, Report No. PGR 94023, University of Nottingham, United Kingdom.
2. Brown, S.F., Tam, W.S. and Brunton, J.M. (1987) Structural evaluation and overlay design: analysis and implementation, Proceedings of the 6th International Conference on the Structural Design of Asphalt Pavements, Ann Arbor, Vol. 1, pp 1013-28.
3. Bonnell, D.G.R. and Harper, F.C. (1950) Thermal expansion of concrete, Institution of Civil Engineers, 33 (4), pp 320-30.
4. Colombier, G., et al. (1988) Fissuration de retrait des chausées à assises traitées aux liants hydrauliques". Bull. Liaison Lab., des Ponts et Chausées, No. 156, pp 37-66, et No. 157, pp 59-87.
5. SETRA. (1990) Techniques pour limiter la remontée des fissures, Note d' Information du SETRA, n° 57.
6. Colombier, G. and Marchand J.P. (1993) The precracking of pavement underlays incorporating hydraulic binders, Proceedings of 2nd International RILEM Conference on Reflective Cracking in Pavements, state of the art and design recommendations, Liege university, Belgium, pp 273-81.

11 Acknowledgements

The authors would like to express their gratitude to the British In-situ Concrete Paving Association (BRITPAVE) for awarding a bursary to the first author and providing assistance during the last two years of the project. A particular debt of gratitude is due to Mr. David York of BRITPAVE for his assistance in organising the site trials.

Special thanks are also due to the following organizations for their support in agreeing to site trials and visits:

1. Balfour Beatty Civil Engineering Limited
2. Atkins / Maunsell Partnership
3. McAlpine Budge Limited
4. Norfolk County Council
5. The Department of Transport

Controlling shrinkage cracking from expansive clay sub-grades

W.S. ALEXANDER
Geofabrics Australasia Pty Ltd, Victoria, Australia
J. MAXWELL
Fisher Stewart Pty Ltd, Victoria, Australia

Abstract
Road pavements built on subgrades of expansive clay soils are effected by volume changes through seasonal wetting and drying cycles. These cracks propagate up through the entire roadbase and asphaltic concrete (A.C.) surfacing, resulting in surface cracks in the A.C. layer. This paper reviews current design and construction practice and discusses the various methods used, to prevent reflective cracking including the use of geogrids and geotextiles placed at subgrade level as well as the use of impermeable moisture barriers placed at the edge of pavement.
Keywords: Reflective cracking, expansive clay subgrade, geogrid, geotextile, impermeable moisture barrier.

1 Introduction

Road construction on the basaltic clay plains of Melbourne's western plains pose special problems to the road engineer. The Western plains of Melbourne are of volcanic origin and exhibit basaltic rock overlaid with residual basaltic silts and clays.

These clays are highly reactive to moisture which result in clays showing significant volume change as a direct result of moisture content variation.

The traditional method of road construction is to remove in situ clays to a depth of 600 mm or greater below road pavement finished surface level and replace insitu material with a stable material which exhibits low Plasticity Index (P.I.) characteristics such as a crushed sandstone. This method of construction has worked well, however, is not cost effective.

Reflective Cracking in Pavements. Edited by L. Francken, E. Beuving and A.A.A. Molenaar. © 1996 RILEM. Published by E & FN Spon, 2–6 Boundary Row, London, SE1 8HN. ISBN 0 419 22260 X.

Another method of road construction is to stabilise the subgrade with the addition of lime which has the primary function of lowering the P.I. of the insitu clays. This method is cost effective, however, it has limited application due to the frequent presence of basaltic rock at the subgrade level which causes considerable damage to the rotorvator plant which mixes lime with subgrade material.

Lime stabilisation of the subgrade has been moderately successful, however, some notable failures to prevent reflective cracking of the A.C. have occurred. This paper examines a case study of one such failure.

2 Test Section

A new residential subdivision, located at New Gisbourne some 50 km north west of Melbourne, has been progressively constructed in four stages over the past two years.

In stage Three of the development, severe longitudinal cracking of the road pavement was observed. The subgrade of the failed pavement had been lime stabilised and construction of the pavement had been carried our during the wet period (winter/spring) of the year. Some six months after the pavement was constructed following a summer period, extensive reflective cracking of the A.C. layer occurred.

The geotechnical investigation revealed a very high moisture content in the subgrade material, however, the insitu clays outside the pavement limits exhibited significantly lower moisture contents. This moisture differential had caused settlement and rotation of the kerb and longitudinal cracking of the road pavement generally within 1 m of the kerb. See Fig. 1.

Fig. 1. Longitudinal cracking of road pavement

In order to eliminate longitudinal cracking of the road pavement, it was apparent that minimal variations in the moisture content of the subgrade material had to be achieved in order to avoid reflective cracking of the A.C. layer.

A number of design options were considered including the following:

- vertical moisture barriers
- subsoil drains
- geosynthetic treatments

They will be discussed in following subsections.

2.1 Vertical moisture barriers

Any subsurface drainage system must a) intercept and dispose of ground water entering from outside of the pavement and also b) dispose of water infiltrating through the pavement surface into the base. Traditional "french" drain designs achieve the first objective but to the extent that it disposes of moisture from the basaltic clay subgrade on both sides of the subsurface drainage line. This results in drying out of the subgrade under the kerb and channel and beyond this into the outer lane of the roadway. Eventually longitudinal cracking of the road surface occurs. To prevent this cracking an impermeable membrane is used to prevent changes in moisture content of the clay subgrade beneath the road pavement.

VicRoads (Ref 1.) reports that the role of a vertical moisture barrier is to stop the seasonal lateral migration of moisture to and from the subgrade beneath the pavement. To effectively stop this moisture migration, the moisture barrier must extend below the depth of cracking. The barrier must also prevent the invasion of plant roots. Thus it preserves a stable suction profile once it has been reached beneath the pavement. This stable suction will be the equilibrium suction that exists in the deeper foundation soils.

The depth of the moisture barriers is based on the zone of seasonal influence, which depends on the soil type and climate. At the site in question, a depth of 900 mm was deemed adequate.

The moisture conditions in the subgrade of the proposed pavement can be critical to the success of this treatment. To reduce the future loss of shape and prevent possible cracking it is important to ensure that the soil suction in the subgrade is at, or near, the equilibrium soil suction. (Ref.4).

At this site 0.2 mm U.P.V.C. sheet was incorporated into the edge of pavement sub-soil drain detail to act as a vertical moisture barrier. See Fig. 2.

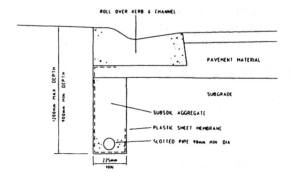

Fig. 2. Combined subsoil drain and vertical moisture barrier

2.2 Subsoil drains

The subsoil detail adopted is shown in Fig. 2. This combined drainage system and vertical moisture barrier was chosen on the basis of effectiveness to form a barrier between the subgrade material and the material external to the pavement where significant variations in moisture content will be encountered. Ease and cost effectiveness of construction were also a consideration. Any moisture that enters the pavement system through the A.C. layer will permeate into the subsoil drainage system as rapidly as possible and then discharge in the underground drainage system. This rapid transfer of moisture out of the pavement system will enable equilibrium of the subgrade moisture content to be maintained. The depth of the trench was chosen to be 1.2 m, below the depth of seasonal influence in this area.

2.3 Geosynthetic treatments

The pavement in this case study was constructed during the wet season when the subgrade material exhibited a very high moisture content and low CBR's. It was not possible to abandon construction works until the subgrade reached its equilibrium moisture content so a geosynthetic treatment was adopted at subgrade level to serve two functions:

2.3.1. Working platform

To provide a platform on which to construct the pavement as the actual CBR values of the subgrade material were less than the design CBR of 3.

2.3.2 Prevention of reflective cracking

Geosynthetic treatments were trialled to test their effectiveness in preventing reflective cracking of the A.C. layer if movement of the subgrade material occurs.

2.3.3 Choice of geosynthetic material

The geotextile used was a non-woven, needle-punched continuous filament polyester of mass 140 g/m² known as Bidim A14. Bidim A14 has a wide strip tensile strength of 8.5 kN/m and CBR burst strength of 1850 N. Roll size is 4 m x 200 m.

The geogrid chosen was Tensar SS1. This material is a polypropylene, biaxial geogrid with a mass of 0.2 kg/m² and roll size of 4 x 50 m. The Q.C. (Quality Control) strength of SS1 is 20.5 kN/m in the transverse direction and 12.5 kN/m in the longitudinal direction. The selection of appropriate geosynthetic material was based on the CBR values of the material encountered which varied over the site.

Soil tests carried out in Stages 2 & 3 of the development indicated in-situ CBR values of 3.0 to 3.5. Whilst CBR values of the subgrade were not tested for Stage 4, the CBR was estimated by visual inspection to be less than 2.0.

Where basaltic rock was encountered at subgrade level, geosynthetic material was not used as rock is a stable material and is not effected by moisture changes. Geosynthetic materials were placed over clay subgrade areas only, which are influenced by moisture changes.

The various geosynthetic treatments are summarised in Fig. 3. below.

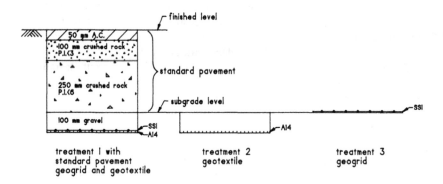

Fig. 3. Trial Pavement Cross-Sections

The area which exhibited the lowest CBR value and highest moisture content was excavated at 100 mm below subgrade and geotextile was laid first, followed by geogrid. A 100 mm layer of locally occurring gravel was then placed to reinstate to subgrade level, see treatment 1. In other areas, 100 mm of subgrade material was excavated and a geotextile only was placed and reinstated to subgrade level, see treatment 2. A third area was treated using geogrid only which was placed at subgrade level without any additional excavation, see treatment 3. Control areas were also constructed, without the use of any geosynthetic material, so that long term performance of the various treatments could be compared and evaluated. See Fig. 4.

Fig. 4. Placing roadbase over geotextile
Note: Moisture barrier at edge of pavement

3 Road Pavement Construction

Upon completion of the subsoil drain/vertical moisture barrier and subgrade treatment works, the pavement was then constructed which consisted of 250 mm depth crushed rock, with a maximum P.I. of 6 being the subbase pavement layer followed by a 100 mm depth crushed rock layer of maximum P.I. of 3 followed by a 50 mm depth A.C. layer. This pavement design is standard for residential roads in this locality and is unchanged from the previous stages which exhibited severe cracking.

4 Performance

At the time of writing this report, April 1996, some 4 months since the pavement construction was completed, there are no signs of kerb rotation and longitudinal cracking of the pavement. This compares favourably to performance of the previous stages of road construction, which failed shortly after construction.

5 Discussion of results

The fact of no cracking to date is a very significant result since this has been observed at the time of the greatest change in moisture content of the in situ soils, over the summer period.

The fact that no cracking has been observed in any section of new roadway including the control sections, may indicate that the vertical moisture barrier, which is present throughout the new constructions may be the major influence on improved performance.

Further monitoring will be carried out and reported at the time of presentation of this paper, October 1996.

6. References

1. VicRoads, Technical Note TN 013, October 1995
2. Evans, R.P. & Holden, J.C. (1994) Use of Geomembranes to Control Pavement Movements from Expansive Soils.   Proc. Ground Modification Seminar No. 3. Geosynthetics in Road Engineering, University of Technology, Sydney. p. 157.
3. Picornell, M and Lytton, R.L. (1988) Behaviour and Design of Vertical Moisture Barriers.   Transportation Research Record 1137, TRB, Washington DC, p. 71.
4. Marmaras, F. Troung, H.V.P. Richards, B.G. and Holden, J.C. (1994) Role of Soil Suction in Cracking of West Gate Freeway Pavements. Proc 17th ARRB Conf; Gold Coast, Vol 2, P33.

Performance of precracked cement treated layers in Spain

C. JOFRE and J. VAQUERO
Instituto Espanol del Cemento y sus Aplicaciones, Madrid, Spain
C. KRAEMER
Departamento de Transportes, Universidad Politécnica, Madrid, Spain

Abstract

Precracking techniques have been used in Spain almost exclusively on roller compacted concrete (RCC) layers. On principal roads, this material is employed in Spain since 1984. By the end of 1995, more than 250 km of two-lane carriageways had been constructed with RCC.

Precracking has also been applied on a dry lean concrete base, as well as on three test sections, two in cement - recycled pavements and the other having a cement - treated base. In spite of the reduced length of these works, compared with those with RCC layers, interesting conclusions have been drawn from them.

In this paper a review of the performance of several sections where precracking has been used is presented. It has been noticed that joint spacing is one of the most important factors. Excellent results have been obtained where joints have been wet-formed every 2.5 to 4 m.

Keywords: Cement treated base, lean concrete, precracking, roller compacted concrete, recycling, Spain

1 Introduction

Precracking techniques have been used in Spain almost exclusively on roller compacted concrete (RCC) layers. On principal roads, this material is employed in Spain since 1984. By the end of 1995, more than 250 km of two-lane carriageways had been constructed with RCC. A significant portion is subjected to the combined effect of a heavy traffic and large thermal gradients.

Precracking has also been used on a dry lean concrete base, as well as on three test sections, two in cement - recycled pavements and the other having a cement -treated base. In spite of the reduced length of these works, compared with those with RCC layers, interesting conclusions have been drawn from them.

Reflective Cracking in Pavements. Edited by L. Francken, E. Beuving and A.A.A. Molenaar. © 1996 RILEM. Published by E & FN Spon, 2–6 Boundary Row, London, SE1 8HN. ISBN 0 419 22260 X.

2 Precracking experiences in RCC pavements

2.1 Introduction

As mentioned above, first applications of roller compacted concrete (RCC) on the Spanish network of main highways and motorways took place in 1984 [1]. Since then, more than 2 000 000 m² of RCC pavements have been constructed as a result of a great number of different types of works: new roads, overlays, duplication and widening of already existing highways, most of them bearing heavy or medium traffic. In 1988, RCC was used for the first time on a motorway section. The main characteristics of some selected projects are given in Table 1.

2.2 Design features

The Spanish Standard for Pavement Sections (Instrucción 6.1 y 2-IC), issued in 1989 [2], provides solutions for all types of new pavements (rigid, semi-rigid and flexible) and all kinds of traffic. Pavement sections with RCC are included in this Standard as one of the seven options admitted for each of the traffic categories considered.

Thickness of RCC varies between 25 cm, for traffic over 2 000 commercial vehicles/day on the design lane during the first year of operation, and 20 cm, for traffic under 800 commercial vehicles/day. For new roads, a soil-cement subbase, 20 or 15 cm thick, is prescribed to provide a support layer rigid enough to obtain the right density on RCC through the compacting process. To correct surface irregularities produced by rolling, an asphalt concrete (AC) layer, 10 cm thick, is specified for traffic over 800 commercial vehicles/day. This thickness can be reduced to 8 cm if measures are adopted to avoid reflection of joints. In some works, however, AC layers totalling a thickness up to 12 or 15 cm have been used. This crack reflection was a potential problem to be solved.

The Standard makes compulsory to place transverse joints in RCC, spaced not more than 7 m, skewed at 1:6 with respect to the carriageway longitudinal axis.

2.3 Construction

2.3.1 Materials

RCC mixes used in Spain have, in general, a continuous grading with a maximum size of 20 mm, and a high content of blended cement (290 to 330 kg/m³), with an important proportion of fly ash (between 36 and 50 per cent of the total weight of binder). An average splitting strength of 3.3 MPa is required at 90 days.

2.3.2 Mixing and laying

RCC has normally been mixed in continuous plants. A great variety of equipment (motor graders, autogrades, asphalt finishers and slip-form pavers with some changes) has been used for spreading, usually across the full width of the carriageway.

In most cases, compaction has been carried out using a heavy vibrating roller, combined with a rubber-tyred roller.

Table 1. Characteristics of selected RCC projects (motorways)

Section	Length of two-lane carriageway (km)	Opening to traffic	Average daily commercial traffic in each direction (1994)	Temperatures (°C) July T_M/T_m	Temperatures (°C) December T_M/T_m	RCC thickness (cm)	Contraction joints Spacing (m) & Construction	Contraction joints Protection	Bituminous layers over RCC (cm)
1	22.5	1988	2 555	34/17	11/2	22-24	15-S 10-S	geogrid strip between asphalt layers (0.85 and 1.7 m width)	• 8 to 12 AC (in 2 layers) • 8 BL + 4 MAC (5% SBS) • 8 BL + 2 GGAM with fibers
2	60 *	1988	6 096	31/17	9/2	25	15-S	geogrid strip between asphalt layers (0.85 m width)	12 to 15 AC (in 2 layers)
3	27	1989	700	27/13	7/0	23	7-S	• sealed (bitumen/rubber product) • SAMI (2.5 kg/m²) • sand asphalt layer, 2 cm thick	• 5 to 8 AC • 5 OGAC (5,6% binder) • 5 cold MAC
4	4.5 *	1990	3 223	32/17	10/2	25	• 6-N • 3.5-N • 2.5-N	• none • sealed (bitumen/rubber product) • asphalt roofing strip • sand asphalt layer, 2 cm thick	12 AC (in 2 layers)
5	25	1990	n.a.	34/17	12/2	22	7-N	none	8 AC (in 2 layers)
6	96	1989-1991	n.a.	34/17	12/2	22	7-S	asphalt roofing strip, 10 cm wide	8 AC (in 2 layers)
7	34	1992-1996	n.a.	34/21	12/6	22	4-Nb	none	8 AC (in 2 layers)
8	9.2	1992	800	26/13	9/3	22	2.5-C	• none • sand asphalt layer, 2 cm thick	4 DAC + 6 to 8 BL

1. N-IV Valdepeñas - Almuradiel
2. N-IV Villaverde - Seseña
3. CL-803 Sanchidrián - San Pedro del Arroyo
4. N-II Alcalá - Meco
5. A-92 Archidona bypass
6. A-92 Salinas - Granada
7. N-321 Jaén - Torredonjimeno
8. A-15 Irurzun bypass

S   Saw cut
N   Notching of fresh concrete
Nb   Notching + bituminous emulsion
C   CRAFT equipment
n.a.   no data available
*   One lane plus shoulder

AC   Asphalt concrete
MAC   Modified AC
GGAM   Gap graded asphalt mix
OGAC   Open graded AC
BL   Binder layer
DAC   Draining AC

T_M Average maximum daily temperature
T_m Average minimum daily temperature

2.3.3 Joint construction

Up to 1987, most of RCC pavements constructed in Spain carried relatively low traffic volumes (less than 500 commercial vehicles per day in each direction). Joints were unusual, and in most cases RCC cracks soon reflected through the thin surfacings of these works (4 to 8 cm or bituminous dressings). Conversely, since that year, where RCC began to be extensively used in motorways, transverse contraction joints are customary, in order to control the cracking process due to shrinkage and thermal stresses. They have been made with different techniques and spacing. Distances between joints have been reduced from 15 m to 2,5 - 4 m in the last works. This slab shortening aims at decreasing stresses and movements at joints, reflective cracking being therefore minimized or even eliminated. Up to 1989 joints were sawed but in that year wet-formed joints began to be used. This technique provides important savings in construction, allowing shorter spacings between joints. It also makes the operation independant of weather conditions which, in case of works done under high temperatures, has resulted in uncontrolled cracks by late cutting.

Wet-formed joints have normally been created by means of vibrating plates with a blade welded to the bottom. In one work the CRAFT equipment has been used. As it is known, it cuts the whole thickness of the uncompacted material, simultaneously spraying a jet of bituminous emulsion between the two blades forming the cutting tool. The emulsion prevents the sides of the joint from bonding again.

In some works, a longitudinal joint has also been provided.

2.3.4 Methods to minimise reflection cracking

It has been already mentioned that joint spacing has been progressively reduced in order to minimize reflection cracking. On joints more than 4 m apart, additional measures have been frequently employed, as for instance:

1. *Discontinous treatments applied over joints:*
- *geogrid strips,* 0.85 to 1.70 m wide, placed between the binder layer and the asphalt concrete wearing course;
- *capping of joints* with rubberized bitumen;
- *asphalt roofing strips,* 5 to 10 cm wide, heated at the time of placing to bond them to RCC.
2. *Continuous treatments:*
- *stress absorbing membrane interlayers (SAMIs),* consisting of a thick film (2.5 kg/m^2) of SBS modified bitumen covered with chippings;
- *sand asphalt layers,* 2 cm thick, with a high percentage of SBS modified bitumen (10%).

2.3.5 Asphalt concrete layers

On high-speed facilities, conventional AC layers are used on top of RCC in order to achieve a suitable surface evenness. In some cases, special mixes have been tested to minimise cracking reflection, e.g.:
- *AC with SBS modified bitumen* (aproximately 5% SBS);
- *gap graded mixes* reinforced with fibers allowing a higher bitumen content.

2.4 Performance of precracking in RCC pavements

When comparing the performance of precracking in the RCC sections built so far, it is difficult to draw a definite conclusion, since there are many differences between them : traffic, cross section, joint spacing, supplementary protective measures, etc. (Table 1). However, it has been noticed that joint spacing is a very influential factor.

In sections 1 and 2, where contraction joints were sawn every 15 or 10 m, cracks reflected through the bituminous layers very quickly, within two years of use. The fact that they are subjected to heavy traffic and that wide fluctuations of temperature occur between night and day, and from winter to summer should be pointed out. Some other factors, as the use of siliceous aggregates or, in the case of section 2, the laying of RCC during a very hot period, have also contributed to increase movements at joints. At present all of them have reflected on the road surface and there is also a considerable number of intermediate cracks. Rocking of slabs under the passage of trucks is also noticed. Moreover, geogrid strips placed over the joints have impaired bonding between the asphalt layers, this resulting in a high ramification and severe deterioration of reflected cracks at the outer lanes.

Conversely, performance of the portions of section 4 where joints are spaced 2.5 and 3.5 m without any supplementary protective measure, is excellent (Table 2). After six winters only 5.3% of joints have reflected. It can be noticed in Table 2 that in this work, four sections with a spacing of joints equal to 6 m were also built. Their performance will be examined later.

In this regard, it is interesting to compare how sections 2 and 4 have performed. Both are widening works near Madrid, with similar design features (Table 3), subjected to a very heavy traffic and relatively close (about 40 km from center to center), that is to say, also with similar climatic conditions. The main difference between them is joint spacing: 15 m in section 2, and 2.5 to 6 m in section 4. After three winters in service, all the joints had reflected in section 2; whereas in section 4, the proportion was only 4% after the same period.

Short slabs have also performed very well in sections 7 and 8, where joints are spaced 4 and 2.5 m respectively. No transverse reflection cracks have been noticed after four years under traffic. With regard to section 7, it has been constructed in three phases, one of them in 1992 and the two others in 1995. In the first phase, no longitudinal joint was provided, and longitudinal cracks have appeared in some parts. To avoid this problem, a longitudinal joint has been sawn in the remainder of the work.

Where joints are 6 to 7 m apart some general remarks can be made:
- heavy traffic does not rock the slabs;
- the amount of reflection cracks depends largely on the thickness of the AC layers;
- additional protective measures adopted to inhibit reflection cracking can also influence the performance, but mixed results have been obtained. For instance, it should be pointed out that in section 4, four stretches with a spacing of joints equal to 6 m were also built (Table 2), as mentioned above. At each of them a different additional measure to prevent reflective cracking was also adopted: capping of joints; asphalt roofing strips on the joints, with different weight (5.2 and 2.2 kg/m^2, respectively); and a continuous sand asphalt layer, 2 cm thick. The stretches with capped joints or asphalt roofing strips have performed very well

(about 5% of reflected joints), whereas, on the contrary, in the one with the sand asphalt layer this percentage is much greater (over 20%).

In section 3, with an AC thickness of 5 cm, all the joints, placed every 7 m, reflected after two winters irrespective of the special techniques adopted to mimimise this phenomena. In a test stretch where the AC was increased to 8 cm, crack occurrence was delayed by two years, but in four winters all the joints had also reflected. The use of siliceous aggregates and the construction in summertime have probably contributed to this generalized cracking.

Table 2. Performance of section 4 (N-II Alcalá - Meco)

Stretch	Reflection control system	Total of joints	Reflected joints					
			April 1992		March 1994		October 1995	
			Total	% (*)	Total	% *	Total	%
1	Joints every 6 m Sealing	152	2	1.3	2	1.3	7	4.6
2	Joints every 2.5 m No protection	360	7	1.9	9	2.5	18	5.0
3	Joints every 3.5 m No protection	257	8	-	8	3.1	7	7.0
4	Joints every 6 m Roofing strips 5.2 kg/m²	100	-	-	4	4.0	7	7.0
5	Joints every 6 m Roofing strips 2.2 kg/m²	50	2	4.0	2	4.0	3	6.0
6	Joints every 6 m Sand asphalt layer	150	15	10	19	12.7	31	20.7
	Total	1 069	34	3.2	44	4.1	80	7.5
	Total without section 6	919	19	2.1	25	2.7	49	5.3

* Percentage of cracks relative to the total number of joints in the stretch

- on the whole, reflected cracks show no ramifications and do not deteriorate under the passage of traffic. Most frequently, their spacing varies from 12 to 20 m. The absence of distresses suggests a correct load transfer. This satisfactory performance can be explained by the fact that, in the beginning, only one joint every two or three opens. This has been observed, for instance, on cores extracted from some jointed RCC layers shortly after spreading. During a certain period, these "working" joints experience relatively high movements, giving rise to reflective cracking. Finally, all joints open, their movements being reduced to an amount that the wearing course can afford with no damage.

Table 3. Compared performance of sections 2 and 4

Characteristic		Section	
		2 (N IV Villaverde - Seseña)	4 (N II Alcalá - Meco)
Type of work		Widening of existing road	Widening of existing road
Average daily commercial traffic (in each direction)	1990	5 029	2 393
	1994	6 096	3 223
Asphalt concrete (cm)		15	12
Roller compacted concrete (cm)		25	25
Soil-cement subbase (cm)		20	20
Joint spacing (m)		15	2.5 - 3.5 - 6
Geogrid strips over joints		Yes	No
Reflected cracks after 2nd winter	Total number	2 700	34
	%	67	3
Reflected cracks after 3rd winter	Total number	4 600	44
	%	115 *	4

* All the joints reflected, plus intermediate cracks

3 Precracking experiences in compacted lean concrete sections

Compacted lean concrete, a material with a cement content (7 - 8%) intermediate between those of roller compacted concrete and cement - treated bases, has been used up to now only in one site in Spain: the Medinaceli bypass, about 9 km long, in the Madrid - Zaragoza expressway [3].

This work is placed on a region with very soft subgrades (layers of clayey muds, up to 40 m thick). To avoid settlements, different treatments of the fills were used (preloading or gravel columns) and a pavement was selected, rigid enough to reduce the stresses on the subgrade, and at the same time able to accommodate to possible differential settlements. After considering several alternatives, the following solution was chosen:

- 15 cm of soil - cement subbase;
- 25 cm of compacted lean concrete base;
- 10 cm of asphalt concrete wearing course.

The lean concrete layer was built with transverse joints 3.5 m apart, skewed at 1:6. They were created in the fresh concrete by a vibrating plate with a keel, and sealed by a injection nozzle pouring bituminous emulsion into the wet - formed groove to ensure joint opening. A longitudinal joint at 4.5 m was also created by means of a similar blade attached to the finisher used to spread the material.

It should be pointed out that, due to opening constraints, it was necessary to lay the compacted concrete and the bituminous layers during the last months of 1991, a period where low temperatures, as well as rains, were often recorded. Therefore, the retention of the bituminous film in the asphalt concrete was probably impaired in some stretches. In the faulty parts, an important reflective cracking has occured, both transversally and longitudinally. Groups of reflected craks 3.5 m apart are frequent, this showing that all the joints have opened; but, due to the stiffness of the AC wearing course, this has been unable to accommodate even to the reduced movements of these close joints. On the contrary, and excepting some distresses due to the ocasional soft spots in the subgrade, in those parts where AC was correctly constructed, joints have reflected with a spacing between 10 and 20 m; that is to say, similarly to what has been observed in some RCC stretches. In the same way, these reflected cracks are performing correctly, showing no distresses.

4 Precracking experiences in cement - treated bases

As mentioned above, up to now precracking of cement - treated bases in Spain is limited to a short test section in one expressway in the North of Spain.

An average daily traffic truck over 2 800 per sense in the opening year was adopted for design. Thus, the following pavement was chosen:
- 20 cm of cement - treated granular material;
- 23 cm of cement - treated base;
- 9 cm of hot - mixed AC, G-25 type, in binder layer;
- 6 cm of hot - mixed AC, S-20 type, in wearing course.

Due to the AC thickness on top of the cement - treated base, this has been allowed to crack naturally, except in a short test stretch, 300 m long, laid in October 1995, where joints were created every 3 m in the fresh material using the CRAFT equipment already described. Tests performed on cores to evaluate the weakening at the joints have shown a decrease about 50% in the splitting strength values (6 MPa at 28 days in jointed cores, compared with 12 MPa in cores without joint).

5 Precracking experiences in cement - recycled pavements

Cement recycling of existing pavements has been only of limited use in Spain up to now (just three works), but the accumulated experience can be considered important, regarding both the length (over 50 km) and the traffic of the involved roads (two of them in the National network managed by the Ministry of Public Works).

The first application, 13 km long, occured between December 1991 and March 1992, in a section of the National Highway N-431 connecting the cities of Huelva and Cartaya [4]. It is placed at the SW corner of Spain, very close to the Atlantic coast, in a region with a mild climate. The existing pavement was milled in a depth of 30 cm and mixed in situ with 4.5% of cement by means of a high - performance equipment. The recycled layer was then covered with an AC wearing course, 5 cm thick. Cement content was adjusted in order to obtain a compressive strength of 3.8 MPa at 90 days, that is to say, that required for a soil - cement subbase when using a cement with a high percentage of active additions. Such type of low - strength material cracks by itself at short distances. However, due to the heterogeneity of the existing pavement, in situ strengths have frequently been much higher (up to 11 MPa, measured on cores). Therefore, it is no strange that, in spite of the favourable climatic conditions, in December 1994, that is to say, two years and a half after opening to traffic the recycled pavement, some very fine cracks appeared, spaced between 12 and 14 m. In June 1995, this distance had reduced to 6 - 7 m, whereas crack width had increased to 0.4 - 0.6 mm.

This fact gives a clear indication that, in recycled pavements, due to their frequent lack of homogeneity, zones with a relatively high strength, therefore prone to reflective cracking, are not unusual. In these cases, it seems very convenient to dispose wet-formed joints at short distances.

With this intention, some precracking experiences were performed in a cement - recycled section, 30 km long, of the National Highway N-630. The recycling took place between March and July 1995. The existing pavement was scarified in a depth of 25 or 30 cm, depending on the different stretches, and mixed with a 5% of cement. The work was executed by half widths. After compaction had been carried out, a bituminous curing seal was sprayed, then covered with chippings. Once a sufficient length had been completed, the recycling plant moved to the other half-width and traffic was allowed to pass on the treated material after 4 hours since milling of the last stretch. An AC wearing course, 10 cm thick, was spread some months later.

In two short test stretches of this work, each 100 m long, transverse joints were created 2.5 or 3.5 m apart, using a vibrating plate provided with a welded knife. In one of these stretches, a vibratory roller was passed at 24 hours on the joints. After rolling, joints were clearly marked on the surface. On cores extracted two months later, it was noticed that joints had opened along the whole thickness of the recycled layer. However, it should be mentioned that all the joints of the stretch where rolling has not been applied on the hardened material were also fully formed. This can be attributed to the passage of traffic (about 600 commercial vehicles per day in each direction) directly on the pavement a few hours after recycling.

In view of all these results, it is advisable for cement - recycled pavements to precrack them at short spacings. On the contrary, it does not seem necessary to try to force the opening of the joints by rolling them or by other means, specially when, as usual, traffic is allowed to pass almost immediately on the uncovered pavement.

5 Conclusions

With regard to RCC layers, it is difficult to draw conclusions from the performance of the several stretches built in Spain, because there are a lot of differences between them concerning their design features, and the traffic they bear. However, it has been noticed that joint spacing is the most important factor influencing the efficiency of precracking, not only in RCC bases but also in other types of cement - treated materials. Pavements with joints 10 to 15 m apart have performed badly. Those with joints every 7 m show some reflective cracking, but up to now these distresses have not deteriorated under trafic. Finally, the excellent results obtained on the stretches where joints have been wet-formed every 2.5 to 4 m makes this measure highly recommendable for all future sections with RCC or cement-treated bases.

It seems interesting to have all the joints opened from the very beginning, and not one every two or three, as it is at present the normal situation. Experiences with rollers or other types of equipment (e.g. pavement breakers) should be continued.

Precracking is also highly recommendable in cement - recycled pavements, specially if they present large variations. In these works, passage of traffic on the recycled material at an early age seems to warrant the opening of all the joints.

6 References

1. Jofré, C., Vaquero, J. and López Perona, R. (1994) *Performance of roller compacted concrete pavements in Spain* in *Proceedings of the 7th International Symposium on Concrete Roads*, Cembureau, Brussels.

2. Ministerio de Obras Públicas y Urbanismo (1989) *Instrucción 6.1 y 2 - IC sobre Secciones de firme (Standard for Pavement Sections)* (in Spanish), MOPU, Madrid.

3. Alberola, R. and Rocci, S. (1994) *Design and construction of a concrete pavement on compressible muds* in *Proceedings of the 7th International Symposium on Concrete Roads*, Cembureau, Brussels.

4. Jofré, C. (1995) *Reciclado con cemento de firmes existentes (Recycling of existing pavements with cement)* (in Spanish). *Carreteras*, No. 77, pp. 60-70.

Finite elements modelling of cracking in pavements

A. SCARPAS, A.H. DE BONDT, A.A.A. MOLENAAR and
G. GAARKEUKEN
Delft University of Technology, Delft, Netherlands

Abstract
Cracking consists one of the main causes of pavement overlay deterioration. Various techniques and products are currently being utilized as possible remedies for delaying the propagation of cracks into and through the overlay. In order to assist the designer with his attempt to determine the factors leading to cracking and in order to enable him to quantify the effectiveness of the various alternative rehabilitation solutions CAPA, a user-friendly, PC based, finite elements system has been developed. A variety of options enable the simulation and the detailed evaluation of the response of rehabilitated pavements.
Keywords: Finite elements, overlay deterioration, pavement rehabilitation, reflective cracking.

1 Introduction

The rehabilitation of cracked bituminous pavements by overlaying is rarely a durable solution since after a while cracking developes rapidly through the new layer. Depending on the situation, cracking can be due to traffic loads, subsidence, thermal gradients and combinations thereof. It consists one of the primary causes of pavement deterioration and as such it imposes heavy financial strains on national and local pavement maintenance authorities.

On a physical basis and according to the theory of fracture mechanics, the discrete discontinuity introduced by the presence of the existing crack tip in the old pavement, results to the development of high tensile stress concentrations at the base of the overlay. These lead to the initiation and propagation of cracking into the overlay at a location directly above the existing crack, Fig. 1(a). In this study this type of cracking (commonly termed reflective cracking in literature) will be characterized as **type A** cracking.

Nevertheless, in some circumstances, depending on geometry and material properties, cracking has been also observed to develop and propagate into the overlay from locations other than the tip of the existing crack. In these cases it is accompanied by debonding between the

Reflective Cracking in Pavements. Edited by L. Francken, E. Beuving and A.A.A. Molenaar. © 1996 RILEM. Published by E & FN Spon, 2–6 Boundary Row, London, SE1 8HN. ISBN 0 419 22260 X.

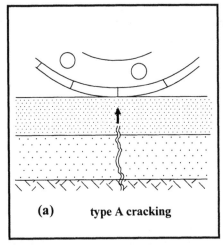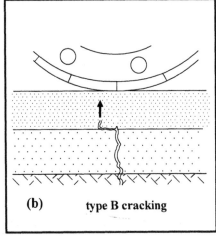

Fig. 1 Types of cracking in pavement overlays

overlay and the old pavement, Fig. 1(b). In this study this type of cracking will be characterized as **type B**.

Because of the multitude of parameters involved and the interplay between them, it is usually difficult if not impossible to determine a-priori, for a given pavement, which of the above two modes of cracking will prevail and what will be its implications on overlay deterioration.

Cracks introduce physical discontinuities in the otherwise homogeneous -for engineering analysis purposes- body of the pavement. Because their stiffness is usually low, if not zero, their presence plays a major role in determining the integrity and hence the load carrying capacity of the pavement. The discontinuous nature of cracks renders inappropriate the application of classical continuum based layered analysis methods for studies of crack initiation and propagation. On the other hand, within the context of the Finite Elements Method (FEM), cracks in the body of the pavement can be easily simulated by disconnecting the nodes of elements on either side of the crack propagation path, Fig. 2.

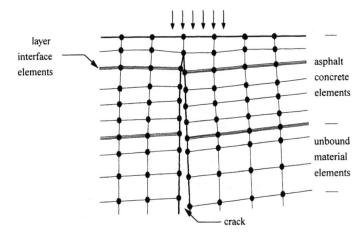

Fig. 2 Finite elements crack simulation

In an attempt to gain some insight on the mechanics of overlayed pavements, the causes and the consequences of cracking and the effectiveness of various rehabilitation techniques and products, CAPA, a user-friendly, PC based, FEM system, has been developed at Delft Technical University. By means of CAPA, extensive studies have been undertaken of the response of overlayed pavements, Bruijn et al. [1], Pajunen [2], Gaarkeuken et al. [3]. In this contribution several features of the FEM simulation of pavement response by means of CAPA will be reviewed. An example of the utilization of the system for the study of an actual overlayed cracked pavement will be presented.

2 Fracture mechanics aspects of FEM crack simulation

Within the context of the FEM, the singularity of strains in the vicinity of a crack tip can be modelled accurately and elegantly if the elements surrounding the crack tip node are substituted by specially formulated "crack tip" finite elements, Fig. 3. Several special elements have been developed over the years. Among the best performers are the elements developed by Barsoum [4]. By correlating the displacements u and v of the nodes at the crack tip computed via the FEM with the theoretical expressions for the displacements, Ingraffea et al. [5] have derived closed form formulae for computation of the stress intensity factors.

For the case of a crack tip simulated by means of Barsoum elements, in the crack local coordinate system (t, n) of Fig. 3 :

$$K_I = \sqrt{\frac{2\pi}{L}} \ \frac{G}{\kappa+1} \left[4\left(v_B - v_D\right) + v_E - v_C \right] \tag{1}$$

$$K_{II} = \sqrt{\frac{2\pi}{L}} \ \frac{G}{\kappa+1} \left[4\left(u_B - u_D\right) + u_E - u_C \right] \tag{2}$$

with : $\kappa = (3-v)/(1+v)$ for plane stress

$\kappa = 3-4v$ for plane strain

G is the shear modulus and v is Poisson's ratio.

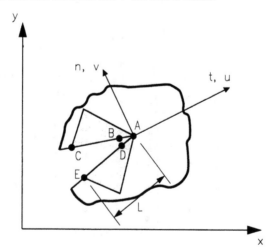

Fig. 3 Crack tip simulation by means of Barsoum finite elements

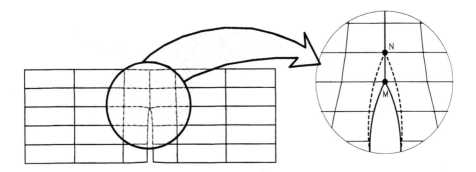

Fig. 4 Discrete crack propagation

Once the stress intensity factors are known, it is possible to compute the number of load repetitions N necessary for the crack in the finite elements mesh to propagate a distance $\Delta c = MN$, Fig. 4, by means of Paris law, Molenaar [6] :

$$\frac{dc}{dN} = A \cdot \left(K_{eq}\right)^n \tag{3}$$

in which A and n are overlay material parameters and K_{eq} represents the equivalent stress intensity factor for mixed mode crack propagation conditions.

Within the context of the FEM crack extension requires a series of successive analyses, in each of which the mesh is progressively modified manually by disconnecting the nodes of elements on either side of the crack propagation path. This is a laborious and error prone task. In CAPA an incorporated remeshing technique completely eliminates this task. Starting from the initial cracked pavement configuration as input by the user, the system automatically propagates the crack into the finite elements mesh computing at the same time all necessary fracture mechanics parameters, Scarpas et al. [7].

3 Simulation of layer interface regions

The successive material layers in a pavement are seldomly, if ever, rigidly bonded to each other. Depending on the nature of the materials and the construction techniques involved, some magnitude of interlayer slip can occur.

The degree of interlayer bonding can influence significantly the overall structural response. This can be demonstrated by means of an example of a simply supported beam subjected to a point load at the middle, Fig. 5(a). The beam is made of two layers of similar material (E=5500. MPa, v=0.35) interconnected to each other by means of a thin layer of bonding material. A parametric analysis was performed with CAPA for several values of bond shear stiffness [3]. The horizontal stresses at the centre of the beam corresponding to the various bond stiffness values are also shown in Fig. 5(b).

It can be seen that for the case of stiff interlayer bonding, the beam responds as an one-material system without discontinuity in the distribution of stresses at the level of the bonding layer. On the contrary, for low bonding stiffnesses, there is a distinct discontinuity in

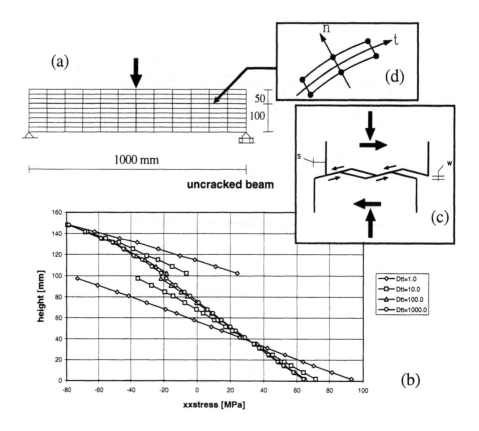

Fig. 5 Layered beam finite elements simulation

the distribution of stresses along the height of the beam. As the degree of bonding between the two layers is decreased, the discontinuity increases. It can be concluded that, by altering the internal distribution of actions in a structure, bonding can influence significantly the response of individual structural components.

Layer interface regions do not constitute actual physical materials in themselves. Instead they represent regions of physical discontinuity between other actual materials. The local response of the contact region between two interacting bodies of material can be described in terms of a relative displacement s and a relative, normal to the interface midplane, displacement w, Fig. 5(c). It is these two displacements that are meant to be modelled by means of an **interface** element, Fig. 5(d). It consists of two pairs of nodes, one pair on each side of the element axis. The thickness of the element in its undeformed configuration can be very small or even zero. One normal n and one transverse t axes of material anisotropy can be associated with the principal local axes of the element. The constitutive relation associating local stresses to relative local displacements can be defined as :

$$\begin{Bmatrix} \tau \\ \sigma \end{Bmatrix} = \begin{bmatrix} D_{tt} & 0 \\ 0 & D_{nn} \end{bmatrix} \cdot \begin{Bmatrix} s \\ w \end{Bmatrix} \tag{4}$$

Damage at the interface may result either as a result of monotonic stresses of magnitude higher than the corresponding strength of the material or due to fatigue cycling at lower stress levels. According to Miner's rule, the damage incurred at the material after N cycles of fatigue loading can be computed as :

$$\eta = \frac{N}{N_{tot}} \tag{5}$$

in which N_{tot} represents the total number of cycles until failure at a given stress level.

4 Calculation of the lifetime of an asphalt concrete overlay

A schematic of the finite elements mesh of a cracked overlayed pavement is shown in Fig. 6. The passage of a wheel load is simulated by placing a normal distributed load of 0.707 MPa at different distances from the crack axes. For a given crack length, the equivalent stress intensity factor K_{eq} is calculated for the various load positions.

During the passage of a wheel, the interface region between the old pavement layer and the overlay is subjected to shear and normal stresses. The exact state of stress depends, among others, on the position of the load and the extent of cracking. Before the stress intensity factors corresponding to a given configuration can be computed, the normal stiffness D_{nn} and/or the shear stiffness D_{tt} of the interface elements must be adjusted so as to reflect the local physical conditions of the interface region over which they span (e.g. $D_{nn} = D_{tt} = 0$ for elements in which tensile fracture has occurred or $D_{tt} = 0$ for shear failure etc.).

From the above distribution of K_{eq} the number of traffic cycles necessary for the crack to propagate over one row of overlay finite elements can be calculated by means of Paris Law. Also, by means of Miner's rule, the number of traffic cycles necessary for fracture of the most critically stressed interface finite element can be computed. The least of these two numbers N_{sim} corresponds to the number of simulated traffic cycles before either mesh modification (by means of crack propagation one element row) or interface element stiffness properties modification (to represent layer interface failure) is necessary. This procedure can be repeated

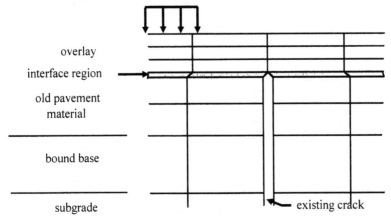

Fig. 6 A50 pavement profile

until all overlay elements are cracked and the economic life of the rehabilitated pavement has been exhausted. It is algorithmically portrayed in Fig. 7.

As mentioned in the introduction, for a given pavement, it is usually difficult to determine a-priori which of the observed two types of cracking will prevail. Because of the stress concentration induced to the bottom of the overlay by the presence of the existing cracks, at the **early stages of overlay service** type A reflective crack initiation and propagation will occur. However, **at latter stages**, and depending on the material characteristics of the interface region

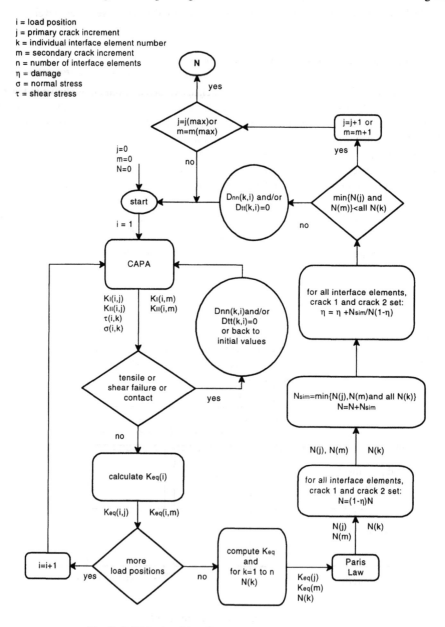

Fig. 7 CAPA algorithm for pavement service life evaluation

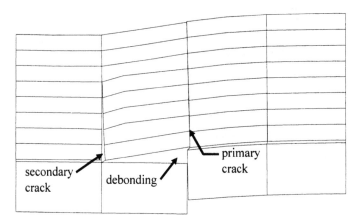

Fig. 8 Primary and secondary crack propagation

adjacent to the existing pavement cracks, debonding of the overlay from the old pavement may occur due to material failure. This may cause high tensile flexural stresses at the bottom of the overlay, Fig. 8, and result to the initiation and development of secondary cracks leading eventually to surface cracking of type B.

CAPA was specially developed to be capable of handling any number of cracks in the body of the pavement. The case of pavement service life evaluation for two simultaneously developing cracks is presented in the algorithm of Fig. 7.

5 Case study

Application of CAPA in pavement rehabilitation studies will be demonstrated by considering the case of an actual Dutch highway pavement, the A50 highway in Friesland, will be considered. A schematic of the cracked pavement profile and the overlay has been shown already in Fig. 6. The specified material properties of the pavement layers are listed in Table 1.

Table 1 Pavement layers properties

	thickness [mm]	E [MPa]	ν	D_{tt} $[(^N\!/_{mm})\,/\,mm^2]$	D_{nn} $[(^N\!/_{mm})\,/\,mm^2]$
overlay	50	5500	0.35	-	-
interface	1	-	-	various	5500
old layer	200	3500	0.35	-	-
bound base	400	10000	0.2	-	-
subgrade	200	100	0.35	-	-

In order to investigate the influence on overall pavement response of the characteristics of the interface region between the overlay and the old cracked pavement two different cases will be considered, a high strength interface material, leading to type A cracking and, a low strength material, leading to type B cracking.

5.1 Type A crack propagation

The algorithm of Fig. 7 has been utilized for the calculation of the number of traffic load cycles N necessary for the old crack at the bottom of the unreinforced overlay to reach to the top [3]. Four different values of interface shear stiffness were utilized so as to cover the whole range of response, Fig. 9. For traffic conditions typically available interface products would normally range between $D_{tt} = 0.5$ and 5.(N/mm)/mm^2. The case of $D_{tt} = 100$. (N/mm)/mm^2 represents almost ideal bonding conditions and has been included mainly for reasons of comparison.

It can be concluded that in case of unreinforced overlays, the presence of a soft interface decouples the two pavement layers and in doing so prevents the energy associated with the existing crack from driving the crack in the overlay.

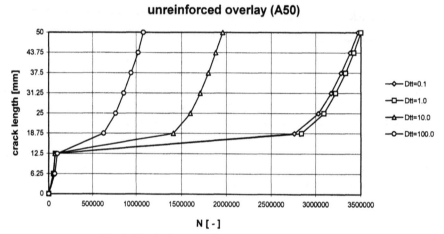

Fig. 9 The influence of interface stiffness on overlay life

5.2 Type B crack propagation

As mentioned earlier, interface failure can create the conditions necessary for type B cracking to develop. This postulate was verified by means of CAPA [3] by repeating the analysis of the pavement after having specified uniaxial tensile strength σ_f and shear strength τ_f values for the strength characteristics of both, the overlay material and the interface, Table 2.

Table 2 Material strength values

	σ_f [N/mm^2]	τ_f [N/mm^2]	D_{tt} [($^N\!/_{mm}$)/mm^2]	D_{nn} [($^N\!/_{mm}$)/mm^2]
overlay	1.0	5.0	-	-
interface	1.0	2.0	1.0	5500.

As expected, because of the stress concentration introduced at the bottom of the overlay due to the presence of the existing crack in the old pavement layers, type A crack initiation and propagation was observed first, however it was accompanied by failure and hence debonding of the adjacent interface region. The resulting double flexure mode of deformation (e.g. see Fig. 8) led to tensile fracture at the bottom of the overlay and the generation of a new crack at a short distance from the existing crack axis. In subsequent increments, this new crack became

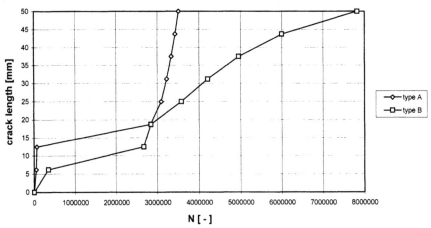

Fig. 10 Influence of cracking type on overlay life

dominant and was the first to reach to the top of the overlay resulting to type B cracking. A comparison of the influence of type of cracking on overlay life is presented in Fig. 10. It appears that for the pavement under consideration, type A cracking will propagate to the top of the overlay at half the number of traffic load cycles N necessary for type B cracking.

6 Acknowledgements

Development of the CAPA system has been made possible by the financial support of the Netherlands Technology Foundation (STW).

7 References

1. Bruijn, N., Scarpas, A. and de Bondt, A.H. (1994) *(Un-)Reinforced Interfaces in Asphalt Concrete Pavements*, Report 7-94-203-19, Division of Road and Railway Construction, Faculty of Civil Engineering, TU-Delft, The Netherlands.
2. Pajunen, M. (1995) *Finite Elements Simulation of Frost Heave Cracking*, Research Report, Road and Transport Laboratory, Faculty of Civil Engineering, Univ. of Oulu, Finland.
3. Gaarkeuken, G., Scarpas A. and de Bondt, A.H. (1996) *The Causes and Consequences of Secondary Cracking*, Report 7-96-203-23, Division of Road and Railway Construction, Faculty of Civil Engineering, TU-Delft, The Netherlands.
4. Barsoum, R.S. (1976) On the use of isoparametric elements in linear fracture mechanics, *Intenational Journal of Numerical Methods in Engineering*, Vol. 10, pp. 25-37.
5. Ingraffea, A.R. and Manu, C. (1978) Stress-intensity factor computation in three dimensions with quarter-point crack tip elements, *International Journal of Numerical Methods in Engineering*, Vol. 12, pp. 235-248.
6. Molenaar, A.A.A. (1983) *Structural Performance and Design of Flexible Road Constructions and Asphalt Concrete Overlays*, Ph. D. Dissertation, Division of Road and Railway Construction, Faculty of Civil Engineering, TU-Delft, The Netherlands.
7. Scarpas, A., Blaauwendraad, J., de Bondt, A.H. and Molenaar, A.A.A. (1993) CAPA: A modern tool for the analysis and design of pavements, in *Proceedings of the 2nd RILEM Conference on Reflective Cracking in Pavements: State of the Art and Design Recommendations*, (ed. J.M. Rigo et al.), E & FN Spon, London.

Forecasting the formation of reflective cracking in asphalt pavements reinforced with glass fiber mesh

B.S. RADOVSKY, V.V. MOZGOVOY, I.P. GAMELYAK, H. SABO
and V.R. SHEVCHUK
Transportation University of Ukraine, Kiev, Ukraine
A.E. MERZLIKIN and O.G. BABAK
State Highway Research Institute, Moscow, Russia

Abstract

The method of forecasting the formation of reflective cracking is offered, which enable to evaluate reinforcement of asphalt pavements with synthetic grids. The method is based on the account of thermo-rheological properties of asphalt concrete. The influence of transport and temperature on the formation of cracks are considered.
Key words: asphalt pavement, forecast of the formation of cracks, synthetic geotextile grid, thermo-rheological properties of asphalt concret.

1 Forecasting the formation of reflective cracking in asphalt concrete pavement

1.1 Solution scheme

As a solution diagram, asphalt concrete pavement is consi- dered as a slab of infinite length and finite width and thickness spanned with the underneath layer of the base having slabs of finite length, witdh and thickness (block base). This layer of the base leans on a multi-layered half-space.

The slabs are considered as double or triple layered. All layers are spanned among themselves. One of layers represents a

Reflective Cracking in Pavements. Edited by L. Francken, E. Beuving and A.A. Molenaar. © 1996 RILEM. Published by E & FN Spon, 2–6 Boundary Row, London, SE1 8HN. ISBN 0 419 22260 X.

composite as asphalt concrete, incorporated with reinforced grid. The other layers consist of asphalt concrete.

The material of the pavement possess thermo-viscoelastic properties (asphalt concrete, polymer-asphalt concrete). The material of the block base possess elastic properties (concrete, reinforced concrete) or thermo-viscoelastic properties (asphalt concrete).

1.2 Limiting condition

For the forecast of crack formation we used the time before the occurrence of cracks on the pavement. With this purpose limi- ting condition of asphalt pavement is devised based on account of climatic character of destruction of its material [ ].

$$\left| M(\sigma_T, t, z_j) \right|^\alpha + \sum_{i=1}^{2} \left| M_i(t, z_j) \right|^{\beta_i} - \left| M_R(t, z_j) \right|^\gamma \leqslant [M] \qquad (1)$$

where $M(\sigma_T, t, z_j)$ - damage index of the structure of asphalt concrete for j-sublayer of pavement from action of tensile horizontal thermo-shrinkage stress; $M_i(t, z_i)$ - also from action of a transport, causing cross-sectional cutting-off forces (i=1) and tensile horizontal normal stress (i=2) over joints or cracks of the block base; $M_R(t, z_j)$ - restored part at a time t earlier than accumulated damage index of the structure; [M] - limiting - allowable value of the damage index; α, β, γ - experimental parameters.

Here under damage we mean formal kinematic concept of irreversible bonding gaps in asphalt concrete, accumulated in a time t under certain character of influence of one of the factors analyzed in (1). Damage index - relative parameter, describing degree of damage of the structure of asphalt concrete under the effect of destroying factor in comparison with maximum possible value of damage.

For damage index of asphalt concrete, accumulated under the action of stress we have the folowing expression

$$M_k(\sigma_T, t, z_j) = \int_0^{t_p} \Phi((t-\tau); n_1; z_j) \varphi(\sigma_T(\tau); T(\tau); z_j) d\psi(\tau), \qquad (2)$$

where Φ - damage function, reflecting the history of damage accumulation in interval $(t-\tau)$; n_i - experimental constants of this function, reflecting nonlinear character of damage accumulation; φ - influence function reflecting thermo-shrinkage stress and temperatures of damage accumulation; ψ - determining variable parameter, in relation to which the damage accumulation is observe.

The expression (2) is applied in case of establishment of adequate expressions for function Φ and on the basis of experimental data when satisfying the condition

$$
\begin{aligned}
&\text{if} \quad \tau \to 0, \quad \Phi \to 0; \\
&\text{if} \quad \tau \to t_p, \quad \Phi \to \text{const},
\end{aligned} \tag{3}
$$

where t_p - time before the brake of bonding gaps in asphalt concrete in all section of j-sublayer.

In individual cases expression (3) gives known criterion of "long" strength of R.Baily, A.A.Iliushin, V.V.Moskvitin and others at appropriate kinds of functions Φ, φ and parameter ψ.

The value [M] is determined from expression

$$
[M] = M_N K_H K_y, \tag{4}
$$

where M_N - nominal value of damage index($M_N=1$); K_H - factor of safety; K_y - coefficient of condition of work.

Time before the formation of reflective cracks under the action of tensile and cutting-off forces is determine as the sum of the time necessary for complete breakage of entity of every sublayer of pavement

$$
t_{tp} = \sum_{j=1}^{m} t_p(z_j), \qquad m=2 \text{ or } 3. \tag{5}
$$

1.3 Forcasting temperature in pavement

When forcasting temperature in pavement, the stratum of the pavement structure and regime of temperature changes on surface during daily and annual fluctuations are considered. For this purpose we use well known theory of hert conduction, on the basis of which we recieve analitical relations, describing temperature changes of i-layer as a function of time and coordinates $T=f(t, x, y, z)$.

For the amplitude of temperature fluctuation of pavement surface we use the relationship between the parameters of temperature changes, air, wind velocity, radiation of the sun and heat-physical characteristics of the surface.

1.4 Stress determination due to temperature fluctuation

When determining thermo-shrinkage stress in asphalt concrete pavement we solve the problem of theory of thermo-viscoelasticity with use of the relationship between stress, strain, time and temperature.

$$\sigma_T(t) = \int\limits_0^t R(\xi(t) - \xi(\tau)) d\varepsilon_T(\tau),$$

$$\xi(t) = \int\limits_0^t [a_T(T(t), Q)]^{-1} dt,$$

where $\varepsilon_T(\tau) = (\varepsilon_T^1(\tau) + \varepsilon_T^2(\tau) + \varepsilon_y^1(\tau) + \varepsilon_y^2(\tau))$ – relative strain; R – function of relaxation; t – time, for which stress is determine, τ – time, previous to t; ξ – time, converted on the basis of the principle of Temperature-time analogy (TTA) to temperature Q, in which the parameters of function of relaxation R(t) are experimentally established; a_T – function of TTA; ε_T^1 – unrealised deformation temperature of pavement; ε_T^2 – pavement deformation, caused by temperature changes of slab length of cracked block-base; ε_y^1 – unrealised pavement deformation during its irreversible shrinkage; ε_y^2 – pavement deformation caused by irreversible shrinkage of slab of the cracked block-base.

As a result analitical relationships were found for determining longitudinal normal thermo-shrinkage stress in pavement $\sigma_T(t)$ relative to time considering the peculiarities of road or airport pavement structure, thermo-rheological properties of materials, regime of temperature changes in relative to parameters, reflecting these parameters: layer thickness h_i; cracked block-base length l_{s1}; gap between slabs δ; length of section with crack-obstructing interlayer λ; length of repaired crack Δ; coefficient of temperature conduction of layers a_i; coefficient of linear temperature deformation of materials $\lambda_i(t)$; coefficient of irreversible linear shrinkage $\lambda_{y1}(t)$; coefficient of friction between the crack block-base and the crack obstructing interlayer f_1; also between the type of base and underneath layer f; relaxation function of material $R_i(t, T)$; functions of TTA $\lambda_{y1}(t)$ and temperature of temperature joints of the the base $T_2^{'}$;

cooling temp of layers K_1; temperature gradient for the time T_1 and others [10, 17, 20, 24, 32].

1.5 Stress determination from the action of transport

During the valuation of the limiting condition of pavement; from the action of transport we determined cross-sectional cutting-off forces and tensile horizontal normal stresses in the pavement. For this purpose we used computer program based on the method of finite elements.

The pavement structure is considered as an elastic multi-layered package, lying on a rigid base. The border of the bottom layer is spanned to the rigid base without horizontal displacement. The layers between themselves are rigidly spanned. Exfoliation does not occur. Between the layers, interlayers are installed deformation properties of which differ from the deformation properties of the layers they are in contact with for about 100 and more times. Thus inside the interlayers shear deformations and friction in accordance with Culon's Law can be simulated. In some adjacent layers a vertical crack of finite size can pierce through them. Load can be applied to the external surface of the package or in the area of the crack. Direction and diagram of the distribution of the load can vary arbitrarily. For modeling of the layers we used isoparametric square finite element.

Interlayer is a structure having a very small thickness. Therefore it is modelled as hexagonal finite element.

In the vicinities of the top of a crack and places of contact of a crack with interlayer the grid is thicknen with the help of a triangular finite element.

2 Thermo-rheological characteristics of the materials of pavement

In carrying out calculations of forecasting the formation of reflective cracking of pavement, we use the following thermo-rheological characteristics: coefficient of temperature conduction a; coefficient of linear temperature deformation $\alpha(T)$; coefficient of linear shrinkage deformation $\alpha_y(t)$; relaxation function $R(t)$; functions of TTA $a_T(T)$; damage function $\Phi(t)$; influence function $\varphi(6, T)$; durability function $t_p{}^*(6, T)$; parameters of limiting conditions. These indexes of properties happen to be thermo-rheological certificate of asphalt concrete. These indexes were determined both experimentally and analitico-experimentally.

3 Glass fiber grid for reinforcing asphalt concrete

Specially for road construction a glass fiber grid is developed from glass-plastic. During the production of the grid the following experiments were conducted:
   - technology for the creation of large squares of the grid;
   - devising special composition that provides water-resistance and adhesion with bitumen;
   - laboratory fatigue tests of grid and layered structure in aggressive medium;
   - devise a method of laying the grid during constraction;
   - devise a method of calculating rational reinforced road and airport pavements.
Laboratory model tests are carried out. Distilled water and salt solutions are used as aggressive medium.
   In devising the method of test of the samples we suggested that one of the worsening signs of the structure of composite materials is the increase of water-absorption and reduction of tensile strength during bending.
   Periodic freezing-melting were carried out with fully saturated water. Samples were cooled at temperature of $-20°C$. Melting of samples in water is carried out up to $+20°C$. After every 5-10 circles samples were choosen for the determination of tensile strength.
   It is established that grids with defferent types of treatment have different tensile strength, and the technology of laying the grid to asphalt concrete layers affect frost-resistance of the composit material (asphalt concrete + grid + asphalt concrete). Treatments that provide the best strength and frost-resistance were found. Grids with these treatments were tested for resistanse to chemical and aggressive solutions and substances: anti-ice-crusted substance, petrochemical products, distilled and polluted water, salt solutions e.t.c.

4 Analysis of the effect of using grid for reinforcing asphalt concrete pavement

As an example we consider asphalt concrete pavement being constructed in Western Siberia.
   Base structure: reinforced concret slabs 14 cm thick, levelling course 3-5 cm, crushed stone stabilised with bitumen 15 cm.
   Calculations of asphalt concrete pavement without grid were carried out for different life span before the appearance of reflective crackings (t_{Tp}). These results show that the value of

crack resistance of the pavement increases with the increment   of the value of $t_{\text{T}p}$ at first,   and later after the life span (six years), the rate of the increase of pavement thickness   gradually decreases.

In accordance   with   our   Standards,   the   minimum   allowable thickness of asphalt concrete pavement for   these   conditions   is 12-15 cm, which correspond to the life span of the pavement befo-re the formation of cracks 2-3 years.   The minimum life span   be-fore   the   formation   of   reflective   crackings is 6 years,   then pavement thickness is equal to 22 cm.   If $t_{\text{T}p}$=18 years then pave-ment thickness is equal to 39 cm.

In order to reduce the thickness of   the   pavement   we   recom-mend reinforcing it with glass fiber grid.

Below are the results of   the   calculations   of   asphalt   con-crete pavement thickness reinforced with glass fiber grid.

Table 1. Asphalt concrete pavement thicknesses reinforced
         with glass fiber grid

$t_{\text{T}p}$, yrs	thicknesses, cm
6	16
12	24
18	31

Results show that reinforcement reduces thickness   by   22-25%. The   reduction   in   thickness   allows   the   rise in efficiency of construction of the pavement and economy of material resourses.

5 References

1.   Radovsky, B.S., Mozgovoy, V.V. (1982) Determination of   thermal stresses in asphalt concrete   pavements considering viscoelastic properties of material, that   changes   as   temperature   changes, *Journal of roads and road construction*, Kiev, No. 31. pp. 48-50.
2.   Radovsky, B.S.,Mozgovoy, V.V. (1986) Thermal stresses in asphalt concrete pavements,   *in   Increasing   of   durability of   road constructions*, Soyuzdornii, Moscow. pp. 29-46.
3.   Mozgovoy, V.V., Tsehansky, O.E. (1988) Influence of   transverse cracks and joints of stabilised base on thermal crack-resistance of road pavements, *Journal of roads and road construction*, Kiev, No. 42. pp. 66-72.

4.   Sunii, G.K., Radovsky, B.S., Merzlikin, A.E., Mozgovoy, V.V. (1988) Automated design of thermal crack-resistant asphalt concrete pavements, *in New methods of road construction design*, Collection of scientific works of Soyuzdornii, Moscow. pp. 5-21.

5.   Radovsky, B.S., Mozgovoy, V.V. (1989) Ways to reduce low temperature cracking of asphalt pavements, *4th Eurobitume symposium*, Madrid, pp.571-575.

Crack resistance of asphalt overlays

Part two

Crack resistance of
asphalt overlay

An approach for investigating reflective fatigue cracking in asphalt-aggregate overlays

J.B. SOUSA
CONSULPAV, Oeiras, Portugal
S. SHATNAWI
Research Program, California Department of Transportation, Sacramento, California, USA
J. COX
Cox & Sons, Inc., Colfax, California, USA

Abstract

Reflective cracking has been one of the major causes of distress in asphalt-aggregate overlays. A new reflective cracking device (RCD) has been developed to operate within the shear testing system developed during the Strategic Highway Research Program. With this device it is possible to simulate in the laboratory the Mode I (opening) and Mode II (shearing) fatigue failure patterns that have been observed in the load associated reflective fatigue cracking. Preliminary results obtained with the device indicate that it has been able to demonstrate that asphalt rubber mixes have greater resistance to reflective cracking than conventional dense graded mixes.
Keywords: reflective cracking, shear device, asphalt rubber, fatigue, finite elements

1 Introduction

The rehabilitation of PCC pavements with bituminous overlays has been a solution adopted worldwide [1]. However, design criteria, test methods and material properties are not readily available and in many cases the cracks propagate through the new overlay during the first few years of service. This mode of distress is traditionally referred to as "reflective cracking" and is a major concern to highway agencies.

1.1 Crack origin and propagation
Three different mechanisms have been identified as the origin of cracks in the overlays of rigid pavements [2]:
- Thermal stresses (or thermal fatigue) occur when the daily temperature variations induce openings and closures of the cracks of the rigid pavement. If the asphalt concrete overlay perfectly adheres on the PCC, the thermal movements induce stress concentrations in the overlay and either a

Reflective Cracking in Pavements. Edited by L. Francken, E. Beuving and A.A.A. Molenaar. © 1996 RILEM. Published by E & FN Spon, 2–6 Boundary Row, London, SE1 8HN. ISBN 0 419 22260 X.

debonding between the layers occurs or a crack will propagate in the overlay from the bottom to the top;

- Thermal stresses may be initiated by a rapid cooling down of the top layer which induces critical tensile stresses that result in cracks [2].
- Traffic Loads: the repetitive traffic loads induce additional distress to the overlay and increase the rate of crack propagation originated by thermal movements.

Authors do not agree on the predominant mechanism by which traffic loads are affecting crack propagation [1, 3]. Some authors argue that the most important effect is obtained by the opening of the cracks (Mode I) as a result of the bending of the pavement structure, while others consider the vertical shear stresses (Mode II) which result from a poor load transfer between the two edges of the crack to be more detrimental. Differential vertical movement (Δ-vertical) at underlayer discontinuities (such as joints or cracks in overlaid PCC pavement) has long been known to be a major cause of reflective cracking in AC overlays [4,5]. Based on these latter considerations this paper addresses traffic induced reflective cracking although the device could be used to simulate also environmental effects.

1.2 Significance of the Problem

Designing a structure and evaluating its cost effectiveness is a current engineering task for which the essential starting elements are:

- The fundamental material properties which can be determined by experimental methods.
- The analytical tools to model system behavior. An essential part of the process is the development of a computational tool to determine the stress-strain distribution in the structure combined with a physical law that describes the evolution of the damage.
- The calibration and/or validation of the system behavior. Development of experimental designs in which the tools and models are used to simulate and predict monitored actual pavement performance. In the absence of perfect predictive tools the development of transfer functions ("shift factors") is a methodology that has been widely used and accepted.

All of these three steps are well defined for the flexural fatigue cracking problem. However, standard test methods and models to evaluate the resistance to reflective cracking are not widely used.

The performance laws used in fatigue cracking predictions, which are based on flexural fatigue laws or crack propagation laws, address the crack opening mode (Mode I). Similarly, for cracking caused by thermal movements only Mode I is considered in the predictive tools. However, in the reflective fatigue cracking problem caused by traffic loads, more complex patterns have been recorded [6] (see Fig. 1). In these cases, a combination of Mode I and Mode II crack openings are simultaneously present. No fatigue law has been developed that specifically addresses the combined effect of these two modes.

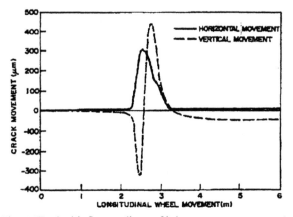

Fig. 1. Typical influence lines of joint movements recorded by the CAM over a joint concrete pavement (6).

Therefore, it is the purpose of this paper to introduce a device capable of simultaneously imposing Modes I and II deflection patterns on asphalt concrete specimens and to also report comparative results showing the retardation improvements due to the use of rubber in asphalt-aggregate mixes.

2 Concepts Behind the Development of a Reflective Cracking Device

The new device was designed so that one day reflective cracking fatigue tests could be executed to obtain predictable fatigue properties. At this stage, only comparative results between mixes is possible, no actual reflective cracking prediction can be made. However "shift factors" could be developed after a research program is undertaken to measure crack movements on typical sections. Systems like the Crack Activity Meter (CAM) [6] can be used to measure displacement before and after the overlay has been placed. The movements recorded by those devices could then be programmed in the RCD. Reflective fatigue tests could be executed at an appropriate testing temperature and a comparison of laboratory fatigue predictions with actual field data could be made.

2.1 Reflective Cracking Failure Criteria
Asphalt aggregate mix stiffness depends on strain level, strain rate and temperature. It also depends on the level of fatigue damage the material has sustained. In fact AASHTO TP8 standard test defines fatigue life of flexural beam specimens when the specimen stiffness is reduced to 50% of the initial stiffness. In the reflective cracking phenomena, due to high stress and strain concentration, there are different rates of damage accumulation within the crack zone. Near the crack edge the rate of damage accumulation is far higher than in zones away from the crack edge. As such, it is likely that some points in the crack zone have their stiffness reduce to zero, while other points are not yet subjected to significant stiffness reduction. However, from a pavement performance viewpoint, it is important to consider the load transfer characteristics over the entire crack zone. A measure of that quantity could be the equivalent stiffness of the overlay over the crack zone. This can be defined by the following two quantities: Equivalent Axial Stiffness of the Crack Zone (EASCZ) and Equivalent Shear Stiffness of the Crack Zone (ESSCZ), as presented below:

Equivalent Axial Stiffness of Crack Zone $= (F_A/ L*H)/ (\delta_A/ w_c)$ Eq. 1
Equivalent Shear Stiffness of Crack Zone $= (F_S/ L*H)/ (\delta_S/ w_c)$ Eq. 2

Where,

        L= crack length (i.e. specimen length (150 mm)), H= specimen height (i.e. 50 mm), w_c= crack width (i.e. 10 mm), F_A=axial force (compression or tension) F_S=shear force , δ_A= axial displacement (opening or closing of the crack -Mode I) and δ_S = shear displacement (Mode II)

The crack zone is defined as the volume of material in the overlay immediately above the crack on the PCC to the surface. Clearly, with a zero crack width the volume would be reduced to zero. However, this situation is outside the scope of this paper. See Fig. 2 for definition of the parameters.

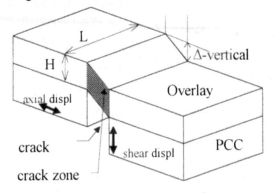

Fig. 2. Schematic representation of crack zone

Initially, before being fatigued, the stiffness of the material portion over the crack zone is identical to the portion away from the crack. As the traffic loads cause relative axial and shear movements between the two edges of the crack, stress concentrations propagate microcracks throughout the crack zone, thus effectively reducing the equivalent stiffness of the material in that zone, and possibly in its vicinity. In this paper, fatigue life was defined as the number of cycles to reach 50% of the equivalent axial or shear stiffness in the crack zone (as defined by equations 1 and 2).

     Similar to flexural fatigue cracking, the reflective fatigue cracking mechanism can be controlled either by the deformability of the underlying layers in thin overlays (strain controlled), or by the stiffness of the overlay for thick overlays (stress controlled). In the failure fatigue criteria adopted in this research, it has been assumed that material characterization would be conducted for conditions representative of those in which displacement control would apply.

2.2 Equipment Description

The loading equipment used for the testing program was the new shear testing system introduced during the SHRP program. The CS7000 shear testing system at Caltrans Transportation Laboratory in Sacramento, California was used as the base testing unit. This system has the advantage of providing a very solid and reliable basis from which many types of tests can be generated. The very rigid axial and shear tables can be simultaneously and/or independently programmed under force or displacement control. This permits the imposition of two independent movements which can be used to simulate Mode I and Mode II movements in the reflective cracking phenomenon.

A Reflective Cracking Device (RCD) was designed and developed to fit in the CS7000 and CS7200 shear testing systems developed by Cox & Sons. This device is described in Fig. 3 and 4. It basically consists of two rigid L-shaped pieces of aluminum to which a specimen is glued. The gap between the two L-Shaped pieces (which simulates the two concrete slabs adjacent to a crack) can be changed from 0 to 12.5 mm. Specimen thickness (simulating overlay thickness) can vary between 0 and 70 mm. The maximum specimen length and width are limited to 175 mm. Specimens can be cylindrical or rectangular in shape.

To minimize unwanted misalignment stresses in the specimens, the specimen is glued to the L-shaped platens so that they maintain their relative position between each other when they are fastened to the axial and shear tables of the testing system. To accomplish this, female V-shaped grooves were machined into the platens and during the gluing stage the matching male V-shaped bars precisely tie the two L-shaped pieces together until the glue hardens (the glue takes about 5 minutes to harden and about 60 minutes to cure) and the RCD is placed in the machine. At this stage, the male V-shaped tie bars are removed.

LVDT holders are mounted on the device to measure the crack opening (Mode I) and shearing (Mode II). These LVDTs can be used as feedback controlling transducers and as such the equipment can impose, on these transducers, displacements that match actual crack activity measured in the field.

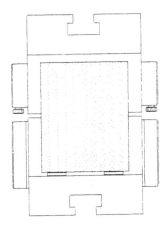

Fig. 3. Diagram of a Front view of the Reflective Cracking Device showing a rectangular specimen and the T slots to mount directly on the shear testing system.

Four pneumatic pistons apply 700 kPa pressure to the specimen simulating tire pressure (see Fig. 4). The four independent cylinders transfer the pressure through two separate aluminum plates and a layer of hard foam rubber. This arrangement minimizes axial movement restraints that would otherwise occur during the pulsing of the axial displacement (Mode I). Although during the tests executed in this research the pressure was held constant at 700 kPa, the system was designed so that the pistons could be controlled by a hydraulic servovalve under the feedback from a pressure transducer which would permit the simulation of the pulse of the tire rolling over the crack zone.

Because the tests were executed at 20 °C, it was considered that permanent deformation would not occur due to the constant pneumatic pressure. However, if reflective fatigue cracking is to be investigated at higher temperatures, it is recommend that a more realistic simulation of the loads applied by the tire be made.

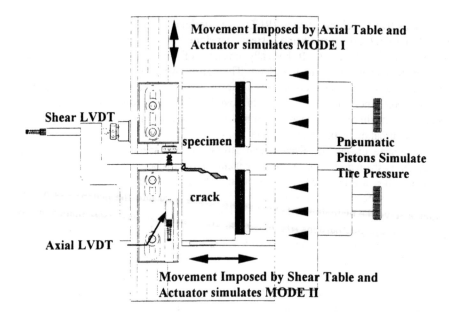

Fig. 4. Side View of the Reflective Cracking Device with Specimen and simulated load associated crack at 45 degrees.

2.3 Finite Element Idealization

A finite element analysis of the test conditions in the RCD was made using SAP90. The SOLID elements were used to simulate plane strain conditions (see Fig. 2 and 5). A 50 mm thick by 150 mm long specimen was placed over two rigid boundaries separated by a 10 mm gap representing the crack width

An axial load and a shear load were applied to one of the boundaries to simulate the test loading conditions. In this case, a 5000 N load was applied in shear (Mode II) and 4000 N was applied in the axial direction (Mode I). Furthermore, a 700 kPa stress was applied to the opposite side of the crack to simulate the tire pressure.

The deformed shape of the specimen under these loading conditions can be observed in Fig. 5.

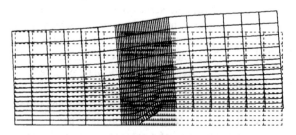

Fig. 5. SAP90 finite element idealization of specimen in the RCD with deformed and undeformed shapes.

A plot of the maximum principal stress (σ_1) occurring in this case is presented in Fig. 6(a). The maximum stress occurs, as expected, near the edge which is subjected to the highest tensile strains. It can be observed that there is a strong gradient of stresses.

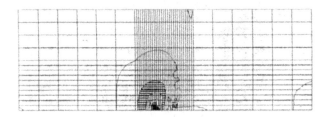

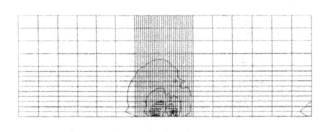

These stresses decrease with distance away from the edge of the crack. From this edge, microcracks will propagate. However, an observation of the Von Mises stress (SVM) (see Fig. 6(b)) defined as:

$$SVM = \{[(\sigma_1 - \sigma_2)^2 + (\sigma_2 - \sigma_3)^2 + (\sigma_3 - \sigma_1)^2]\}^{1/2} \quad Eq.3$$

where σ_1, σ_2 and σ_3 are the three principal stresses, indicates that within the crack zone high stress concentrations occur specifically at each crack edge.

Fig. 6. Stress Contours (a) - Maximum Principal Stress (b)- Von Mixes Stresses

3 Testing Program

3.1 Experiment Design
To achieve the objectives of this research two types of mixes were tested on the RCD, a conventional dense graded asphalt concrete (DGAC) mix and an asphalt-rubber hot mix - dense graded (ARHM-DG) mix. The variables considered included two strain levels and two replicates.

3.2 Materials
The aggregate source used was Teickart Perkins located in Sacramento, California. The gradation was 12.5 mm maximum size aggregate for Type B Medium per Caltrans standard specifications. The asphalt contents were 5.0% and 7.6% for the conventional DGAC and the ARHM-DG mixes, respectively, which are optimum binder contents per Caltrans mix design practices. The asphalt used in both mixes was AR-4000. The rubber content in the ARHM-DG mix was 19% by weight of asphalt. The rubber was a product of automobile ground tires. The specimens were fabricated in the laboratory and cut to the appropriate specimen dimensions (40 mm height by 150 mm in diameter).

3.3 Testing
The reflective cracking fatigue tests were executed under a displacement control mode. The tests were executed by applying relative movements between the two edges of the

crack (see Fig. 1). The total pulse duration was 0.2 seconds. However the actual pulse lasted only 0.1 seconds. The width of the crack was set to 10 mm [7]. The shape and the duration of the waves is typical of crack movements on jointed concrete pavements [6]. This wave was used to simulate the relative movement between the crack edges. The magnitude of those displacements was maintained constant throughout the test. However, it was rescaled to impose two different displacement levels and tensile strains where imposed instead of the compressive strains to expedite testing. The displacement levels used were 0.02 and 0.07 mm for axial displacement and 0.05 and 0.14 mm for shear displacement. These values are consistent with those reported in the literature [8]. It was verified that the shear and axial loads necessary to impose those displacements decreased with the number of cycles (see Fig. 7) indicating distress propagation. The ATS software was used to simultaneously and synchronously generate and control the waves in each direction. Tests were executed at 20 °C +/- 1 °C. This was the room temperature at the Caltrans Transportation Laboratory.

3.4 Data Analysis

The shear and axial loads and displacements applied to the specimens were recorded. For each of the recorded cycles, the equivalent axial and shear stiffness in the crack zone was computed using equations 1 and 2. The traces of the shear and axial stiffness were plotted versus the number of cycles (see Fig. 7), and a determination was made to the number of cycles at which the shear and axial stiffness values had their value reduced to 50% of the initial. The fatigue curves, based on shear fatigue life, for each of the two materials tested (DGAC and ARHM-DG) are presented in Fig. 8. It can be observed that at high displacement levels, the fatigue life is shorter than at low displacement levels. These comparative results indicate that ARHM-DG has a greater resistance to reflective fatigue cracking than the conventional DGAC mixes.

It was noted that the number of cycles needed to reach 50% of the initial equivalent shear stiffness in the crack zone was, in most cases, identical to the number of cycles to needed reach 50% of the initial equivalent axial stiffness.

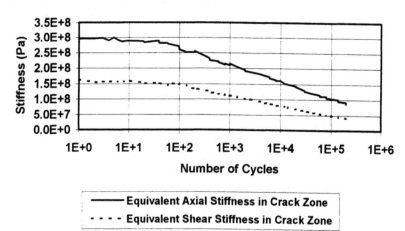

Fig. 7. Typical variation of Equivalent Shear and Axial Stiffness in the crack zone with number of cycles

This may be attributed to the fact that the magnitude of the displacements imposed in Mode I were identical to those imposed in Mode II. For example, at high strain amplitudes the shear displacement in one direction was 0.07 mm (0.14/2 mm). The corresponding axial displacement was 0.05 mm. It was also noted that the repeatability of the device was good. This may be attributed to the long (150 mm) crack edge under a uniform stress distribution. Longer crack initiation zones permit the test to be less sensitive to particle and local specimen non-uniformities, thus providing for more consistent results.

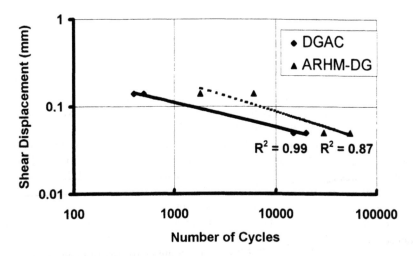

Fig. 8. Comparison of Fatigue Curves for Conventional (DGAC) and Asphalt Rubber (ARHM-DG) Mixes.

4 Summary and Conclusions

A new testing device capable of performing reflective fatigue tests on asphalt-aggregate mixes has been developed. It can simulate the reflective cracking phenomena observed in overlays that are placed over Portland Cement Concrete. The Reflective Cracking Device, when used in combination with the shear system developed during the Strategic Highway Research Program, can simulate Modes I and II (opening and shearing) loading patterns. This device, together with the shear testing platform within which it operates, opens new avenues to develop a standard test to characterize reflective fatigue crack resistance. It can be a useful tool in evaluating mix resistance to thermal, as well as, load associated reflective cracking.

Based on preliminary test results, it can be concluded that asphalt-rubber mixes resist reflective cracking better than conventional dense graded asphalt concrete mixes. RCD has a potential use as a tool to obtain fundamental material properties and to assist in reflective cracking comparative studies. With the RCD it is also possible to execute tests to identify the usefulness of different types of paving reinforcing fabrics or stress-absorbing layers, thus providing the means to rapidly select reflective crack mitigation strategies. The RCD is easy to use and, as such, it could be considered for standardization.

Further research is needed to identify representative loading patterns for each type of pavement condition, effect of crack width, crack edge shape, appropriate testing temperatures, and representative shift factors. Bias and precision studies need to be undertaken.

Acknowledgements

The authors acknowledge the contributions of Mr. Jeff Rush of The California Department of Transportation.

References

1. Reflective cracking in pavements: state of the art and design recommendations: proceedings of the second international RILEM conference, organized by the Belgian Research Centre for Plastics and Rubber. 1st ed. London; New York: E & FN Spon, 1993. Series title: RILEM proceedings ; 20

2. Nunn, M. E. An investigation of reflection cracking in composite pavements in the United Kingdom / M.E. Nunn (Transport and Road Research Laboratory). IN: Reflective cracking in pavements. Liege, Belgium : C.E.P.-L.M.C., State University of Liege, 1989.

3. Reflective cracking in pavements: assessment and control / edited for the conference held in Liege, Belgium on March 8, 9 and 10, 1989; a conference organized by C.E.P., Belgian Research Centre for Plastics and Rubber, The University of Liege, Belgium: C.E.P.-L.M.C., State University of Liege, 1989.

4. McGhee, K.H. Efforts to reduce Reflective Cracking of Bituminous Concrete Overlays of Portland Cement Concrete Pavements. Virginia Highway and Transportation Council, 1975

5. Roger D. Smith. Laboratory Testing of Fabric Interlayers for Asphalt Concrete Paving". California Department of Transportation, Sacramento, California Report no. FHWA/CA/TL-84/06. June 1984

6. Rust, F. C.. Load Associated Crack Movement and Aspects of the Rehabilitation of reflection Cracking in Cemented Pavements. Faculty of Engineering, University of Pretoria, South Africa. 1987.

7. Coetzee and C.L. Monismith. Analytical Study of Minimization of Reflection Cracking in Asphalt Concrete Overlays by Use of a Rubber-Asphalt Interlayer. IN Transportation Research Record no.700, pp. 100-108.

8. H. McGhee. Attempts to Reduce Reflection Cracking of Bituminous Concrete Overlays on Portland Cement Concrete Pavements. IN Transportation Research Record no.700, pp. 108-114

On the failure strain of asphalt mixes produced with SBS-modified binders

T. TERLOUW, C.P. VALKERING and W.C. VONK
Shell Research and Technology Centre, Amsterdam, Netherlands

Abstract

Where an increase in viscosity at high service temperatures is realised with the polymer modification of binders to obtain high stability asphalt mixes, the higher failure strain in tension allows applications of such binders to crack bridging systems and bridge decks.

In this paper we report on the failure strains of asphalt mixes with SBS-block-copolymer modified binders. Failure strains were measured in different configurations or tests and with binders modified with different concentrations of polymer. In addition to single loading to failure in bending, compression and indirect tension, repeated loading was used to determine the fatigue strains. The strain may be taken as an indicator for the ability of the asphalt mix in an overlay to sustain the high strains at cracks in the old structure.

Depending on the test method and temperature, an increase in the failure strain is found with SBS modification. Particularly when the mix formulation is adapted, this specific feature can be exploited.

Keywords : Failure strain/stress, polymer modification, crack-propagation rate, indirect tensile test, fatigue bending

Reflective Cracking in Pavements. Edited by L. Francken, E. Beuving and A.A.A. Molenaar. © 1996 RILEM. Published by E & FN Spon, 2–6 Boundary Row, London, SE1 8HN. ISBN 0 419 22260 X.

1.   Introduction

Polymer modification of road binders is increasingly applied in many countries to cope with the ever increasing severity of traffic loadings. The first objective of the modification is to improve the mechanical stability or resistance to rutting of asphalt layers at higher service temperatures. However, there are more benefits to be gained, provided that the right formulations are used.

This paper reports on the effect of modification with polystyrene-polybutadiene-polystyrene (SBS) copolymers, and specifically on the flexibility of the binders and asphalt mixes. The effects on the low temperature properties, particularly at higher concentration levels, have given this polymer a dominant position in the modification of bituminous compounds for roofing applications in many countries.

From the above it is obvious that there is an opportunity to formulate bitumen/SBS binders which are optimal in terms of (base) bitumen grade and SBS content and which provide a longer and cost-effective service life. Over many years these principles have been applied in the research and development of the Shell proprietary polymer-modified bitumens CARIBIT* and CARIPHALTE*.

The modification of road binders with SBS has such a strong effect on the mix stability that the base bitumen used in these modified binders is generally softer than the standard grade conventional bitumen. Such softer grade base bitumens not only contribute to the additional advantage of low temperature performance, but also to an improvement of the service life of asphalt mixes in terms of binder hardening.

In this paper we report on the properties of asphalt mixes with modified binders as determined in destructive testing with single and repetitive loadings. Particular attention is given to the deformation at failure which is considered to be directly related to crack propagation/arresting under road pavement conditions. Specifically with overlays, the flexibility or stress/strain at break may be taken as indicator for the ability of the mix to sustain the high stress/strain values at cracks in the old structure.

Part of the test series concern binders with different concentrations of SBS, but formulated in such a way that the penetration is kept the same. In this series, on which single and repetitive three point bending (fatigue) tests were carried out, with increasing polymer content, the base bitumen is softer. The composition of some of these formulations are quite similar to the polymer-modified bitumens, CARIBIT* and CARIPHALTE*. Further, results of fundamental crack propagation tests on mixes with SBS-modified binders are reported.

*CARIBIT and CARIPHALTE are Shell trademarks

2. Single load tests

Two single-load tests were used to study the effect of SBS modification on asphalt mixes: a three-point bending test (carried out at different temperatures) and an indirect tensile test (ITT; carried out at 0°C). The ITT has been proposed as a test that gives an indication on the resistance to crack propagation of an asphalt mix [1], [2].

2.1. Bending tests

The bending tests were performed on 230x30x22 (mm) specimens that were deflected at a constant rate using an Instron tensile tester. At each temperature six specimens were tested with test temperatures ranging from -70°C to +20°C and the tensile stress and strain were recorded.

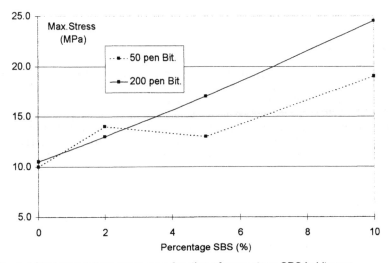

Figure 1 Maximum failure stress as a function of percentage SBS in bitumen

Two binders have been included in these tests; both of the base bitumens (one 200 pen and one 50 pen bitumen) were blended with 3, 5 and 10% Kraton D-1101-C* resp. and the resulting decrease in penetration was corrected by the addition of a heavy flux, so that each of the modified binders had a penetration equal to that of its base bitumen. Kraton D1101-C* is a linear SBS of medium molecular weight. We note that the polymer concentrations reported have been calculated for the base bitumen including the flux added. Although the binders thus obtained are not necessarily practical binders, the total series covers effects that can be expected in practice. As can be seen in Fig. 1, which is a plot of the maximum (failure) stress versus the polymer content, the SBS content has a distinct effect: the higher the polymer content, the higher the strength. Note that the maximum stress reported in Fig. 1 is encountered at different temperatures and, as shown in Fig. 2, the temperature with the maximum strength value decreases with increasing polymer content.

* KRATON D is a Shell trademark

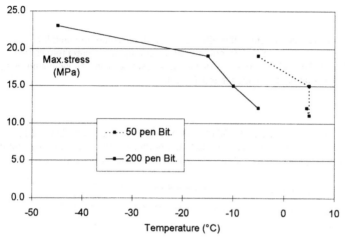

Figure 2. Maximum failure stress vs. temperature of testing.

Similarly, the strain at failure increases with the polymer concentration as shown in Fig. 3 for the results obtained at -20°C. Figures 2 and 3 indicate that the major effects are obtained with polymer concentrations of 5% and higher: the effectiveness of the modification in increasing the failure strain and in reducing the temperature at which maximum strength is obtained depends on the base bitumen and, in particular on the penetration grade. Hence part of the results may be attributed to the effect of softening obtained by adding the flux, but in general it is common practice to use softer base bitumens with polymer-modified binders.

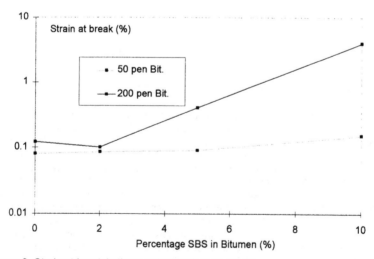

Figure 3. Strain at break in three-point bending at -20 °C

2.2. The ITT tests

60 mm high cylindrical asphalt specimens were subjected to a compressive diametrical loading with a displacement rate of 0.85 mm/s. The force versus strain plot was recorded and the following characteristics have been determined:

the strength (force peak-value/cross-sectional area through centre of sample);
the fracture energy (area under curve till fracture point);
the total energy (total area under curve) and
the strain at break.

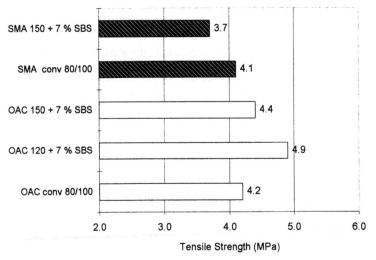

Figure 4. Tensile strength in ITT on OAC and SMA test specimens

Tests were carried out on open asphaltic concrete (OAC) and stone mastic asphalt (SMA). In the OAC, two formulations containing 7% (m/m) SBS (one based on a 120, the other on a 150 pen base binder) were compared to an 80/100 reference binder.

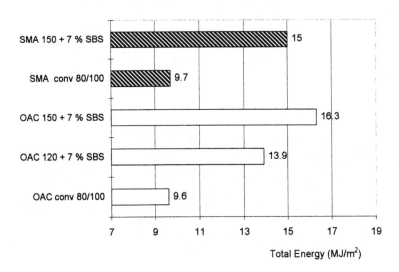

Figure 5. Total energy in ITT on OAC and SMA test specimens

In the SMA, only the reference and the softer modified binder were compared. Binder characteristics are listed in Table 1. The results as given in Figs. 4 and 5 clearly show that the tensile strength in this test is only slightly influenced by the presence of the polymer, but that the total energy and the energy to break are significantly increased. The higher force that is required for the unmodified asphalt mix reflects the higher stiffness of the binder at 0°C: the modified binder is based on a 150 pen bitumen and the SBS modification does not necessarily lead to an increase in stiffness at this temperature. The energy is also markedly dependent on the hardness of the base bitumen. Note that, to a certain degree, the comparison of the binders is influenced by the temperature of the test.

2.3. Single load tests discussion

From the results above one firm conclusion may be drawn: the addition of SBS has a positive effect on the resistance to cracking of asphalt mixes as determined in single-load tests, particularly in combination with somewhat softer base bitumens and at contents ≥5% (m/m) of SBS. That a minimum content of SBS is required is understandable: at service temperatures, all bitumen/polymer binders consist of two-phase structures with the morphology depending mainly on the quality of the base bitumen. An essential feature is the (dis)continuity of the polymer-rich phase. SBS can absorb up to 9 times its own weight of bitumen constituents and may become the continuous phase. The SBS-rich phase is essentially extended SBS and hence highly flexible and elastic. Crack initiation and propagation are strongly dependent on phase continuity, although a discontinuous polymer-rich phase will also have a positive effect as a crack arrester, as is the case in e.g. toughening plastics with rubbers. From the literature on the latter subject it is known that rubber content and the particle size of the dispersed phase influence the effectiveness of the modification on impact and crack-growth resistance. Although not part of the current study, the differences observed in the three-point bending tests between the binders of equal penetration, but from different sources, are likely to be due to differences in the phase morphology.

Table 1. Properties of the binders used in the stone mastic asphalt mixes

Binder	Conventional 80/100	120 pen + 7 % SBS	150 pen + 7 % SBS
$Pen_{25°C}$, 0.1mm	92	75	80
$Pen_{40°C}$, 0.1mm	407	155	177
$T_{R\&B}$, °C	46.5	97.5	75
PI (Pen/Pen)	-0.5	4.6	4.0
PI (Pen/$T_{R\&B}$)	-0.6	(7.6)	(5.0)

3. Tests with repetitive loading

3.1. Fatigue bending tests on dense asphaltic concrete mixes.
The fatigue performance was measured in the 3-point bending test for the series of modified binders investigated in the single load bending strength test and reported in section 2.1. Tests were carried out in the constant strain mode at a frequency of 50 Hz and at temperatures ranging from -30 to +25 °C. Mix stiffness as reported in the Figs. 6 and 7 indicates that at fixed penetration of the binders and under constant test conditions the stiffness decreases with the addition of polymer and of the compensating flux to the binder. This effect is also significant at the very high polymer concentration with the low penetration base bitumen, which suggests the continuity of the polymer-rich phase.

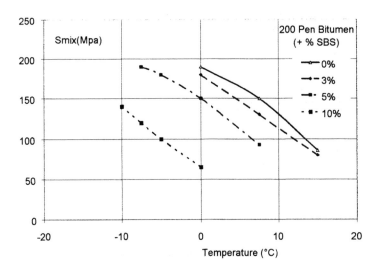

Figure 6. Mix stiffness versus temperature for different SBS contents in binders of 200 pen

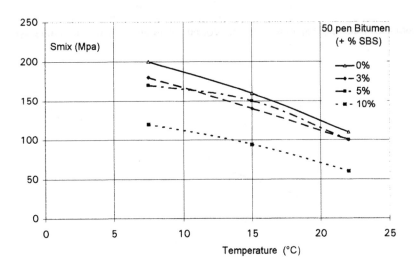

Figure 7. Mix stiffness versus temperature for different SBS contents in binders of 50 pen

As an example, the number of fatigue cycles as a function of the initial strains in the bending test is shown in Fig. 8 for the tests at 0 °C for the series with one of the 200 pen base bitumens. Fatigue life increases as a function of the SBS concentration: in particular, the 10 % SBS concentration shows a large increase. An effect of increasing polymer percentage on the slope of the fatigue curve is clearly found. In other test series this change of slope has also been observed at polymer contents of 5 % m/m.

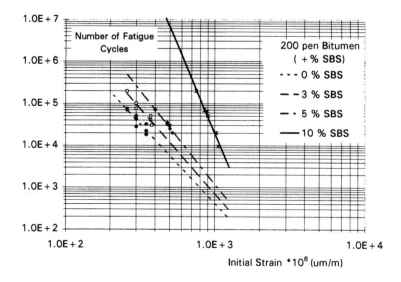

Figure 8. Mix fatigue life versus initial strain at 0 °C and a frequency of 50 Hz for different SBS contents in binders of 200 pen

From Figs. 6 and 7 temperatures corresponding to a mix stiffness of 100 MPa were determined for the different binders applied. Subsequently the strain values corresponding in the tests to a fatigue life of 100.000 cycles were determined from the individual fatigue performance plots (e.g. Fig. 8) at that specific temperature and plotted against the polymer content in the binders in Fig. 9.

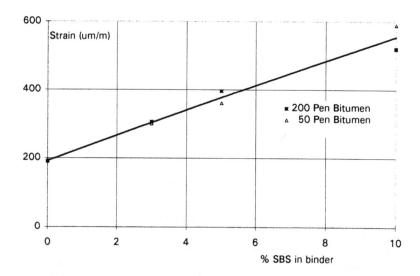

Figure 9. Strain versus percentage of SBS added to the binder for the different base bitumens used.

In this way it is possible to identify the effect of the percentage of SBS in the base bitumen in Fig. 9. It is clear that irrespective of the base bitumen the allowable strain increases with increasing percentage of SBS in the binder

3.2. Fatigue bending tests on stone mastic asphalt mixes
A specific application for stone mastic asphalt mixes could be on steel orthotropic bridge-decks. Because of the higher bitumen content in this type of mix they will better sustain the considerable strains as a result of traffic loadings. For this present series of tests a rather low binder content of 6.2 pha was chosen, and therefore the obtained results will be on the conservative side compared to more standard binder contents of 7 pha. Fatigue bending tests were carried out with stone mastic asphalt mixes at a temperature of 10 °C to compare the fatigue performance of the mix with a 150 pen +7 % SBS and a conventional 80/100 binder. The tests were carried out at constant strain and a frequency of 40 Hz. The constant strain mode is considered to be the relevant mode for testing of mixes on such steel bridge configurations.

The fatigue results obtained for the SMA mixes with the different binders are presented graphically in Fig. 10. The 150 pen + 7 % SBS binder has a considerably better fatigue performance than the conventional 80/100 binder. The increase in fatigue life at a given strain for the modified binder is more than a factor 9.

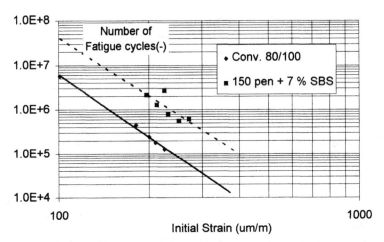

Figure 10. Fatigue life as a function of initial strain for SMA mixes with conventional 80/100 and 150 pen + 7 % SBS binders.

3.3. Crack Propagation Tests

The propagation of cracks is generally considered as one of the phases in the fatiguing of asphalt mixes up to failure. In some cases, e.g. with reflection cracking, more emphasis is given to the investigation of crack propagation than to the preceding crack initiation.
As part of a special study, carried out by Netherlands Pavement Consultants (NPC), stress-controlled 4-point bending tests were performed with different binders in open asphalt concrete base course mixes to assess the relative performances of these binders in terms of crack-propagation. The binders applied were a conventional 80/100 as a reference and binders modified with SBS. Propagation of the cracks in the test samples was followed by means of crack-foils attached to either side of the beams (dimensions: 70*45*600 mm). From this type of test it is possible to determine the parameters of the Paris cracking law :

$$da/dN = A \cdot \Delta K^n$$

In which :

da/dN = Crack-propagation rate (mm/load)
ΔK = Stress-Intensity at crack tip (N/(mm$^{3/2}$))
A and n = Constants in the Paris equation depending on material properties

In Fig. 11 the crack-propagation rates for two of the binders tested are given as a function of the stress intensity. The binder with 7 % SBS modification performs significantly better than the conventional binder in the open asphalt concrete mix .

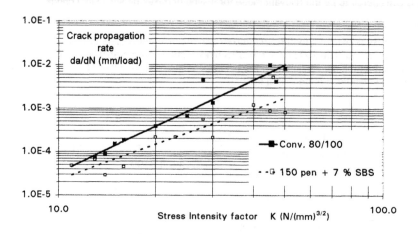

Figure 11. Crack propagation in 4-point bending tests with different polymer-modified binders and a conventional 80/100 binder in an open asphalt base course mix.

4. Conclusions

The flexibility of an asphalt mix in terms of strain or deformation to failure at low temperatures increases as a result of the modification of the binders; this was found in both single and repetitive loading tests.
Similarly, the stress or loading to failure in these tests does increase considerably when polymer-modified binders are applied. The positive effects on failure stress and deformation increase with increasing level of polymer concentration in the binder. The actual effect is also dependent on the type of base bitumen used.

In fatigue bending tests with repetitive loading, polymer modification of the binders increases the strain to failure; this effect is proportional to the polymer content in the binder.
In crack-propagation tests at 0 °C it has been demonstrated that the rate of crack propagation is reduced by polymer modification of the binder.

References

1. A.A.A.Molenaar, A.v.d.Meent, N.C.Mul, N.Nataraj, W.Don, "Design of polymer-modified asphalt mixes for Amsterdam's Schiphol airport", 9th AAPA International Asphalt Conference, November 1994, Australia.

2. E.T. Harrigan, R.B.Leahy and J.S.Youtcheff, " The SUPERPAVE Mix Design System Manual of Specifications, Test methods and Practices" SHRP-A-379 Report, March 1994.

Semi-circular bending test: a practical crack growth test using asphalt concrete cores

R.L. KRANS
Road and Hydraulic Engineering Division, Ministry of Transport and Public Works, Delft, Netherlands

F. TOLMAN
Netherlands Pavement Consultants, Utrecht, Netherlands

M.F.C. VAN DE VEN
Department of Civil Engineering, University of Stellenbosch, Stellenbosch, South Africa

Abstract

The Semi-Circular Bending Test (SCB) has been applied to asphalt mixes, to test the crack growth and strength parameters. In the SCB, diametrically halved cores were loaded in a three point bending set-up, both with cyclic and with static loads.

After treatment of the material behavior, various crack growth tests are described, and compared to the SCB: direct and indirect tensile tests and three and four point bending tests, both with and without initial cracks. Then results are presented of cyclic direct tensile, four point bending, and SCB tests. Finally static indirect tensile tests and static SCB tests are compared. In all experiments asphalt concrete specimens have been used.

The preliminary conclusion is that the SCB test is suitable to determine the crack growth properties of bituminous materials. Extra research has to be carried out to confirm this. A large advantage is that drilled cores can be used for design (functional specification), and also for quality control, for instance using a Marshall press.

Keywords: Asphalt concrete, cores, crack growth, notch, Paris, SCB, semi-circular bending test, tests.

1 Introduction

In pavement design, the use of fracture mechanics to calculate layer thicknesses becomes more and more familiar. Combination of parameters from laboratory tests with a modeling of the structure enables calculation of the number of load repetitions until failure. This is illustrated in the flow chart of figure 1.

A new type of test has been developed for testing asphalt concrete. The rationale of this test is twofold: a quick, cheap, and reliable quality control test is requested for road construction and mix production, and furthermore a mix design test is need-

Reflective Cracking in Pavements. Edited by L. Francken, E. Beuving and A.A.A. Molenaar. © 1996 RILEM. Published by E & FN Spon, 2–6 Boundary Row, London, SE1 8HN. ISBN 0 419 22260 X.

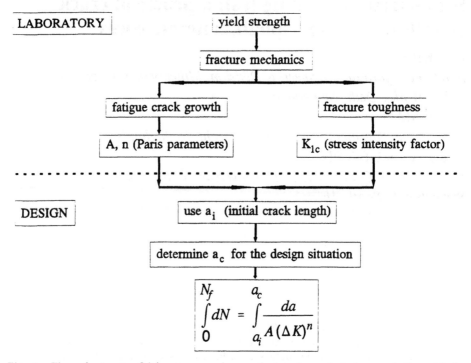

Fig. 1. Flow chart: use of laboratory crack growth data in road design. a_c is the critical crack length, at which failure occurs. N_f is number of cycles to failure from initial crack a_i.

ed when specification of asphalt concrete mixes based on mix composition is replaced by functional specification.

Our research is dedicated to crack growth, fracture toughness, strength and fatigue. Crack growth tests are performed, using specimens with notches that act as starter cracks. Fatigue and strength tests are performed using unnotched specimens.

Previously, several crack growth tests have been applied to asphalt concrete using cyclic loading, each having its own typical drawbacks [1,8-12]. Some tests, like the direct tensile test (DT) with external starter notches and the center cracked tensile test (CCT) with central starter notches, require gluing of the specimen, which is time consuming. Furthermore the DT will bend to one side once one crack has grown. The CCT does not have this drawback, but its specimens are large and difficult to make. (Almost) pure bending tests, like the three and four point bending tests (3PB and 4PB), require relatively large specimens. The indirect tensile test (IT) is only suitable for materials having a compressive strength considerably larger than the tensile strength. Asphalt can not satisfy this requirement at most test conditions. To produce a starter notch in the center of the specimen is also a problem.

In order to overcome these problems we have chosen to load a diametrically halved core in a three point bending set-up (figure 2). This test is called the Semi-Circular Bending Test (SCB), and has previously been applied in stone and ice mechanics [2-4]. Advantages are that no gluing is needed and that specimens can be fabricated from cores. A notch can simply be sawn in the specimen. Comparison of the SCB with the IT demonstrates that a specimen already fails in the SCB at a quarter of the force needed in the IT. At these lower force and stress levels, local

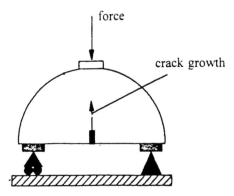

Fig. 2. The Semi Circular Bending test set-up, with diametrically halved specimen.

shear distortions hardly occur. As a mix design test a cyclic load can be used; for the quality control test a static load is sufficient.

This paper is organized as follows: first the material behavior is discussed, then the crack growth tests are treated, and finally the results of crack growth experiments are presented and discussed.

2 Material behavior

2.1 General
Different parameters can be used to describe failure, depending on the characteristics of a construction. In table 1 an overview of several employed parameters is given.

As a pavement is loaded alternatingly, fatigue resistance is a primary design criterion. However, at very low temperatures, temperatures can induce a static load, which can result in fracture (either on its own or in combination with cyclic loads from traffic). At the surface the interaction of the material with tires can be very complex, and cracks can occur locally.

The prediction of fatigue life is by no means straightforward: Fatigue life comprises both crack initiation and propagation stages. Furthermore, an exact definition of the transition from initiation to propagation is generally impossible. At low stress levels, crack initiation and micro-crack growth usually account for most of the fatigue life.

Table 1. Parameters for characterization of crack tests

	static load	cyclic load
unnotched	$\sigma_{fracture}$ (strength)	N_f
notched	K_{1c}	A, n

2.2 Linear elastic behavior
At low temperatures c.q. high loading speeds, the behavior of asphalt concrete is mainly linear elastic. Under these conditions, stress and strain distributions in the specimens can be calculated, using well-known linear elastic formulas; analytical or finite element methods can be used.

Geometries with a crack can also be calculated linear elastically. Especially the stress field in the crack tip area is important here. This can be described by the stress intensity factor, K. The definition of K is:

$$K = \sigma\sqrt{\pi a} * Y(a/W) \tag{1}$$

with σ   = stress
      a   = crack length
      W   = specimen width
      Y   = geometry factor for K, a function of a/W (figure 3)

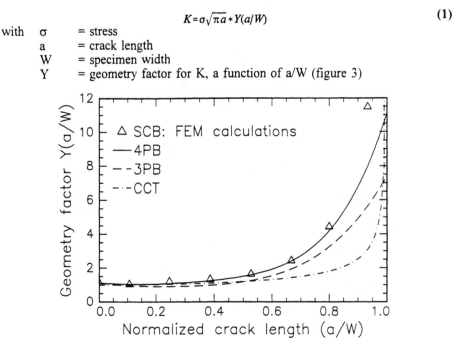

Fig. 3. Geometry factors Y(a/W) from equation (1).

We will describe the failure behavior in tests with a cyclic load signal with ΔK and the Paris equation [5]:

$$\frac{da}{dN} = A(\Delta K)^n \tag{2}$$

with ΔK = K_{max} - K_{min}
      A, n   = regression constants from the Paris equation

The dominating crack growth mode in our tests is the K_1-mode (crack growth in tension). K_{1c}, the critical K_1 value, is the value for which crack growth becomes instable. For tough materials this value is high.

2.3 Deviations from linear elasticity

Bituminous mixes display viscous effects. For example, the stiffness of bituminous materials depends on temperature and loading speed. The viscous effects may have a large influence on the tests, especially at higher temperatures and for asphalt mixes that are readily deformed. Therefore, for the evaluation of experiments, the relative importance of linear elastic and viscous effects should always be considered. The description we use is always linear elastic, even when viscous effects are predominant. Non-linearity is not included. It can become considerable in case of large deformations and crack growth.

3 Description of crack growth tests

3.1 Existing tests

In the introduction, §1, several crack growth tests have been mentioned, with their common abbreviations. A division can be made in direct tensile tests, the indirect tensile test, and bending tests. In this paragraph the tests are discussed in more detail.

3.1.1 Direct tensile tests

Two direct tensile tests have been investigated: the DT with external starter cracks, and the CCT with internal starter cracks.

In the DT the stress distribution is uniform, at least in principle. In practice, however, undesirable disturbances appear, because of edge effects, of material inhomogeneities, and of the influence of the starter notch. Bending moments may occur once a crack starts to grow from one side of the specimen.

To reduce the disturbances caused by external starter cracks, the CCT has been developed. In this test two cracks grow from a central starter notch. If one crack has exceeded the length of the other crack, growth of the first crack will decrease until both cracks have comparable lengths (or K_1) again. Both types of tensile tests require gluing, a cumbersome process. Making the starter notch in a CCT specimen also is an involved operation.

3.1.2 Indirect tensile test (IT)

In the IT, a cylindrical specimen, either with or without a central starter notch, is diametrically loaded in compression, resulting in homogeneous tensile stresses in the plane perpendicular to the direction of loading. Especially at elevated temperatures the stress distribution in the specimen and the length of the evolving crack is strongly influenced by shear cracks (mode II) arising below the loading strips, which makes interpretation of the results difficult.

3.1.3 Bending tests

Two bending tests have been investigated: the 4PB and the 3PB.

In The Netherlands much experience has been gained with the 4PB for fatigue experiments. The stress distribution in the unnotched specimen is simple (Bernoulli strain distribution). If the specimens are sufficiently slender the loads and therefore the stresses near the loading points are small. The stresses are influenced by specimen and loading geometries.

The 4PB can also be used for crack growth experiments. In that case a notch is fabricated in between the central loads, to act as a starter crack. In the crack tip area the stresses are sufficiently uniform. Major drawback of the 4PB is that it is expensive to obtain 45 cm long beams from a road construction.

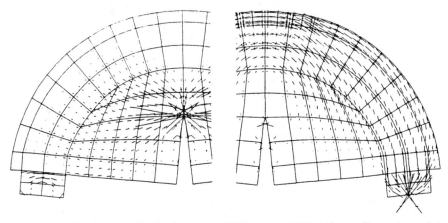

Fig. 4. Calculated stress distribution in an SCB set-up. SCB left: tensile stresses, right: compressive stresses.

The stress distribution in the 3PB is more complicated because no zone exists where shear stresses may be neglected. Shear stresses become more important when the span/height ratio decreases. As a rule of thumb, pure bending may be assumed when this quotient is larger than 10. In cement concrete standards, however, a factor 4 is prescribed.

3.2 The Semi-Circular Bending test
The SCB is depicted in figure 2. In this test a diametrically halved cylinder is loaded in a three point bending set-up. Cylinders can easily be obtained, as drilled cores from road constructions or slabs, and fabricated by Marshall or gyratory compaction. For crack growth, a notch (starter crack) is sawn in the middle of the flat side of the semi-circle, which is straightforward to accomplish. The SCB-test can be used in all set-ups where the indirect tensile test is performed, with minor adaptations. All these factors make the SCB test a feasible test for the asphalt concrete practice.

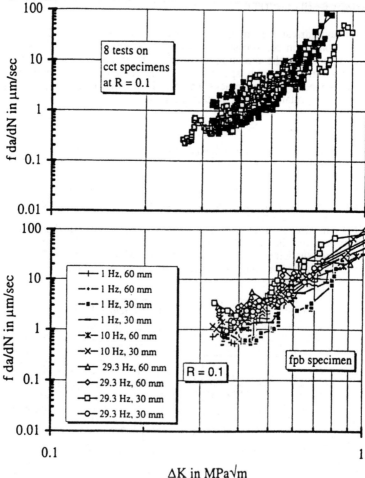

Fig. 5. Paris equation plots for CCT (upper figure) and 4PB (lower figure) tests on micro-DAC [11]. For both tests the loading frequency ranged from 1 to 30 Hz. The R factor is the ratio of minimum over maximum applied stress of a (shifted) sine signal. E.g., for a real sine R = -1, for a haversine R = 0, and for a constant signal R = 1.

The stress and strain distributions for both uncracked and cracked geometries have been determined using the Finite Element Method-program CAPA. The results of calculations for an SCB configuration with a crack length of 29 mm are presented in figure 4. In figure 3 the geometry factor for the SCB is shown as calculated with CAPA. This demonstrates that, for linear elastic behavior, the stress situation can be described at any length of the crack, using finite element analysis. Also, the stress situation at the supports can be analyzed, in order to investigate the influence of the external forces locally.

4 Results of crack growth experiments

4.1 Existing tests
In our search for a suitable crack growth test for asphalt concrete, we started with the DT and the 4PB tests [1,8,9]. Tests have been carried out on different types of asphalt concrete. One material was a sand asphalt with a continuous aggregate gradation, similar to dense asphalt concrete, but with 2 mm as largest grain size. We call this material: micro-dense asphalt concrete (micro-DAC). The advantage of this material is that the inhomogeneities (aggregate grains) are smaller, resulting in less scatter in the experiments. Using micro-DAC for CCT and 4PB tests yielded much information on crack growth behavior of asphalt concrete. The most important results are presented in figure 5 and 6 [11]. In figure 5, measurements using loading frequencies of 1, 10 and 30 Hz are displayed, showing that for both CCT and 4PB tests, the product of crack growth per cycle and frequency (*i.e.*, crack growth per time) hardly depends on the actual loading frequency. Thus much of the crack growth of these experiments is similar to creep crack growth. In fact, a more detailed analysis has shown that creep crack growth accounts for approximately two third of the total crack growth in these experiments [11]. In figure 7 it is shown that measurements with R-values (the R factor is explained in the caption of figure 6) can be combined by plotting them against ΔK_{eff}, the effective ΔK as defined in the figure. In other words, a master curve for crack growth at different loading frequencies and for different R-values has been constructed in figure 7.

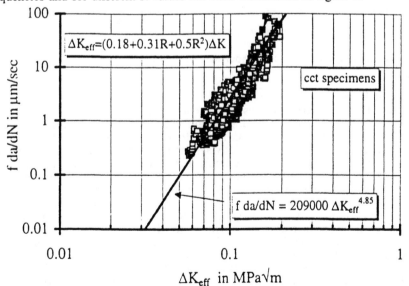

Fig. 6. Master curve of crack growth in CCT tests [11].

This research has been accompanied by 4PB tests on dense asphalt concrete (DAC) with 45/60 pen bitumen and a maximum aggregate size of 16 mm [8.9]. On the same material well-instrumented DT tests have been carried out, in which the force, displacement, crack opening displacement (COD), and crack length have been recorded. Variations in excitation (force or displacement controlled) and signal (sinus combined with different static loads, including zero) were carried out. The results have not been used to characterize the material but to evaluate the tests. All tests could be carried out well, but were rather cumbersome, especially when loading plates had to be glued.

Furthermore a number of IT tests on specimens with different types of initial starter notch have been carried out [10]. The IT test proved to be unsuitable for asphalt concrete at temperatures of 0 °C and above, because of failure of the specimen due to compressive stresses near the loading points, besides or even instead of crack growth in the center of the specimen. Experiments aimed at improvement of the IT test finally led to the application of the SCB test.

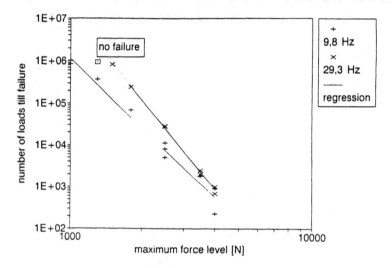

Fig. 7. Results of cyclic SCB tests without starter notches on DAC [12].

4.2 The SCB test

SCB tests have been carried out, using a cyclic signal and notched specimens [10]. Several results have been obtained: The deflection caused by viscous effects is neglectable at a temperature of 0 °C and a frequency of 30 Hz (testing time 30 minutes). The loads can be chosen relatively small (a factor of 2.3 to 3 smaller than in the IT), thereby reducing the undesirable effect of local failure near these loads and letting the investigated phenomena appear in a pure form. The specimen dimensions (core diameter 100 mm in this case, but 150 and 200 mm cores can also easily obtained and applied) and the length of stable crack growth (two third of the specimen height) are sufficiently large in relation to the grain size of Dutch mixes (11 to 32 mm) to obtain reasonable results. The ΔK range of the measurements is large enough to determine A and n in the Paris law. In the SCB test there are no instabilities from alternating crack growth, as opposed to DT tests.

Next, static and cyclic SCB tests have been performed at T = 20 °C [12], on 25 mm thick specimens without starter notches. The material was DAC, from a different batch than mentioned before. Despite the lack of the notch, cracking almost always started centered between the supports, as desired. In figure 7 the number of loads to failure are presented for the cyclic SCB tests. The static SCB tests yielded an average failure force of (4.7 ± 0.4) kN, with a force per unit width of 190 N/mm.

Finally static IT and SCB tests have been performed at 0 and 30 °C on porous asphalt [13]. The results are summarized in table 2. When the temperature is raised to 30 °C the decrease of the maximum force per width is a factor of 3 larger for the IT than for the SCB. This correlates to more damage occurring at 30 °C near the loading strips for the IT than for the SCB. For both test types, using porous asphalt, no damage is visible near the strips at 0 °C. In figure 8 two SCB force displacement curves are given, with viscous behavior visible at 30 °C, and not at 0 °C.

Table 2. Comparison of static IT and SCB tests (coefficient of variation of F/W is 20% for both tests). Porous asphalt from test sections in A12 highway, Zoetermeer, Netherlands. σ_{max} is the maximum stress, F_{max}/b is the ratio of maximum force and thickness of the specimen, and E/b is the ratio of total energy dissipated in the SCB test and thickness of the specimen

		IT tests '93, 3 cores		SCB tests '94, 5 half cores	
Type of binder modification	Temperature (°C)	σ_{max} (Mpa)	F_{max}/b (N/mm)	F_{max}/b (N/mm)	E/b (J/m)
Inorphil	0	1.5	242	98	48
Multiphalte		2.0	317	98	39
no modification		1.7	269	100	41
Inorphil	30	0.14	22	22	153
Multiphalte		0.32	51	46	115
no modification		0.15	25	30	127

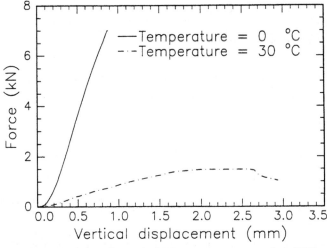

Fig. 8. Force displacement curve for static SCB tests on porous asphalt [13].

5 Conclusions and recommendations

1. The SCB test appears to be a good test to determine crack properties of asphalt concrete. It is promising for design (functional specification) and for quality control. For design, tests with both static and cyclic loading have to be carried out. The crack growth parameters can be quantified with the cyclic test; the static load test can be used to determine when stable crack growth changes into instable crack growth. For quality control, displacement controlled tests can be carried out using a Marshall press.

2. The following experiences demonstrate the value of the SCB test:
- specimens are easy to obtain
- if the loads are sufficiently small no failure near these loads occurs
- stable crack growth occurs
- the test set-up is relatively simple
3. Comparison with other test types shows the following advantages of the SCB test:
- advantage over tests using (bending) beams: SCB-specimens can be obtained by coring
- advantage over IT tests: loads can be chosen 2.3 to 3 times smaller, thereby reducing the undesirable effect of local failure near these loads and letting the investigated phenomena appear in a pure form
- advantage over DT tests: in the SCB test the instabilities from alternating crack growth, typical for DT tests, do not occur; furthermore gluing is avoided

Finally we remark that mode II crack growth can also be investigated with the SCB, by making oblique notches [3]. Because mode II cracking occurs in asphalt structures, it is important to be able to investigate this mode. Recommendations for the near future are:
- fixed positioned roller bearings are recommended both to prevent the change of span and to minimize friction between specimen and bearings
- a calibration instrument is to be devised in order to evaluate the test.

6 References

1. Krans, R.L. and Ven, M.F.C. van de (1994) Vergelijking van splijtproef met vierpuntsbuigproef voor asfalt; *CROW*, Wegbouwkundige Werkdagen.
2. Lim, I.L., Johnston, I.W., Choi, S.K., and Boland, J.N. (1994) Fracture testing of a soft rock with semi-circular specimens under three-point bending, part 1 and 2. *Int. J. Rock Mech. Min. Sci & Geomech. Abstr.*, Vol. 31, No. 3, p. 185.
3. Lim, I.L., Johnston, I.W., and Choi, S.K. (1994) Assessment of mixed mode fracture toughness testing method for rock. *Int. J. Rock Mech. Min. Sci & Geomech. Abstr.*, Vol. 31, No. 3, p. 199.
4. Adamson, R.M., Dempsey, J.P., and Mulmule, S.V. (in press) Fracture analysis of semi-circular and semi-circular-bend geometries; *Int. J. Fract.*
5. Ewalds, H.L. and Wanhill, R.J.H. (1984) *Fracture Mechanics*, DUM, Delft, The Netherlands.
6. Tan, S.-A., Low, B.-H., and Fwa, T.-F. (1994) Behavior of Asphalt Concrete Mixtures in Triaxial Compression. *JTEVA*, Vol. 22, No. 3, p. 195.
7. Planas, J., Elices, M., and Guinea, G.V. (1992) Measurements of the fracture energy using three-point bend tests. *Materials and Structures*, Vol. 25, p. 305.
8. Krans, R.L. and Ven, M.F.C. van de (1994) Crack growth experiments on asphalt concrete beams; SHRP congress, The Hague.
9. Ven, M.F.C. van de and Versluis, A. (1993) Mechanisch scheurgroeionderzoek in de vierpuntsbuigproefopstelling; NPC 92138-2 en 3.
10. Ven, M.F.C. van de and Versluis, A. (1994) Mechanisch scheurgroeionderzoek cyclische indirecte-trekproeven, NPC 93105-III, en mechanisch scheurgroeionderzoek cyclische directe-trekproeven, NPC 93105-IV.
11. Kleemans, C.P., Zuidema, J., Krans, R.L., Molenaar, J.M.M. and Tolman, F. (1995) Fatigue and creep crack growth in asphalt pavement materials, ESIS Conference (no proceedings).
12. Versluis, A. (1994) Orienterende driepuntsbuigproeven op halve kernschijven; NPC 94140.
13. Kooij, Jochem van der (1995) eenmalige SCB proeven bij 20 °C aan DAB, ongemodificeerd, met Multiphalt en met Inorphil; internal memo, RHED, Delft.

Fracture behaviour and bond strength of bituminous layers

E.K. TSCHEGG
Inst. of Applied and Tech. Physics, Technical University of Vienna, Austria
S.E. TSCHEGG-STANZL
Inst. of Met. and Physics, University of Agriculture Vienna, Austria
J. LITZKA
Inst. of Road Construction and Maintenance, Tech. University of Vienna, Austria

Abstract
One of the best known testing methods to determine the bond strength of bituminous layers are the pull-off test and the shear test. Both testing procedures, however, yield strength values (adhesive tensile strength or shear strength), which do not allow predictions on the fracture behaviour of an actual bond of bitumen layers and are not sufficient to numerically calculate and simulate the deformation and fracture behaviour of road constructions. Therefore a new fracture testing procedure has been developed, which allows to use cores from road constructions. In addition it yields fracture mechanical characteristic values besides the usual mechanical values. First experimental results, which have been obtained with laboratory specimens as well as with specimens from road pavements, are reported and the resulting new understanding of bond strength and fracture behaviour of bituminous layers is discussed.
Keywords: Bond strength, interface cracks, mode I fracture behaviour, specific fracture energy, temperature dependence of fracture mechanical values, splitting test.

1 Introduction

Formation and propagation of reflection cracks in bituminous pavements are unsolved problems in theory and practical application until today. It is necessary to characterize the mechanical behaviour of the basic components in order to be able to calculate, model and simulate reflection cracks according to Francken [1]. As an important information, the fracture behaviour and bond strength of bonds between the asphalt layers, has to be determined. Relevant guidelines for the testing of such adhesion bonds, as well as test equipments were developed in the past. pull-off and shear test are the best known experimental methods. Both test procedures supply strength values (adhesive tensile strength and shear strength) which, however, make no statements on the fracture behaviour of the bonds [2]. Testing methods and equipments to characterize the fracture mechanical

Reflective Cracking in Pavements. Edited by L. Francken, E. Beuving and A.A.A. Molenaar. © 1996 RILEM. Published by E & FN Spon, 2–6 Boundary Row, London, SE1 8HN. ISBN 0 419 22260 X.

characteristics of bonds between bituminous layers of road constructions are not reported in the relevant literature so far.

Several studies, however, exist on the fracture mechanical characterization of cement bonded materials [3-6, e.g.]. The influence of the pretreatment of the old concrete surface on the adhesion and the fracture mechanical behaviour of old-new concrete bonds have been investigated. It showed that the conventional fracture mechanical concepts (like stress intensity concept, J-integral, and COD concepts) are not appropriate for a characterization of these materials. There are two main reasons for this: i) Exact measurement of crack lengths is absolutely necessary in order to apply the above mentioned concepts. It is very difficult or even impossible, however, to measure the crack length in heterogeneous materials (consisting of matrix and aggregates) exactly enough, as the crack tip is placed between process zone in front of the crack tip and the grain bridging zone behind the crack tip. The transition between these two areas is continuos and cannot be recognized and defined exactly. ii) The stress distribution around the crack tip differs from that of ideally elastic materials, being assumed for these concepts. These two restricting reasons are likewise effective for bonds between bituminous layers. In addition, specimen size and shape may influence the experimental results in conventional fracture concepts. Asphalt layers and bonds with thicknesses between 10 and 100 mm are critical in this respect and make fracture testing difficult.

A simple and efficient concept to characterize the fracture behaviour of disordered materials is the "total work of fracture" concept [7], which has been successfully applied to concrete, asphalt, wood, fiber reinforced polymers and bonds between old and new concrete [2-6]. Crack propagation is stable during the whole experiment until final fracturing takes place, so that the complete load displacement diagram (F-δ curve) may be recorded. This curve contains all informations to fully characterize the fracture behaviour of bonds.

The pull-off test is not appropriate to determine load displacement curves, as separation of the layers takes place mostly with unstable crack propagation. The maximum load can be measured and from this the bond strength may be derived. These values, however, do not allow to decide whether a material bond has broken in a "ductile" or "brittle" manner and therefore do not render relevant fracture mechanical data. First fracture mechanical tests on bonds of bituminous layers are reported in [9]. Cubic specimens of bond material were broken with the wedge splitting technique according to Tschegg [10] and the load displacement diagrams were recorded. The basic difference between adhesive strength and crack growth resistance of bituminous asphalt bonds has been demonstrated very effectively in this study for the first time. The investigations are continued in the present paper. The wedge splitting method was further developed and improved, employing core specimens as used for other tests of bituminous pavements. Systematic fracture tests were performed on laboratory specimens as well as on cores of road pavements with typical asphalt bonds in a temperature range of -22° to 15°C.

2 Principle of the testing method

Principle of the wedge splitting test according to [10] and its handling are extensively described in [11,12] and therefore treated only shortly in this paper. The principle of the testing method is shown schematically in Fig. 1. Core specimens (see Figure 2) are placed on a narrow linear support in a compression testing machine. The specimen has a rectangular groove with a starter notch in the interface (bond between die bituminous layers) at the bottom of the groove, where the wedge splitting facility is placed. Wedge, starter notch, interface and linear support area are positioned in the same vertical plane.

The wedge transmits a force F_M from the testing machine to the specimen. The slender wedge exerts a large horizontal component F_H and a small vertical force component F_V on the specimen. The horizontal component splits the specimen along the interface, similar as in a bending test.

Splitting must take place during stable crack propagation until of the specimen is complete separated. This is possible with a stiff loading system only (wedge-loading system + testing machine). The wedge loading device itself is extremely stiff, i.e. the force flux is short and direct and so little deformation energy is stored. Therefore, such experiments can be performed for example with usual mechanical spindle driven machines together with the described loading device.

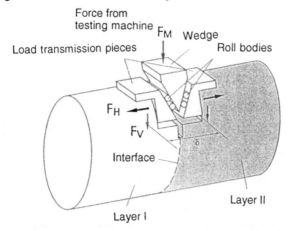

Figure 1: Principle of wedge splitting method with cylindrical specimen according to [10].

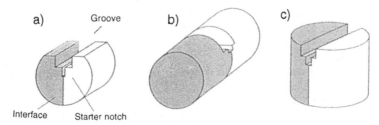

Figure 2: Specimen shapes for the wedge splitting test to determine fracture mechanical values of bonds between bituminous layers.

The force F_M is determined with a load cell in the testing machine. Knowing the wedge angle (the wedge angle is 7.5° in these tests), the force F_H in horizontal direction can be simply calculated. The vertical force is small and does not influence the results [11]. Friction between wedge and load transmission pieces is minimized by use of roll bodies (or needle bearings) and thus may be neglected (mean error approximately 1%) [11].

The load displacement δ is measured on both ends of the groove on the specimen surface with two electronic displacement gauges. The two measurements of δ serve to obtain mean values of δ on one hand and to be able to detect whether the crack front has propagated parallel to the starter notch in the direction of linear support area on the other hand. If the resulting values (δ_1 and δ_2) differ by more than 20 % these measurements are not evaluated any more. Force F_M and displacement δ_1 and δ_2 are recorded by an electronic data logger; from this, the load- displacement curve (F_H-δ curve) is obtained.

3 Specimen, material, experimental details and data evaluation

In this study, cylindrical specimens as shown in Fig. 1 and Fig. 2b have been used exclusively. The diameter of the cores was 150 mm and the total length 160 mm. The bond area was placed in the center of the specimen, so that the thickness of the layers aside the interface was approximately 80 mm. The groove was introduced symmetrically into the interface with a stone saw and was 40 mm wide and 20 mm deep. The starter notch was cut with a 3 mm thick saw blade. Its depth was 55 mm. Three different types of specimens have been used, from which two were produced in the laboratory (called L1 and L2 in the following) and one has been taken from a bituminous pavement of motor way A21 in the east of Austria (called A21 in the following).

3.1 Specimen preparation of L1 and L2
From a bituminous base layer sample of type BT I 16 (crushed aggregates, maximum grain size 16 mm, binder content 5 % of 100 pen-bitumen ; according to RSV 8.05.14 [13]) plates of 80 mm thickness and with dimensions of 400 x 1500 mm were produced. Compaction was performed with a vibration roller. Thus the asphalt aggregate mixture and production of the layer was similar to that, which is used in road construction. The plate surfaces were treated with a water jet until all binders and fine aggregates were removed from the surface. In this way a road with traffic was simulated. Then, a second, 80 mm thick layer of BT I 16 was placed on this surface without any tack coat as test series L1. For series L2, a polymer modified bitumen emulsion (HB-60 K/PM with a bitumen content of 60%, polymer is styrene butadiene styrene SBS, approximately 3% by weight of binder content) was sprayed on the surfaces. The amount of this tack coat was 0.2 kg/m^2. Then, a 80 mm thick layer of BT I 16 was applied and compacted. Cores with a diameter of 150 mm were taken from these plates.

3.2 Specimens of A21
Cores of road A21 were taken from the area between the two wheel-tracks. They had the following structure: Two layers of 90 mm thick BT I 22 (crushed aggregates, maximum grain size 22 mm, binder content 5 % of 100 pen-bitumen ; according to RSV 8.05.14 [13]) were connected with the classical emulsion Colas V40 (amount: 0.2 kg/m^2) as a tack coat . The road was frequented by heavy traffic during several years, so that some rutting appeared in the first track. Few specimens were taken directly from the ruts.

3.3 Testing conditions
To test the influence of temperature on the quality of the interface bonding, six temperatures haven been chosen. The specimens were stored about 30 hours in cooling chambers at -22°, -10°, -5°, 0°, 5° and 10°C. The accuracy of the chamber temperature control was +/- 1°C.

Testing was performed with a mechanical spindle machine with a load capacity of 100 kN, and a cross head speed of 2 mm/sec [2]. As the whole testing procedure (mounting plus testing) did not last more than 5 - 7 minutes, the specimens did not warm up substantially during testing, though no cooling chamber in the testing machine was provided.

3.4 Data evaluation

The area under the load displacement curve is directly proportional to the absorbed energy G to fracture the specimen. Division of this energy by the nominal (projected) fracture area A, renders the material specific value G_f, called specific fracture energy:

$$G_f = G/A$$

G_f is independent of specimen shape and size, if a minimum specimen size is provided. It characterizes the resistance against crack propagation. It has been demonstrated in [2] that a specimen diameter of 150 mm is sufficient, so that the "size-effect" does not play a role.

A nominal notch tensile strength σ_{ns} may be obtained from the maximum load F_{max} of the F-v- diagram [2], which is in some way similar to the maximum load of the pull-off test.

$$\sigma_{ns} = F_{max}/A + M/W$$

with M being the applied moment of force and W the moment of resistance of the specimen.

Characteristic fracture values may be obtained from the F-δ diagram for FE-simulations and modelling was performed with simple calculation procedures in a similar way as for asphalt aggregate mixtures [2].

In order to determine the scatter of the results, at least 3 to 4 specimens of each specimen type were tested at each testing temperature . The mean standard deviation of the σ_{ns}-values was 12% and of the G_f-values 9%.

4 Experimental results

4.1 Typical load - displacement curves

The load displacement curves of the L2 series (laboratory specimens with polymer modified binder) are shown in Fig. 3. These diagrams demonstrate that the results are completely different for different temperatures, which means that they reflect different fracturing behaviour quite well.

Linear elastic behaviour is observed for the initial part of the load - displacement curve. The "elastic line" is steeper for low temperatures which is caused by a higher modulus of elasticity. With decreasing temperature, the displacements δ corresponding to F_{max} decrease and the region where the maximum force, F_{max} effective, becomes smaller and smaller. Plastic behaviour (flat elongated curve, extensive plastic deformation) of the bonding area is typical for temperatures around 10°C. Mechanisms being responsible for this typical behaviour of bonds between bitumen layers are extensively discussed in [2].

Crack propagation starts after the maximum of the tensile load F_{max} is obtained. At low temperatures, the decreasing part of the load displacement curve is very steep, which corresponds to brittle separation of the material. Contrary to this, the curve is flatter at higher temperatures after the maximum load is obtained which points to ductile material separation.

4.2 Maximum horizontal force - temperature dependence (F_{max} vs. T)

The maximum horizontal forces, F_{max} resulting from wedge splitting tests at different testing temperatures are shown in Fig. 4. The F_{max}-T curves and σ_{ns}-T curves,

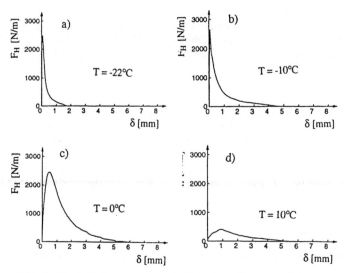

Figure 3: Typical load-displacement curves (splitting force F_H versus displacement δ) of
bond BT I on BT I with 0.2 kg/m² polymer modified binder at testing temperatures
between -22° and 10°C.

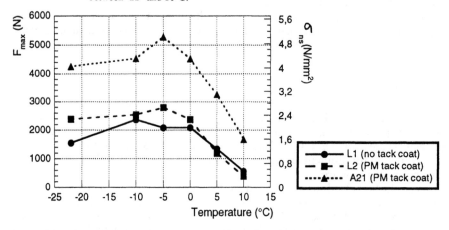

Figure 4: Splitting force F_{max} and nominal notch tensile strength σ_{ns} versus testing temperature for
bonds without and with tack coat, as measured with laboratory specimens (L1, L2) and cores
from roads (A21).

respectively, are in principle similar for all three specimen types. High values are obtained at low temperatures, and the curves are almost horizontal in the low temperature
region. At temperatures between -5 and 0°C, a more or less pronounced maximum is attained, and after this, the curves drop down steeply at a testing temperature of approximately 10°C. The bond looses much of its strength in this temperature range. An increase of the strength by a binder is obtained for laboratory specimens at temperatures
below approximately 0°C only.

The absolute F_{max}- and σ_{ns}-values from laboratory specimens and specimens from the road differ very much, though the structure (asphalt of layers and binder) of the three specimens is not too different. This result needs some discussion.

4.3 Specific fracture energy - temperature dependence (G_f vs. T)

The specific fracture energy value G_f is more appropriate to characterize the adhesive power and the resistance against crack growth than the maximum load value F_{max}.

All G_f vs. temperature curves (Fig. 5) show an increase of the G_f values from -22°C to approximately 0°C, where a maximum is obtained. The curves decrease towards higher temperatures and obtain similar values at 10°C as for -22°C. A polymer modified binder as tack coat increases the resistance against crack propagation by a factor of 2 in comparison with a bond without tack coat in the temperature range of -22° to 5°C. The curve of road specimens is higher than of laboratory specimens, and the difference is especially high in the temperature range between -5°C and +10°C. Reasons for this result are discussed in the following chapter.

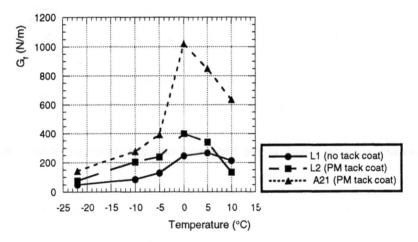

Figure 5: Specific fracture energy G_f versus testing temperature for bonds without (L1) and with (L2) tack coat, as measured with laboratory specimens and cores from roads (A21).

5   Discussion and conclusions

5.1   Testing method and accuracy

Advantages of the wedge splitting technique are its easy handling, its simplicity, favorable specimen shape and relatively easy specimen preparation. Drawbacks are, that deeply notched specimens may be damaged easily if they are transported or stored or mounted into the testing machine in an inappropriate manner.

The standard deviation of the experimental results of approximately 10% is acceptable in comparison to other mechanical tests and may be attributed to inhomogenities of the specimen of the bond, as has been studied and discussed extensively in [2].

5.2 F_{max}- / σ_{ns}- temperature- and G_f-temperature   relationship

The results of this and the study of [2] all show F_{max}- or σ_{ns} vs. temperature, as well as G_f vs.temperature curves, which are typical for laboratory as well as for core specimens with different bonds. A schematic curve is reproduced in Fig. 6. The F_{max} or σ_{ns} curves start with high values at low temperatures and drop steeply at a temperature between

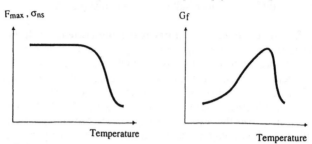

Figure 6: Schematic curves of a) max. splitting force F_{max} and nominal notch tensile strength σ_{ns} and b) spec. fracture energy G_f versus testing temperature of bonds between bituminous pavement layers.

-5°C and 10°C. Contrary to this, the G_f curves start with low values at low temperatures, increase then and obtain a maximum value at around 0°C, from where they drop steeply if the temperature is raised to 10°C. Depending on the tack coat agent and composition of the two layers, this maximum may be shifted to somewhat lower or higher temperatures [2]. Figure 6 clearly shows, that the F_{max}- and G_f vs. temperature-curves are in principle neither identical, nor similar and therefore do not represent similar properties of the tested bond material. Therefore no statements on characteristic fracture mechanical properties of the adhesive bond can be derived from the measured F_{max}- or. σ_{ns}- values. The wedge splitting method reveals load displacement curves and resulting characteristic values (e.g. G_f and strain softening curves) [2], which describe the fracture behaviour extensively and serve as basic data for FEM simulations and numerical analysis. A simple case of reflecting crack formation in a two layer asphalt road pavement has been described and simulated in [2] with the aid of an FEM calculation (considering the measured fracture characteristic values of the asphalt and the bond). These studies show new and interesting results for scientists and engineers, though many simplifying assumptions were necessary.

5.3   Fracture behaviour of bonds in laboratory and core specimens
The absolute values of the F_{max}- and G_f-curves of laboratory specimens and such, which were obtained from road pavements, are different. This observation proved right for specimens, which were tested in this study, as well as in [2]. In [2], specimens from the outer road parts (without traffic) were tested and similar results, as for laboratory specimens were obtained i.e. lower F_{max} and G_f values, than for traffic loaded road specimens. It is assumed that the reason for the higher values of the specimens from pavements of roads with traffic load are better compaction and compression of the bond by the acting traffic load. This assumption is based on two mechanisms:

(i) Experience in glueing technique [2] tells, that the bonding agent is squeezed better into the pores of a fine structured (rough) surface of two layers, if the contact pressure is high, so that the effective area for the adhesion between binder and layer is increased.

(ii) Higher pressure between the two layers favors indenting of the aggregates in the layers in the close vicinity of the bond area. Thus the energy dissipation for crack formation in the interface is increased by the energy needed to break out the aggregates ("grain bridging" effect). In addition, the resistance against a displacement of the layers during shear loading is increased. This interpretation is supported by the result of a practical study [14], saying that bonds of layers with crushed aggregates are more resistant against damage than layers with natural gravel.

Though the laboratory specimens were compacted by a vibrating roller, the traffic load obviously has a favorable effect on the bond quality. If, however, ruts are formed by the traffic load, the bond is damaged, as has been found in studies [2] of cores of the A21 road. This study shows that the G_f values of the ruts are certainly lower at temperatures below 0°C than of the areas between the wheel paths. It has to be mentioned, however, that these results have been obtained from a few number of tests only. At temperatures above 0°C, on the other hand, the G_f values, i.e. the resistance against crack propagation, was found to be approximately the same for the two areas.

5.4 Improvement of bonds

Fracturing is much influenced by structural changes and their energy consumption (micro cracking and bridging mechanisms, as well as debonding and plastic deformation) in the bond and adjacent areas of the layers [2]. An improvement of the "quality of the bond" may be obtained by increasing the energy consumption of these structural changes during crack formation and propagation.

Considering an ideal bond (the interface is not considered as an inhomogeneity towards the inner structure of the layers), adhesion tensile strength and resistance against crack propagation are the same as for the layer itself. The crack does not find a pregiven area for its propagation, is deflected by aggregates and forced to run in a tortuous way, thus consuming a maximum fracture energy. This energy consumption is increased by roughening the contact surface of the asphalt layers [2].

Tack coates increase the adhesion of layers only slightly compared to mechanical procedures; improvements of layer bonds are mainly obtained by mechanical procedures (roughening), by forming such geometries of the interfaces, as being effective in an ideal bond.

Promising equipments have been developed in recent times to roughen bituminous contact areas. With a special high pressure water jet technique, roughening of the bond area of several millimeters and subsequent careful cleaning of the bond area is performed [16], which are important parts of the whole procedure. An improvement of the quality of the layer bond adhesion in pavement constructions against tension and shear loading may be expected by the use of such promising procedures in comparison to the conventional use of spraying tack coats.

5.5 Comparison of pull-off test and splitting test results

The pull-off test is discussed in comparison to the results obtained with the described wedge splitting technique, as a lot of data and experience from pull-off tests at 0°C and 10°C exist in Europe already.

Cores have been taken from roads with different adhesive bonds in [17]. These bonds were tested with the wedge splitting technique at 0°C and 10°C. It showed that the F_{max} values of the splitting tests are always proportional to the maximum load values of the pull-off test, though the absolute values turned out to be different. If the typical F_{max}/σ_{ns} vs. temperature dependence (Fig. 6) is applied to the pull-off test, the following features are recognized: Most test instructions for pull-off tests provide testing temperatures of 0°C and 10°C. The test results at 0°C characterize the properties at low temperatures, as the curve in Fig. 6 drops steeply only at temperatures above 0°C. According to this, too low temperatures in the pull-off test cause small measuring errors only (the curves are almost horizontal at temperatures below 0°C), which is in contrast to the temperature regime above 0°C (steep drop of the curves). The higher temperature regime is characterized by measurements at 10°C, where the curve becomes horizontal again (this part is not drawn in Fig. 6). The choice of 0°C and 10°C as testing temperatures for the pull-off test therefore characterizes two typical and interesting regimes of the bond strength of bituminous material. The testing temperatures of 0°C and 10°C should be controlled exactly and deviations tolerated only, if they are below 0°C for the nominal 0°C measurements and above 10°C for those at nominal 10°C.

7. Acknowledgment

The authors thank MR. Dipl.-Ing. Fichtl and MR. Dipl.-Ing. Dr. H. Tiefenbacher for stimulating discussions. This publication is a part of a research project which is supported by the Federal Ministry for Economic Affairs, contract no. 3.125.

8. References

1. Francken L.(1993) Reflective Cracking in Pavements, ed. by J.M. Rigo, R. Degeimbre and L. Francken, RILEM, E&FN Spon, London, pp 75-99
2. Tschegg E.K., Tschegg-Stanzl S.E. and Litzka J., Charakterisierung der Hafteigenschaften von Asphaltschichten, Schriftenreihe Straßenforschung, Heft 456, BMWA, Wien, 1996
3. Tschegg E.K. and Tschegg-Stanzl S.E., Adhesive power measurements of bonds between old and new concrete, J. Mat. Sci., 26, 1991, 5189-5194
4. Tschegg E.K. and Tschegg-Stanzl S.E., Adhesive power of bonded concrete, Proc. RILEM/ESIS Conf. Fracture Processes in Brittle Disordered Mat.: Concrete, Rock, Ceramics, J.G.M. Mier, J.G. Rots, and A.Bakker, E&FN, London, 1991,809-818
5. Harmuth H., Investigation of the adherence and the fracture behaviour of polymer cement concrete, Cement and Concrete Res., Vol. 25, No.3, 1995, pp 497-502
6. Tschegg E.K., Tan.D.M., Kichner H.O.K., Tschegg-Stanzl S.E., Interfacial and sub interfacial fracture in concrete, Acta Metall., 41 (2), 1993, pp 569-576
7. Nakayma J., Bending Method for Direct Measurements of Fracture Surface Energy of Brittle Materials, Jpn.Appl.Phys. 3 (7), 1964, 422-423
8. Tan D.M., Tschegg E.K., and Tschegg-Stanzl S.E., Fracture Mechanical Characterization of Asphalt Aggregate Mixtures at Different Temperatures, Proc. of "Localized Damage III" Computer-Aided Assessment and Control, ed. M.H.Alisbadi et al., CPM Southampton Boston, 1994, 217-224
9. Tschegg E.K., Kroyer G., Tan D.M., Tschegg-Stanzl S.E. and Litzka J., Investigation of Bonding between Asphalt Layers on Road Constructions, ASCE J. of Transportation Engineering, Vol 121, No4, 1995, 309-316
10. Tschegg E.K., Equipment and appropriate specimen shapes for tests to measure fracture values AT No. 390328, Austrian Patent Office, Vienna,
11. Tschegg E.K., New Equipments for Fracture Tests on Concrete, Materialprüfung (Materials Testing) 33, 11-12, 1991, 338-342
12. Tschegg E.K., Tschegg-Stanzl S.E. and Litzka J., New Testing Method to characterize Mode I Fracturing of Asphalt Mixtures, Proc. of the 2nd RILEM Conf. "Reflecting Cracking in Pavements", C.R.R., Liege, Belgien, ed. J.M. Rigo, R. Degeimbre and L. Francken, F&FN SPON , 263-270
13. RSV 8.05.14, Directions and prescriptions for road constructions, "Oberbauarbeiten (ohne Deckarbeiten), bituminöse Tragschichten im Heißverfahren" Forschungsges. für das Verkehrs- u. Straßenw., Vienna, Austria, 1994.
14. Fenz F., Gregori H. and Krzemien R., Die Verklebung von Asphaltschichten, Schriftenreihe Straßenforschung, Heft 316, BMWA, Wien, 1987
15. Tschegg E.K.,W..Zikmunda and Tschegg-Stanzl S.E., Improvements of New-Old Concrete Bonds in Road Constructions -Procedure and Testing Methods, Proc. of the 7th Intern. Symp. on Concrete Roads, Vienna, Vol. 2/3, 1994, 51-56
16. Egle K., FRIMOCAR, Haag, Schweiz, private communication, Jan. 1995,
17. Krzemien R and Tschegg E.K., Schichtverbund, Asphaltprüfung und Qualitätsabzüge, Schriftenreihe Straßenforschung, Heft 444, BMWA,Wien, 1995

Mixed mode fatigue crack propagation in pavement structures under traffic load

J. ROSIER, CH. PETIT, E. AHMIEDI and A. MILLIEN
Civil Engineering Laboratory, University of Limoges, IUT, Egletons, France

Abstract
This paper is about mixed mode fatigue due to traffic load. Last works assume parameters of Paris law in opening mode in order to predict life time of pavements structures. A new fatigue crack test in mixed mode has been conceived. The obtained results indicate that mixed mode parameters differ from those in opening mode. Finite element models of cracked pavements structures under traffic load show that they give a lower crack velocity than mode I parameters. This methodology contributes to better modelize fatigue of pavement structures under traffic loads.
Keywords: Crack propagation, Mixed mode, Fatigue, Bituminous concrete, Finite element model.

1 Introduction

All pavement structures exhibit almost pronounced damage. As the most expensive maintenance costs are due to fatigue crack propagation from the base through the rolling layers, the main problem is now "how to predict and how to retard the crack propagation velocity".

Our objective is to better evaluate the life time of pavement structures. We consider pre-existing transverse cracks in base layers of rigid or semirigid structures. For these structures, the lack of knowledge of the experimental intrinsic fatigue parameters does not allow accurate numerical lifetime predictions [1][2][3]. It is the reason why we have conceived a new testing fatigue apparatus [4], in order to evaluate precisely the fracture mechanic characteristics of each layer and of their interfaces.

We present here experimental results of crack fatigue in mixed mode [5], which are introduced in a finite element model of pavement structures under traffic load. A

Reflective Cracking in Pavements. Edited by L. Francken, E. Beuving and A.A.A. Molenaar. © 1996 RILEM. Published by E & FN Spon, 2–6 Boundary Row, London, SE1 8HN. ISBN 0 419 22260 X.

generalized Paris law is proposed, taking into account coupling effect. The influence of mixed mode is identified and modelized.

2 Mechanisms of fatigue crack propagation

Two phases of fatigue damage can be identified whatever the solicitation (Fig. 1):
• an initiation phase also called damaging;
• a fatigue propagation phase, studied in this paper within the frame of the fracture mechanics theory (Paris law).

Although these two phases have been observed through our experimental work, we restrict our study to the second one.

Pavement structures are subjected to two types of mechanical solicitations (thermal and traffic load). Most of the authors consider the solicitation scheme of pavement as follows:
• mixed mode often considered as pure shear mode [1][2][3] (Fig. 2a, 2c);
• pure opening mode illustrated in Figure 2b (under traffic load, this mode is often neglected) and in Figure 2d (thermal effect).

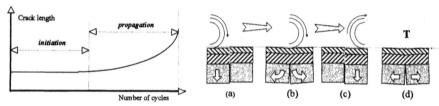

Fig. 1. Phases of fatigue damage   Fig. 2. Crack pavement solicitations

3 Experimental process and results

3.1 Fatigue crack testing apparatus and samples

We have conceived a new crack test, which enables us to reproduce opening mode and mixed mode in order to identify the intrinsic parameters of bituminous materials in both solicitation type [4]. This apparatus allows investigations in the present limits:
• bituminous materials whose granular maximal dimension is less than 10 mm;
• frequency range of mechanical solicitations included between 1 Hz and 10 Hz;
• test temperature included between -40°C and +40°C.

This test machine COLAREG 001 is now under European certification. Figures 3 and 4 present a schematic view of its main features. Choice and optimisation of sample and test types proceed of several elements [4].

Concerning opening mode, after a review of the literature, as our objective is to obtain accurate experimental results, we decide to use mechanical simple tests and specimen geometry. So, a three point bending test is adopted, and stress intensity factor K_I is calculated by finite element method with a plane strain assumption [6]. Specimen geometry and dimensions are presented in Figure 5.

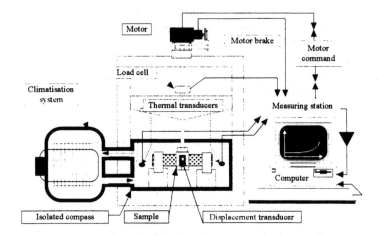

Fig. 3. COLAREG 001 testing machine

Opening Mode   Mixed Mode

Fig. 4. Mode I and II test configurations of COLAREG 001

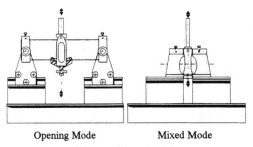

A L = 300 mm; h = 100 mm;
e = 50 mm

B L_c = 100 mm; L_e = 50 mm

Fig. 5. Three point bending (A) and modified compact shearing (B) samples

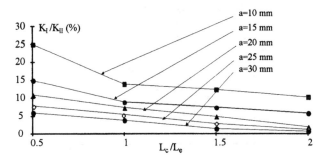

Fig. 6. Parametric evolution of solicitation modes ratio

On the other hand, for shear mode, we have selected a modified compact shearing concept. In order to minimise opening mode effects, the dimensions of the samples are optimised by the mean of a parametric finite element study. Our numerical calculations show that the opening mode contribution decreases as the ratio L_c/L_e increases and as the length of the cuts a increases (Fig. 6). This optimisation process leads us to the specimen geometry and dimensions presented in Figure 5.

For both modes, the length of the initial cuts is chosen as 25 mm, in order to allow the use of crack propagation gages.

It is worthwhile to point out that, according to these dimensions, opening mode and shearing mode can be investigated with the same experimental device.

3.3 Exploitation process of experimental tests

The mechanical fatigue tests presented in this paper are performed, for both pure opening mode and mixed mode investigation, at the constant temperature $T = 0°C$ and at the solicitation frequency $f = 1Hz$. Two bituminous materials are studied: a sand asphalt and a bituminous concrete.

3.3.1 Pure opening mode

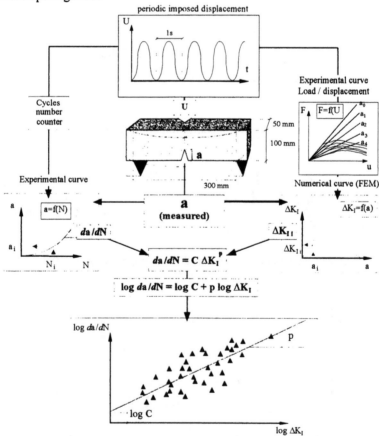

Fig. 7. Exploitation process of a fatigue crack test

The fatigue behaviour description of bituminous materials starts, for the opening mode, with the Paris law, which relates the crack propagation velocity and the stress intensity factor K_I as follows :

$$\frac{da}{dN} = C_I \left(\Delta K_I\right)^p \tag{1}$$

where : a is the crack length, N the number of cycles, $\Delta K_I = K_I^{max} - K_I^{min}$, and C_I, p are fatigue parameters, depending on the solicitation frequency and the temperature.

We present in Figure 7 the opening mode identification protocol used to determine the intrinsic coefficients C_I, p of the Paris law (1) for the sand asphalt and the bituminous concrete.

3.3.2 Mixed mode

Most of the mixed mode fatigue theoretical investigations consider that fatigue cracks always propagate in pure opening mode direction (defined as K_I*) [7].This assumption enables the user to restrict experimental studies to pure opening mode .

In this paper, we propose to investigate in a generalised Paris law [8][9] in mixed mode :

$$\frac{da}{dN} = C \left(\Delta K_{eff}\right)^p \tag{2}$$

where ΔK_{eff} is an effective stress intensity factor variation depending on ΔK_I and ΔK_{II}.

In mixed mode, the determination of C and p parameters must be based on coupled processes. The main difficulty of the exploitation of the mixed mode experimental results is due to the different histories of the four crack propagation paths.

So, a specific finite element model must be associated to each experimental test. The experimental follow up of the test history enables us to simulate by finite element models the propagation's process in order to identify ΔK_I and ΔK_{II} against crack length a. The Figure 8 presents an example of the rebuilding of a mixed mode propagation test: the experimental propagation paths as function of the numbers of loading cycles and the associated finite element mesh are presented respectively in the Figures 8a and 8b.

The analysis of experimental results in mixed mode allows us to assert that the parameters C and p of the generalised Paris law (2) depend on the proportion of the shearing. So they are not intrinsic material parameters. The following expressions are obtained with a best fitting (at least square sense) of the experimental results:

$$\Delta K_{eff} = \left(\Delta K_I^4 + 3.5\ K_{II}^4\right)^{1/4} \tag{3}$$

$$p = 3.5 \frac{\Delta K_I}{\Delta K_I + \Delta K_{II}} \tag{4}$$

$$C = 10^{-3} \frac{3.53\ 10^{-6} \Delta K_I + 4.67\ 10^{-3} \Delta K_{II}}{\Delta K_I + \Delta K_{II}} \tag{5}$$

These first results are based on a few tests performed on the bituminous concrete. The parameters (3) (4) (5) are in reasonable agreement with all results except for pure K_{II} mode ($K_I/K_{II} < 10\%$) [6].

An automatic propagation procedure has been implemented in Castem 2000 in order to always have an optimum mesh particularly at crack tip. The fracture mechanic computation has been treated with conservative laws such as path independent integrals (G_θ ([10][11][12], Integral A [13]). These methods are very interesting for the very accurate results obtained particularly for multilayered structures [14][15] and in mixed mode cases because in most of situations, we ought to have the separated mode K_I and K_{II} to supply crack propagation law. Effectively, we propose here to solve the modified Paris law in order to take into account propagation in mixed mode.

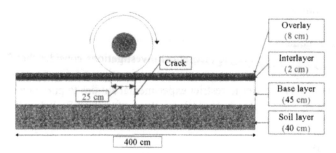

Fig. 9. Road structure studied

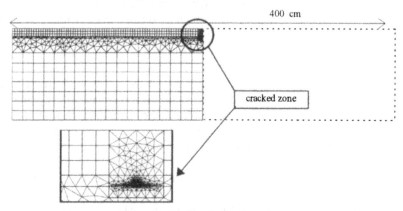

Fig. 10: Finite element mesh.

4.2 Numerical results

The evolutions of the stress intensity factors ΔK_I and ΔK_{II} with the crack length a are presented respectively in Figures 11a and 11b. ΔK_{eff} is calculated with equation (3). We can observe that the mode I is lower than mode II but not null such as most of the studies before [2]. So according to the modified Paris law proposed with our experimental results, we can say that mode I probably has a great effect on fatigue crack propagation velocity. ΔK_{II} is rather constant in the bituminous concrete layer except in the last two centimetres, near the top. This view confirms the large effect of mode II in the last centimetres of propagation in the upper layer. This conclusion has already shown by some authors before. The curve K_{eff} is very near the curve K_{II}, according to the experimental identification given in equation (3).

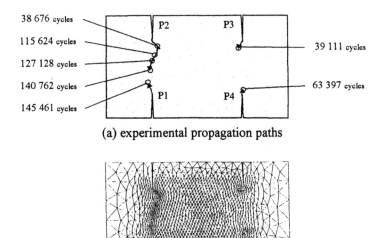

(a) experimental propagation paths

(b) finite element mesh

Fig. 8. Crack propagation paths in mixed mode

4 Application to fatigue crack propagation in shear mode in pavement structures

4.1 Presentation of the pavement structure

The structure is composed of four layers represented in Figure 9; we can see a soil layer under a cement treated layer. We consider an existing crack in this layer, due to shrinkage behaviour. Then, two intact upper layers made of sand asphalt and bituminous concrete are considered. The finite element method is used to study the mechanical behaviour of the structure. In this work, we assume a linear elastic behaviour (Table 1) of the four layers components and consider plane strain hypothesis.

Table 1. Elastic parameters of the different layers

Layers types	E (MPa)	ν
Soil	100	0.25
Cement grave (base layer)	20 000	0.25
Sand asphalt (interlayer)	6 700	0.35
Bituminous concrete (overlay)	9 200	0.35

In order to apply shear mode experimental results, we study the crack propagation under traffic load only. The load is located in the maximum K_{II} according to previous works [2]. In fact, the pressure is just before or just after the crack (Fig. 9) and we apply a pressure of 0.67 MPa on 25 cm width area.

A finite element model has been used in order to determine displacement, stress and strain fields. The finite element mesh is represented in Figure 10.

150 *J. Rosier* et al.

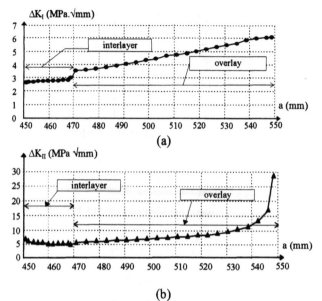

(a)

(b)

Fig. 11: Evolutions of ΔK_I (a) and ΔK_{II} (b) with crack length

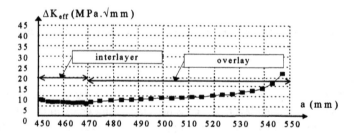

Fig. 12 : Evolution of ΔK_{eff} with crack length

The different parameters of modified Paris law (2) are then identified : the stress intensity factors are computed by numerical integral A with finite element results, then the previous parameters p (4) and C (5) of both materials are introduced in equation (2). Finally, the number of cycles at failure is given by numerical integration of this equation.

The parameters p and C are obtained from the experimental crack fatigue tests. In this work, several results are available in mode I, so the obtained values are significant. However, only few tests have been made in mixed mode. Currently, we can propose an approximate life time computation of the structure under traffic load only. This load is modelized by an uniform pressure at the location defined before.

In the present work, we assume the fatigue parameters presented in Table 2. Mixed mode parameters have been determined from (4) and (5) equations assuming a ratio K_I/K_{II} constant through each layer ($K_I/K_{II} \approx 0,65$ is, for example, verified in almost 7 cm of the overlay according to Figure 11).

Table 2. Experimental fatigue parameters

	Mode I		Mixed mode	
	p	C	p	C
Bituminous concrete 0/10	3.50	$3.53\ 10^{-9}$	1.38	$2.83\ 10^{-6}$
Bituminous sand 0/6	1.24	$3.24\ 10^{-6}$	/	/

The experimental results given in Table 2 enable us to estimate the number of cycles at failure (≈ 270 millions).

The life time computed here correspond to a structure only submitted to traffic load. This value is obviously high, as confirmation that other effects such as thermal ones must be taken into account in order to determine realistic life time of pavement structures.

5 Conclusion

In this paper, we want to go on a methodology already began in the two last conferences in Liege [2][3][4] to evaluate the real life time of pavement structures. Few authors have already studied fatigue crack propagation by finite element model with thermal load effect and traffic load effect uncoupling. Most of them consider the Paris law and use experimental results in mode I. The hypothesis is valid in the thermal effect case because the assumption of pure mode I is verified in this case. But, under traffic load most of authors consider mode II but assume mode I experimental values.

So, in this work, we have chosen to investigate in a mixed mode crack fatigue apparatus in order to identify a fatigue law of asphalt materials. First tests have shown that mode I Paris law is not valid in mixed mode; it is the great conclusion of this paper in order to modify some finite element models, which consider only mode I parameters. In a second time, we propose a generalised Paris law according to Bilby [8]. The identification of the parameters shows that this generalisation leads to non intrinsic parameters, functions of the stress intensity factors in mode I and II. Otherwise, we have a better evaluation of mixed mode crack fatigue law. It enables us now to go on mixed mode tests and to identify a more significant fatigue law.

In perspective, in order to obtain more representative life times, we have to take the following points into account:
- these tests show us that coupling mode I and II is existing and must be taken into account (traffic load K_{II} and thermal effect K_I);
- low levels of K_I can have an important effect (p is higher in mode I than in mode II), so the pressure given by traffic load must be moved to obtain maximum K_I and not necessary only maximum K_{II};
- advances in finite element modelling with more mechanical realistic experimental results enable us to get optimum pavement structure with different interlayers in order to increase life time/cost ratio.

Today we envisage to study properties of the interface between two layers to complete our mechanical approach.

6 References

1. Jayawickrama, P.W., Smith, R.E., Lytton, R.L. & Tirado-Crovetti, M.R. (1989) Development of asphalt concrete overlay design program for reflective cracking, *First int. Conf. on Reflective Cracking in Pavements*, RILEM Conference, J.M. Rigo and R. Degeimbre, Liège, pp.164-169.

2. Vergne, A., Petit, C., Zhang, X., Caperaa, S. & Faure, B. (1989), Simulation numérique de la remontée d'une fissure dans une structure routière, *First int. Conf. on Reflective Cracking in Pavements*, RILEM Conference, J.M. Rigo and R. Degeimbre, Liège, pp.182-189.

3. Petit, C. & Caperaa, S. (1993) Influence of modulus ratio on crack propagation in multilayered pavements, *Second Int. Conf. on Reflective Cracking in Pavements*, RILEM Conference, E & FN Spons, London, pp.220-227.

4. Caperaa, S., Ahmiedi, E., Petit, C. & Michaut, J.P. (1993) A new test for determination of fatigue crack characteristics for bituminous concretes, *Second Int. Conf. on Reflective Cracking in Pavements*, RILEM Conference, E & FN Spons, London, pp.193-199.

5. Ahmiedi, E., Bressolette, Ph., Rosier, J. & Vergne, A. (1995) Comportement de structures multicouches fissurées sous sollicitation de fatigue - Application aux structures routières, in *Des géomatériaux aux ouvrages - Expérimentations et modélisations*, (ed), Hermès, Paris, pp.221-239.

6. Ahmiedi, E. (1994), *Etude de la propagation des fissures en milieu viscoélastique*. Applications aux enrobés bitumineux. Thesis, Université de Limoges.

7. Amestoy, M., Bui, H.D. & Dang Van, K. (1979) Déviation infinitésimale d'une fissure dans une direction arbitraire. *C.R. Acad.. Sci. Paris*, vol. 289B, pp.99-101.

8. Bilby, Cottrel & Swiden. (1963) The spread of plasticity from notch. *Proc. Roy. Soc.*, vol. A 272, pp 304-314.

9. Molenaar, A.A.A. (1993) Evaluation of Pavement Structure with Emphasis on Reflective Cracking, *Second Int. Conf. on Reflective Cracking in Pavements*, RILEM Conference, E & FN Spons, pp.21-48.

10. Bui, H.D. & Proix, J.M. (1985), Découplage des modes mixtes de rupture en thermoélasticité par des intégrales indépendantes du contour, 3ème colloque *Tendances actuelles en Calcul de Structures*, Bastia, pp.631-643, nov.1985.

11. Suo, X.Z. & Combescure, A. (1992), On the application of G_θ method and its comparison with De Lorenzi's approach, *Nuclear Engineering and Design*, vol. 135, pp.207-224.

12. Petit, C. (1994) *Généralisation et aplication des lois de mécanique de la rupture à l'étude des structures et matériaux fissurés*, Thesis, University of Limoges.

13. Petit, C., Vergne, A., Zhang, X. A comparative Numerical Review of Cracked Materials, *Eng. Fract. Mech.*, to be published, May 1996.

14. Petit, C. (1990) *Modélisation de milieux composites multicouches fissurés par la mécanique de la rupture*, Thesis, University of Clermont-Ferrand.

15. Ishikawa, H. (1980), Finite element analysis of stress intensity factors for combined tensile and shear loading by only a virtual crack extension, *Int. Journal of Fracture*, vol.16.

Predicting cracking index in flexible pavements: artificial neural network approach

N.O. ATTOH-OKINE

Florida International University, Department of Civil and Environmental Engineering, Miami, Florida, USA

Abstract

Cracking is one of the important components of surface distress variables used in pavement performance modeling in Pavement Management Systems (PMS). It is also a major distress mode that causes premature failure in flexible pavements. Models for predicting cracking, especially change in cracking, have been developed using age of the pavement, structural number and cumulative standard axle loads, etc. A major characteristic of the models is that they are formulated and estimated using a statistical approach. This paper uses artificial neural networks (ANNs) in predicting the area of indexed cracks in flexible pavement. The index crack is a weighted measure of area of cracking and crack width. With reference to cracking prediction, the neural networks can achieve modeling more parsimoniously than the statistical approach, and it can also identify which type of variables can be used to predict cracking. The networks are trained using past pavement data (structural number, age, ESAL, environmental factors, etc). A suitable learned network has the ability to generalize when presented with inputs no‡ appearing in the training data, and is capable of offering real-time solutions.

Keywords: Neural networks, cracking, pavement management, flexible pavements.

1   Introduction

Pavement distress, especially cracking, can affect pavement serviceability, structural capacity, and appearance. Predicting the cracking progression during pavement design life is very important for pavement management decision making, pavement design and evaluation, and road pricing. There are various types of cracking. Cracking in flexible pavements includes: (a) alligator (fatigue) cracking; (b) block cracking; (c) edge cracking; (d) longitudinal cracking; (e) reflection cracking at joints, and (f) traverse cracking. In rigid pavements there are (a) corner breaks; (b) durability "D" cracking; (c) longitudinal cracking,

Reflective Cracking in Pavements. Edited by L. Francken, E. Beuving and A.A.A. Molenaar. © 1996 RILEM. Published by E & FN Spon, 2–6 Boundary Row, London, SE1 8HN. ISBN 0 419 22260 X.

and (d) transverse cracking.

In rigid pavements, excessive crack openings can lead to of load transfer, causing flexing of the concrete slab under traffic (with resultant spalling of the concrete). They can also lead to punchouts, steel rupture, the infiltration of incompressible material (causing spalling and blowups), and water seepage that can reduce roadbed support and that can cause rusting of the steel [1].

Paterson [2] presented the characteristics of cracking and other existing measurements. Cracking is characterized by these five attributes:

(a) *Extent - the area of pavement covered by cracking technically defined by the perimeter bounding all of the area covered by a set of cracks and expressed in units of area or of percentage of total area of pavement.*

(b) *Severity - usually the average width of crack opening at the pavement surface and expressed in terms of either level (high, low or wide, narrow, etc) or the average width dimension itself.*

(c) *Intensity - length of cracks per unit area expressed, for example, in meters per meter, squared foot (per foot squared) and sometimes alternatively expressed as crack spacing.*

(d) *Pattern - the orientation and interconnectedness of the cracks, usually expressed by cracking type (alligator, block, longitudinal, transverse, etc); and*

(e) *Location - identifies which part of the pavement is cracked, for example, wheelpath, between wheelpath, edge, joint, midslab, or random.*

Various combinations of these attributes are used in the existing measure of cracking.

Sood et al., [3] derived models in change distress as a function of current surface condition, traffic volume and loading, pavement age, pavement strength, and maintenance inputs. The models permit prediction of cracking over time. Knowledge of pavement cracking prediction allows several activities to occur:

- prediction of the future conditions (cracking) of the pavements;
- prediction of the future funding requirements to keep the pavement network at certain level; and
- comparison of the effects of various conditions or funding on cracking performance.

Artificial neural networks (ANNs) have been shown to offer a number of advantages over traditional statistical methods. They are capable of making generalizations and of offering real-time solutions to complex predictions problem because of their massive parallelism and strong interconnection. Because ANNs learn from pavement data historically, no human expert, specific knowledge, or developed models are needed.

The aim of this paper is to evaluate the capabilities of ANNs in predicting cracking progression in flexible pavement from modified structural number, incremental traffic loadings, and environmental mechanisms.

2 Artificial Neural Networks

Artificial neural networks are used in task domains ranging from concept learning to function approximation. While the concept of learning has many features in common with statistical models, large number of algorithms used in statistics can represent neural networks [4]. ANNs, or simply neural networks, are computing systems made up of a number of simple and highly interconnected elements that process information by its dynamic-state response to external inputs [5]. ANNs are non-parametric classifiers. The statistical nature of cracking and cracking prediction, makes the application of artificial neural networks to cracking prediction very appropriate.

The basic unit of an ANN is a processing element (PE). It combines the inputs and produces an output in accordance with a transfer function. The output of one processing element is connected to the input paths of other processing elements through connecting weights. General artificial neural network models draw primarily on these features: massive parallism, non-linearity, and processing by multiple layers of neural units, and finally dynamic feed-back among units [6]. An important characteristic of neural networks is its ability to "learn." The neural networks generate their own rules by learning from examples. Learning or being "trained," which is analogous to being "programmed" in computing, is achieved through adaptive process. When presented respectively with the inputs and the desired output under which falls "supervised learning", a neural network organizes itself internally, gradually adjusting to achieve desired input and output mapping.

Currently, there is no formal way of computing the network structures as a function of the complexity of the problem. It is manually selected by a trial and error process which can be relatively time consuming. The selection algorithm usually involves two methods; destructive and constructive [7]. In the destructive methods, one starts with a large network size and deletes unnecessary units/connections until the "optimum" architecture is obtained, based on prediction and processing time. Constructive methods imply that one starts with a relative simple network and adds units/connections to reduce the error.

A generalized adaptive neural network architecture [8] is used for the analysis. In the generalized neural network architecture the learning paradigm for the neural network constructs network architecture by example. These networks grow to fit the problem at hand and do not require a priori specification of the number of layers, number of nodes, etc. New layers are added provided they offer improved performance based on a selected criteria. The communication flow is unidirectional with no feedback loops permitted. The adaptive neural network architecture can be best described as hybrid, and uses a combination of supervised and self-organized learning. The generalized adaptive neural network is based on the following principles:

(a)   Processing element in the network must be non-linear, easily computed and capable of computing non-trivial functions with minimal number of processing elements.

(b)   The neural network must allow for minimally sufficient interactions between its processing elements.

(c)   The architecture of the network (number of nodes in each layer, number of layers, connection patterns) can and should be learned by example, through the presentation of associated input and output patterns.

3 Development of ANNs models

Cracking indices were generated using the RODEMAN, a menu driven PC version of the Road Deterioration and Maintenance Submodel of HDM-III. The approach utilizes a full empirical simulation model to generate cracking index data. Table 1 is an example subset of data generated. The confidence in the generated data stems from the extensive verification of the empirical simulation model and the comprehensiveness of the interactive form of the primary parameters. Data were generated for an array of two pavement types (asphalt concrete and surface treatment flexible pavements), and three primary variables (primary strength, annual traffic loading and environment).

Autonet, a generalized adaptive neural network software system was used for the analysis. In the analysis, the original data was used as input without any normalization. A total of 100 observations were used: 65 for training and 35 for testing. The minimum sum of squared error was set to be equal to 0.005 and the percentage improvement was fixed at 10 percent. New layers were added provided they offered improved performance, measured in squared output error.

Figure 1 shows the generalized adaptive neural network obtained. The network consists of 8 inputs and 1 output in terms of cracking index. It appears that out of 8 inputs, only pavement age, potholing, patching and pavement thickness were needed to predict the cracking index using the generalized adaptive neural network.

Figure 2 shows the scatter diagram of predicted and actual cracking indices. There were 100 discrete data items.

4 Concluding Remarks

The paper demonstrates the application of ANNs in cracking index prediction in flexible pavements. The present studies show that adaptive neural network approach is more accurate in predicting cracking index. Furthermore, the adaptive neural network approach can be used as a sensitivity analysis tool to identify most significant variables needed to predict cracking index. This approach can applied to various forms of surface defect prediction in both flexible and rigid pavements.

Table 1. Subset of Generated Data

EASLY	Age	Ravel Area	Potholing	Rut Depth	Patching	M	Thickness	Cracking Index
0.01	9	0	0	7	0	0	30	16
0.1	12	0	0	11.6	0.26	0	30	74
1	15	0	0	15	5.42	0	50	100
0.03	14	0	0	2.8	0	0	80	43
3	7	0	0	5.3	0	0	80	38
0.01	18	0	0	8.8	0.12	0	30	100
0.01	14	0	0	6	0.13	0	50	80

EASLY	=	Annual Equivalent Standard Axle Load per Year
AGE	=	Age of Pavement since Resurfacing (Yr)
Ravel Area	=	Area of Ravelling (%)
Potholing	=	Area of Potholing (%)
Rut Depth	=	Mean Rut Depth (mm)
Patching	=	Patching (%)
M	=	Environmental Factor
Thickness	=	Thickness (mm)
Cracking Index	=	Area of Index Cracking (%)

5 References

1. Suh, Young-Chan and McCullough, B. F (1994) Factors Affecting Crack Width of Continuously Reinforced Concrete Pavement. *Transportation Research Record,* No. 1449, pp 134-140. National Research Council, Washington DC.
2. Paterson, W. D. (1994) Proposal of Universal Cracking Indicator. Transportation Washington on 1994) Research Record, No. 1455, pp 69-74. National Research Council, DC.
3. Sood et al., Pavement Deterioration Modeling in India. Third International Conference on Managing Pavements, Volume 1, pp 47-54, San Antonio, Texas.
4. Frohlighaus T. et al., (1994) Hierarchical Neural Network for Time Series and Control Networks, Vol 5, pp 100-116.
5. Kamarthi et al., (1992) Neuroform-Neural Network System for vertical Form work Selection. Journal of Computing in Civil Engineering, Vol 6, No. 2, pp 178-199.
6. Kuan, C. M and White H (1994) Artificial neural Network: An Econometric Perspective. Econometric Review, Vol 13, No. 1, pp 1-91.
7. Nabhan, T. M. And Zomaya, A. Y (1994) Toward Generating Neural Network

N.O. Attoh-Okine

Structures for Function Approximation. *Neural Networks*, Vol 7, No. 1, pp 89-99.

8. Cogger, K (1992) A Learning Paradigms for Neural Network Architecture. Working paper No. 237, School of Business, University of Kansas, Lawrence, KS.

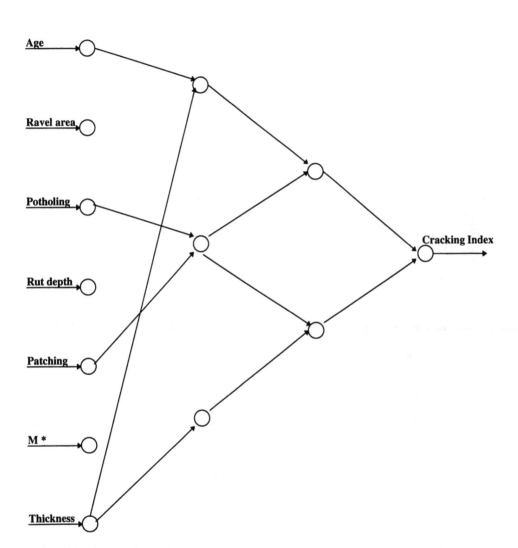

M * = Environmental Factor

Figure 1 : Adaptive Neural Network (Cracking Index)

INDEX CRACKING

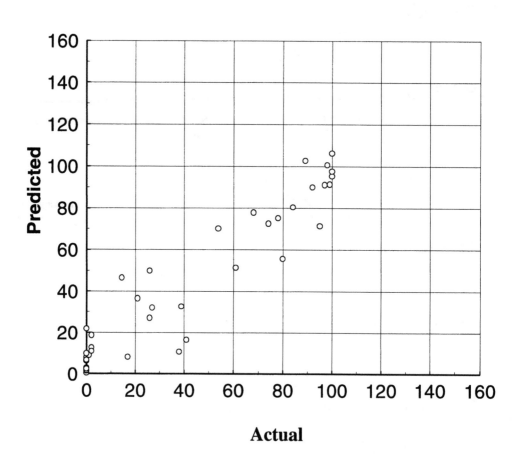

Figure 2 : Actual and Predicted values of Cracking Index.

The stress–strain conditions in road pavements with a cracked roadbase

A.E. MERZLIKIN
State Road Research Institute, Balashiha, Russia

Abstract

In analyzing the problem, the road pavement structure was schematized as an elastic three-layered system laid over the rigid base. The study was limited by plane deformations. The finite element method has been applied. Computations have shown that reducing the crack spacings results in an increase of the deflection not more than by 44%. In the pavement structures with the cracked roadbase, the concentration of horizontal stresses in the surfacing above the cracks does not practically depend on the block length. The paper describes a method for the practical application of the coefficients of stress concentration in the automated design of the road pavements.

Keywords: cracked roadbase, stress-strain conditions, finite element method.

1 Introduction

Structures consisting of asphalt concrete surfacings laid over relatively rigid base courses are called combined or semirigid pavements. As rigid bases those bases may be considered that consist of materials and soils treated with inorganic binders, of monolithic concrete, of reinforced concrete slabs that earlier served as the surfacing and of the old cracked asphalt concrete. A shortcoming of such structures is the probability of developing cracks that reflect joints and cracks in the base.

Reflective Cracking in Pavements. Edited by L. Francken, E. Beuving and A.A.A. Molenaar. © 1996 RILEM. Published by E & FN Spon, 2–6 Boundary Row, London, SE1 8HN. ISBN 0 419 22260 X.

Cracking in the bituminous surfacing is caused by the disturbance of thermal and mechanical stress fields near the roadbase joints and cracks. In studies of the detrimental action of daily and yearly temperature variations on the asphalt concrete surfacing it is found [1] that the thermal crack-resistance of the surfacing increases with reducing the distance between the cracks. This results in loss of the total stiffness of the roadbase and thus, in the development of higher deformations at the bottom of the pavement under mechanical loading. The negative effect of the crack spacing in relation to the type of loading (mechanical or thermal) on the reflection of cracks through the wearing course has required close consideration of this problem.

As is known from composite material mechanics and works of road researchers [2], failure of the layered system may occur according to two schemes. The first one is when the strength of the layer interface zone is not sufficient; this results in segregation and loss in the total stiffness. The second scheme is when the bond between the layers is strong, in such a case the cracks propagate from one layer into the adjacent ones. This report deals with the latter sceme.

2 Statement of a problem

When formulating the problem, the road pavement structure is presented as an elastic three-layered sandwich laid over the rigid base. Even taking into consideration that equations of the elasticity theory are considerably simplified when solving the problem, using a plain strain approach, the interface conditions that reflect the presence of some vertical cuts in one of the layers do not allow to solve the problem analytically, therefore the finite element method has been applied. An isoparametric square element, most efficient from the viewpoint of the accuracy and computation time, was used as the base one.

In building a finite element grid, a fine mesh has been provided at the tops of the cuts, for which purpose the triangle elements are employed along with the quadrangle ones. An algorithm for solving the problem as well as a Fortran program have been developed by a team under the quidance of Ju.A.Tchernyakov, Cand. Ph.-Maths Sc.

3 Results of calculations

To illustrate the results obtained in a numerical experiment, an analysis was carried out of a road pavement structure consisting of a 100 mm thick bituminous surfacing over a 200 mm thick cement-soil roadbase.

The roadbase was divided into blocks by means of 5 mm wide cuts, subsequently called cracks. The block length was varied from 750 to 1870 mm. A load that was distributed through a 350 mm wide strip was placed above the cracks at the edge and the middle of the block. From Fig.1, which presents a scheme of the load application and the computation

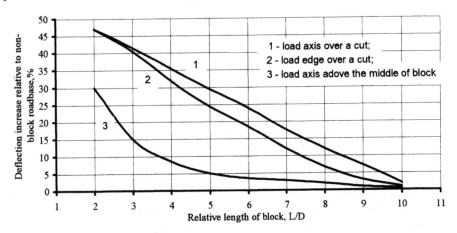

Fig.1. Effect of the block length (L) and load (width D) placement relative
to a cut on the pavement deflection

results, it may be seen that reducing the block lengths results in an increase of the deflection not more than by 44%. Thereby, the effect of load location relative to the cracks becomes lower. Numerous experimental data of different authors [3] on average indicate the same relation between the deflections, and this is one of the evidences in support of the suitability of the solution obtained for practical application.

It has been found that the crack width has a little effect on the deflection, for instance, at a ten-fold reduction of the crack width, a decrease in the deflection does not exceed 7%. To describe the stress concentration near the crack, a coefficient of the horizontal stress concentration was applied. This ratio is defined as a ratio between the maximum horizontal stresses σ^{cr} in the surfacing above crack and the horizontal stress σ in the surfacing laid over the base with the same elasticity modulus but without a crack.

Consideration of the stress conditions in the pavement during the spring period, which in the most unfavourable period since the asphalt concrete stiffness is rather high (E_1=5 000 MPa), has led to the following results (Fig. 2).

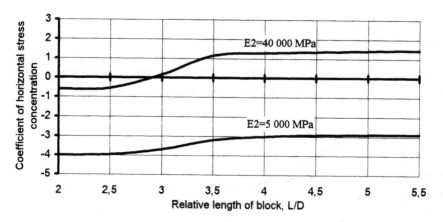

Fig.2. Coefficient of stress concentration in the pavement versus block
length and material stiffress of the cracked roadbase

In the structures with the cracked bases, the horizontal stress concentration
in the surfacing above the cracks are almost independent on the block
length of 1.5 m or more. The value sharply raises when the block length is
reduced from 1.5 to 0.9 m. Such tendency is characteristic of the structures
having the roadbases made of cement concrete (E_2=40 000 MPa) and
cement-soil (E_2=5 000 MPa). In the last case, a decrease in the length of the
base blocks results in about a 30% increase of the maximum tensile stresses
in the asphalt surfacing of thickness H_1=100 mm.

A general idea of the relationship between the coefficient of
horizontal stress concentration ($K=\sigma^{cr}/\sigma$), on one hand, and the material
stiffness of the cracked base and surfacing on the other can be obtained by
analyzing Fig. 3. This Figure presents computation results for the structures
with base courses consisting of blocks more than 1 m long.

As can be seen, the stiffness of the base material has a considerable
effect on the tensile stress concentration in the surfacing only during the
spring period. Thus, above cracks in the cement-soil roadbase K= -4 while
above cracks in the cement concrete roadbase K= -0.5. During the hot
summer months, when the asphalt stiffness is much lower (E_1=500 MPa),
the tensile horizontal stresses transform into the compression ones, and the
coefficient of stress concentration does not depend on the base material.

According to the Road Pavement Design Manual applied in Russia,
cracks (thermal, shrinkage, etc.) in the monolithic layers are taken into
account in computations by underestimating several times the elasticity
modulus of layer material. Thus, no matter what strength design criterium
is accepted, in computations, the elasticity modulus of material is decreased

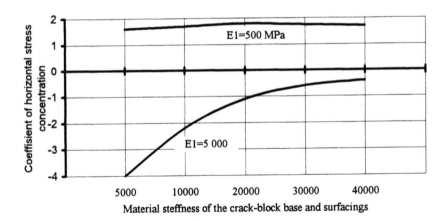

Fig.3. Coefficient of stress concentration versus material steffness
of the cracked base and surfacings

stone course and 25 times for that including a cement-treated loam course. However, computations performed for the structure with the cracked roadbase have shown that dividing the monolithic roadbase into blocks and developing the cracks in it unequally influence various components of the stress tensor. This can be illustrated by the example of flexible road pavement structure with the cement-soil base (see Fig. 3, E_2=5 000 MPa) that is cracked into blocks 1 m long.

An equivalent surface deflection of the cracked pavement can be obtained if the elasticity modulus of the uncracked roadbase material is reduced by a factor of 2.5, whereas to reach equal-in-value maximum horizontal stresses in the surfacing, a 150-fold decrease of the modulus is required.

In order to be able to use elastic layered theory for pavements with cracked layers reduced moduli values for the cracked layers have to be used. To determine the design elasticity moduli of the cracked base E_2, the following formulae are proposed:

for computations where the criterium is the maximum deflection of pavement surface, E_2 is calculated using:
$$E_2 = 17\ 200\ exp(2.37 \cdot 10^{-6}\ E_{m2});$$

for computations where the criterium is the tensile stresses in the surfacing above the cracked layer, E_2 is calculated using:
$$E_2 = 250\ exp(6.58 \cdot 10^{-6}\ E_{m2}),$$

where E_{m2}=modulus of elasticity of the cracked base material.

It is known that an efficient method to reduce the tensile stress concentration in the surfacing above the cracks is to lay an interlayer between the surfacing and the cracked base, which has a lower modulus of elasticity, as compared with that of the adjacent layers. The computer program developed has made it possible to represent the interlayer as a course, the thickness and deformation characteristics of which are equal in all directions. Considering an approximating character of such modelling of the geotextile interlayers [4], an analysis has been carried out of damping properties of the latter. Computations have shown that the interlayer thickness has an effect on the stress concentration: the thinner the interlayer, the lower the coefficient K. The interlayer that possesses the deformation properties of glass fibre geogrid (E_{eff}=600 MPa) and has a thickness of 1 mm allows a 40-10-fold reduction of the coefficient of tensile stress concentration, while thickness of 10 mm gives - only a 5-3-fold reduction. Similar conclusions have been made in paper by N.F.Coetzee and C.L.Monismith [5].

A program is worked out which simulates the interlayer properties by an assemblage of elementary deformed plates hinged to each other. Such approach is based on the fact that a real interlayer does not take the compression forces in the horizontal direction. A new version of the program may be considered as "plane analogue" of our ARMOGR program where an axi-symmetric problem has been solved [6].

4 Realization of results

Numerous computations of the coefficients of tensile stress concentration have permitted to produce compact tables in the form of a subprogram or data base for their subsequent application in the automated design systems. Computations were carried out for structures with the roadbases of blocks 2.68D (940 mm) in length and with a crack spacing of 0.0143D (crack width 5 mm) between them. Table 1 presents primary characteristics from the data base.

The range of parameter variation includes parameters most widely applied in the construction of semi-rigid and flexible road pavement structures, which are tested in spring.

Thus, in the automated design an improved method for determining the tensile stresses in the asphalt surfacing over the crack-block base has been realized, which is as follows:
- determining maximum horizontal stresses in the surfacing with the application of the elasticity moduli of the material of individual blocks, instead of the underestimated elasticity modulus of the base;

Table 1. Arguments, their ranges and gradation of tabulated relationships for determining the coefficients of stress concentration

Input parameter	Parameter variation range	Minimum spacing in parameter variation
Relative thickness:		
surfacing H_1/D	0.1-0.6	0.1
base H_2/D	0.2-1.0	0.2
Relative modulus of elasticity:		
surfacing E_1/E_3	10-100	20
base E_2/E_3	100-1000	200

Note: E_3=modulus of elasticity of the underlying course.

- finding the coefficient of stress concentration in the data base by interpolating;
- transforming into the real tensile stresses by means of the coefficient of maximum horizontal stress concentration.

5 References

1.   I.-P.Marchland, H.Goacolon   Fissuration dans les couches de roulement. - *Papp. lab. Min. uzban. et log. Min. transp. const.   routiere,* 1983, #2, 133-171

2.   A.E.Merzlikin, V.Yu.Gladkov *Strengthening the road granular base courses as a way of increasing their strength.* - M: VPTITransstroy, 1988, 30 p.

3.   *Road Pavements with Base Courses of Stabilized Materials/* Yu.M.Vasiliev, etc - M: Transport, 1989, 191 p.

4.   G.K.Syunyi,   B.S.Radovsky,   A.E.Merzlikin,   V.V.Mozgovoy Automated design of the thermal crack - resistance of asphalt pavements.- In book: *New in the design of road pavement structures.Trudy Soyuzdornii,* M., 1988, p. 5-21

5.   V.D.Kazarnovsky, A.E.Merzlikin, V.Yu.Gladkov On the efficiency of strengthening a road pavement structure performed in the elastic stage.- In book: *Synthetic textile materials in the road structures. Trudy Soyuzdornii,* M., 1983, p. 117-122

6.   N.F.Coetzee, C.L.Monismith   Analitical study of minimization of refection cracking in asphalt concrete overlays dy use of rubber- asphalt interlayer. - *Transp. Res. Res.,* 1979, #700, p.100-108

Fatigue shear and fracture behavior of rubber bituminous pavement

YE GUOZHENG

South China Construction University, Guang Zhou, China

Abstract

This paper presents the fatigue shear and fracture test equipment. The characteization of material as well as the physical and mechanical behavior of rubber bitumen and of general bituminous mixture are demonstrated. Their relationship of the shear stress-fatigue life and of the fatigue fracture ratio (dC/dN)-stress intensive factor (K) are summarized.

The rubber bituminous mixture have many excellent pavement property. The shear fatigue life and fracture fatigue life can be prolonged about eight to ten times as compared with the general asphalt mixtures.

We can use the rubber asphalt mixes to prevent the reflective cracking in pavement. The satisfied results of rubber asphalt pavement have been obtained. This paper the presents calculating analysis of reflective cracking failure in bituminous pavement. It might be useful to pavement design when the fatigue fracture and the fatigue shear designed subsystem are taking accounted.

Keywords: fatigue shear, fracture, reflective crack, rubber.

1 Introduction

The prevention of reflective cracking in pavement have been paid more attention on the whole world. In recent ten years a great number of rigid pavement have been paved in China. Over 12,000 Kilometers of concrete are being operated in Guangdong province. But the concrete pavement behave transverse joints, it's the cause of vehicle jumping and speed reducing as well as felling uncomfortable in travel.

Reflective Cracking in Pavements. Edited by L. Francken, E. Beuving and A.A.A. Molenaar. © 1996 RILEM. Published by E & FN Spon, 2–6 Boundary Row, London, SE1 8HN. ISBN 0 419 22260 X.

Since 1986 the prevention research of reflective cracking in labouratory and insite road have been performed in China.   The goetextile,stress absorted layer and rubber asphalt mixture have been used in bituminous surface layer construction.   The rubber type involves botodiene styrene, polythylene, chloroprene and the reclaimed rubberetc. The rubber bituminous pavement met with good success.

This paper describes the physical--mechanical behavior; the shear fatigue and the fracture fatigue test of rubber bituminous pavement mixture.   The fatigue expression have been summarized.   Both design of the shear   fatigue and the fratigue fatigue subsystem have been accounted.

2 Shear fatigue test

In order to research the shear fatigue property of bituminous mixes a shear fatigue machine had been used (Fig.1).   In fatigue test the repetitive frequency 4--6 Hz was used.   The load of application varied from 0.5--4 KN.

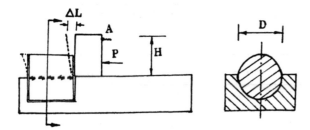

Fig.1.   Schematic graph of shear fatigue machine

The shear stress and the elastic modulus of specimen is determined by:

$$\tau = \frac{P}{Dh}$$
$$\gamma = \tan \gamma = \frac{\Delta L}{H} \tag{1}$$
$$G = \frac{\tau}{\tan \gamma}$$

where: D, h -- the diameter and height of specimen (70mm),

p ----- horizonal shear load,

Δ L --- horizonal elastic displace of point A,

H ------ height of point A

The characterization of bitumen and rubber bitumen are listed in table 1.

The composition of bituminous mixes and rubber bituminous mixes are listed in table 2.

Table 1. Property of bitumen and rubber bitumen

type	ductility (cm)		penetration	soft point	stiffness	elastic deformation
	25 ℃	0 ℃	(0.1mm)	T_{RB} (℃)	-10 ℃(MPa)	(%)
bitumen	71	4.8	87	50	203	68
rubber bitumen	59	8.9	71	54	167	82

Table 2. Composition of bituminous mixes and rubber bituminous mixes

percent passing in sieve		(mm)						
25	15	5	2	1	0.5	0.25	0.15	0.074
100	74	50	32	20	17	12	9	6

rubber bitumen 5 %   (by weight)
rubber/bit =   4 %   (by weight)

The main property of two mixes are listed in table 3.

Table 3. Propery of two mixtures

item	bituminous mixes			rubber bituminous mixes		
	20 ℃ (50 ℃)	0 ℃	-10 ℃	20 ℃ (50 ℃)	0 ℃	-10 ℃
shear strength $\sigma = 0.5$(MPa)	0.86	1.39	1.88	0.96	1.64	2.13
split strength (N)	1.62 0	3.40	4.82	1.58	2.86	3.82
tensile strength (MPa)	2.06	4.40	5.86	1.83	3.92	4.98
stiffness (MPa)	1108	2172	4160	908	1823	2610
Marshell stabitity (KN)	(8100)	---	---	(10240)	---	---
flow (100-1cm)	(36)	---	---	(34)	---	---
dynamic stability DS (cycle/mm)	(2428)	---	---	(5262)	---	---

From the test results it's clearly shown that the rubber bituminous mixture behave many excellent pavement behavior.   Such as higher viscosity and soft temperature, higher deformation capacity and elasticity as well as flexibility and anti-fracture property in low temperature.

In the fatigue shear test, the shear modulus have been calculated by the shear stress and strain as a function of time.   The shear modulus depend upon the numbers of load application.   It's almost unchanged in a large part of fatigue life.   When the shear modulus is decreased to 50% of initial modulus, the numbers of load application is accounted as the fatigue life of specimen.

The shear fatigue relationship show in Fig.2 and may be presened by:

For the bituminous concrete:

$$K = \frac{S}{\tau} = 0.29 \, N^{0.164}$$ (2)

$$R = 0.971$$

For the rubber bituminous concrete:

$$K = \frac{S}{\tau} = 0.25 \, N^{0.152}$$ (3)

$$R = 0.964$$

where: S -- shear strength

τ -- shear stress of repetitive application,

K-- safety coefficient of structure,

R -- relative coefficient,

N -- numbers of stress application.

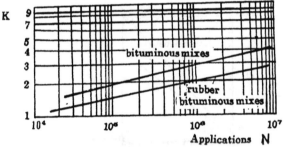

Fig.2 Relationship btw K *VS* N

From the test results it can be seen that the rubber asphalt mixes can prolong about nine times of shear fatigue life as compared with to the asphalt mixes.

3 Fracture fatigue property

In order to study the fracture behavior of pavement the fracture fatigue test had been performed by slab fatigue machine (see Fig.3a). The size of slab is 150×100×10cm. The loading tire was rolling on the slab repetitively. The repetitive frequency 0.5Hz was used. The load application carried from 0.3 to 5KN.

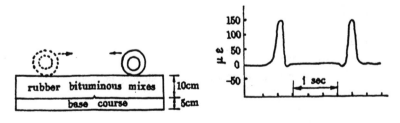

(a) Equipment   (b) Strain-time history

Fig.3 Slab fatigue machine

The elastic modulus of bituminous concrete layer had been computed by the strain and he stress. The example of the strain-time history curve is given in Fig.3b.

The ratio of cracking propagation is　expressed as: (Fig.4)

$$\frac{dC}{dN} = aK^b \tag{6}$$

For bituminous mixes

　　　$a=3.44 \times 10^{-6}$

　　　$b=2.32$

For rubber bituminous mixes

　　　$a=2.86 \times 10^{-6}$

　　　$b=1.94$

Where: K--- stress intensive factor　　$(MPa.cm^{1/2})$

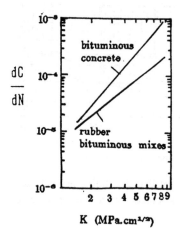

Fig.4　Relationship　btw dC/dN　*VS*　K

From above expressions the ratio of cracking propagation of rubber asphalt mixes is decreased from 50-150 % as compared with the asphalt mixes.　The stress intensive factor is also decreased.

The fracture fatigue life is given by:

$$N_f = \int_{co}^{cf} \frac{dc}{ak^b} \tag{4}$$

where: co — initital cracking length,

　　　cf — thickness of layer.

the stress intensive factor may be determined by the following equation:

$$K = P \sqrt{\frac{E}{2(1-\mu^2)h} \frac{\partial L}{\partial C}} \tag{5}$$

where: P —load;

μ --poisson coefficient;

h --layer thickness;

L --compliance i.e. converse of elastic modulus;

$\dfrac{\partial L}{\partial C}$ --- increment compliance/propagation of unit cracking length

4 Vertical shear fatigue sub system

The preventive design of reflective cracking in pavement involve two parts: the vertical shear fatigue subsystem and fracture fatigue subsystem. In shear fatigue design many data will be required as the traffic condition, material property (cohission, friction angle and shear strength etc.) and the different deflection at both of the joint in the old pavement.

In order to calculate the vertical shear stress the ratio of loading transfer may be introduced:

$$R = \frac{\Delta D}{\Delta Y} \qquad (7)$$

The schematic diagram of different deflection of cracking pavement is shown in Fig.5.

If the designed wheel load is L, then the vertical shear load is L (1-R). The designed ratio of loading transfer RD may be obtained by the measure of differential deflection from the cracking (Joint) pavement.

$$RD = R - Z\sigma \qquad (8)$$

where: R--the mean of loading transfer ratio;

σ--mean-square deviation;

Z--dependability coefficient; Z=1.5 when the design level of 93.3% .

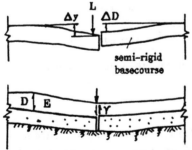

Fig.5   Schematic diagram of different deflection

The vertical shear stress is expressed as:

$$\tau = \frac{L(1-RD)}{DL} \qquad (9)$$

where: D--thickness of (rubber) bituminous layer;

L--width of one traffic lane.

The vertical shear stress must be satisfied in following expression:

$$\tau \leqslant \frac{S}{K} = [\tau] \qquad (10)$$

where: S--shear strength.

The fatigue life (cumulatic applications of standard axle load) may be estimated by:

$$N = (\frac{4S}{\tau})^{\frac{1}{0.152}} \qquad (11)$$

5 Fracture fatigue subsystem

The flow chart for fracture fatigue subsystem is developed as Fig.6. It may be applied by computer program according to the traffic data, material evalution and structure parameters by means of the layers elastic system theory and the fracture mechanical law.

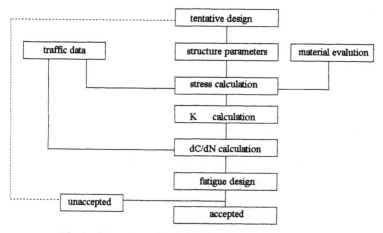

Fig.6   Flow chart for fracture fatigue subsystem

6 Conclusion

It's clear to known that the rubber bituminous mixture behave many advantage. It not only improves the viscosity of bitument,but also enhances the stability in high temperature,increases the low temperature ductibility and the shear or the fracture fatigue life of bituminous mixtures.

The good serve quality had been verified by smooth and stabilizing surface, more elastic   behavior and comfortable to traval on rubber bituminous pavement. It's obvious that the   cracking have been reduced greatly in pavement. It's an excellent material for the prevention of reflective cracking in pavement.

7 References

1. Fan Tongjiang: construction technique for anti-cracking pavement with SBR modified asphalt, Conf.of modified bitumen.proc.PP48-53 1992.10, China.
2. Li Zimin, Yu Xuemin: A study on preparing bitumen butadiene styrene rubber by using mother soluation technique and the charateristics Conf.of modfied bitumen Proc.PP38--42 1992.10, China.
3. Zhang Nanlu: The reseach of complex-modified asphalt. Asphalt, No3.1995, China.
4. Luis.f.Da stlva & Juan A confre:modeling of Geotectiles and other membranes in the prevention of reflection cracking in asphalt resurfacing. Public Road, winter,1994.
5. J.verstraten, V.Vereka & L.Franken: Rational and practical design of asphalt pavement to avoid cracking and rutting 5th Int.Conf.on the struct.Des.of Asp.Pav.proc.PP45--58.

Enduit superficiel aux fibres: Colfibre

J.P. MICHAUT

Direction de la Recherche et du Développement, Société COLAS, France

RESUME

L'ingénieur routier est souvent confronté aux problèmes de fissuration des chaussées, fissures qui proviennent soit des matériaux d'assises traités avec des liants hydrauliques, soit de la fatigue de la chaussée. Différentes techniques sont utilisés pour répondre à ces problèmes. L'objet de cette communication est la présentation d'un revêtement superficiel armé de fibres de verre, Colfibre de Colas, pour lutter efficacement contre la remontée des fissures dues aux graves hydrauliques ou à la fatigue. Après une présentation du procédé, les suivis réalisés sur les chantiers retenus dans le cadre de la charte à l'innovation de la direction des routes françaises sont présentés, permettant de tirer des conclusions intéressantes sur le comportement de ce procédé.

Un certificat, établi par le directeur du SETRA (Service d'Etudes Techniques des Routes et Autoroutes, organisme dépendant de la Direction des Routes) a été délivré soulignant le caractère probant de ce procédé vis-à-vis de la remontée des fissures d'une part sur les graves traitées aux liants hydrauliques pour des fissures peu dégradées et d'autre part sur les chaussées en enrobé pour des dégradations de type faïençage et fissuration anarchique. Les suivis toujours en cours permettront de confirmer ces premiers résultats.

INTRODUCTION

La présence de fissures en surface de chaussées supprime l'étanchéité du revêtement, favorisant ainsi la présence d'eau dans les couches de structures. La résistance mécanique de ces couches s'amoindrit et réduit ainsi, peu à peu, la durée de vie de la chaussée. De nombreuses techniques sont proposées pour lutter contre la remontée de la fissuration, certaines faisant appel à des liants modifiés dans des enrobés ou dans des enduits, d'autres à des géogrilles incorporés ou à des fibres

Reflective Cracking in Pavements. Edited by L. Francken, E. Beuving and A.A.A. Molenaar. © 1996 RILEM. Published by E & FN Spon, 2–6 Boundary Row, London, SE1 8HN. ISBN 0 419 22260 X.

mélangés dans un revêtement.

Il nous est apparu intéressant de cumuler les avantages des liants modifiés et des fibres dans un seul procédé pour lutter plus efficacement contre cette fissuration, en cherchant principalement à retarder l'apparition en surface de celle-ci.

C'est un procédé destiné à la lutte anti-fissure quel qu'en soit l'origine: fissures liées au retrait thermique, au retrait de prise ou à la fatigue de la chaussée. Ce procédé a pour objectif:

- de ralentir, en forte proportion, la remontée des fissures,
- de restaurer et d'assurer l'étanchéité du revêtement,
- de réaliser une couche de roulement aux bonnes caractéristiques.

1. DEFINITION - DOMAINE D'EMPLOI

Ce procédé consiste en la réalisation d'un enduit superficiel armé de fibres qui peut servir directement de revêtement gravillonné et notamment sur les revêtemnts fissurés par fatigue de la chaussée ou bien de technique intermédiaire avant l'application d'une couche de roulement.

Sa réalisation se fait par interposition de fibres de verre coupées en place et noyées entre deux couches de liant appliqué sous forme d'émulsion, de préférence élastomère. Sa mise en oeuvre est réalisée à l'aide d'une répandeuse spéciale. Ce complexe est ensuite gravillonné.

Ce procédé est facile à mettre en oeuvre, il s'applique comme un enduit superficiel. Son intérêt repose sur son efficacité, sa simplicité de mise en oeuvre, sa rapidité d'exécution, et par conséquent son faible surcoût par rapport à un enduit superficiel classique.

Il est actuellement employé selon 2 techniques :

1) Selon la technique **S.A.M.** (Stress Absorbing Membran), c'est à dire que l'enduit armé constitue la couche de roulement provisoire ou définitive. Dans ce cas:
    - la structure de l'enduit est choisie dans la gamme des structures d'enduits existantes pour satisfaire l'objectif à atteindre:
          ➲ Monocouche,
          ➲ Monocouche Double Gravillonnage,
          ➲ Bicouche,
          ➲ Bicouche prégravillonée.
    - il s'applique sur des chaussées dont la classe de trafic sera $\leq$ T1, exceptionnellement T0, en fonction de la structure retenue.
    - son domaine d'emploi concerne les travaux d'entretien par renouvellement de la couche de surface.

2) Selon la technique **S.A.M.I.** (Stress Absorbing Membran Interlayer), il est interposé entre le support à traiter et une nouvelle couche de roulement,

généralement en enrobé. L'enduit armé assure en plus l'étanchéité sous le revêtement définitif. Dans ce cas :

  - une structure classique telle le monocouche est choisie,
  - cette technique est employée aussi bien en construction neuve qu'en travaux d'entretien et sur tout type de trafic, T0 inclus.

C'est un procédé simple, économique et efficace, comme l'ont montré les essais de laboratoire réalisés d'une part au LRPC d'AUTUN, et d'autre part à l'université de NOTTINGHAM; ceux-ci confirmés par les chantiers réalisés.

2. FORMULATION - COMPOSITION

COLFIBRE est réalisé comme un enduit superficiel (figure 1). Une répandeuse équipée d'une double rampe à liant applique simultanément :

  - une première couche de liant sous forme d'émulsion,
  - une nappe de fibre de verre coupée,
  - une seconde couche de liant sous forme d'émulsion.

La mise en place des fibres se fait à partir de bobines de grande longueur, d'où le fil est tiré puis coupé à grande vitesse. Leur répartition au sol est assuré par un dispositif à air pulsé.

SCHEMA DE PRINCIPE

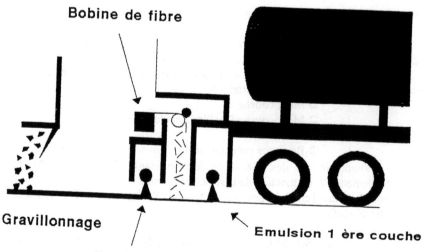

2.1 Emulsions

Les émulsions employées sont des émulsions cationiques à forte teneur en bitume pur ou en bitume élastomère dans les cas de trafics forts ou de sollicitations importantes: COLACID R69, R70A ou POLYCOL X ou P, ces deux dernières étant des émulsions élastomères utilisant du styrène butadiène styrène.

Par rapport à un enduit superficiel classique de structure équivalente, il y a lieu de prévoir un dosage en émulsion très légèrement supérieur, de l'ordre de 10 à 15 %, ceci étant due à l' absorption de l'émulsion par les fibres. Les corrections de dosage liés aux paramètres de chantier tels que trafic, rugosité, environnement etc... sont applicables et identiques aux enduits superficiels classiques.

2.2 Fibres

Les fibres sélectionnées sont des fibres de verre. Elles sont coupées en place à grande vitesse à l'aide d'un dispositif spécial sur une longueur de 5 à 6 cm. La méthode d'application assure une répartition au sol omnidirectionnelle, gage d'efficacité. Le dosage appliqué est généralement de l'ordre de 70 grammes au m², il peut être modulé entre 50 et 100 grammes au m², selon la technique choisie et l'importance de la fissuration.

2.3 Gravillons

Lorsque les gravillons sont soumis directement au trafic, technique **S.A.M.**, ils doivent être conformes aux spécifications en vigueur relatives au choix des granulats pour la réalisation d' enduits superficiels : norme P 18 101 et fascicule 23. Les dosages en granulats sont identiques à ceux couramment appliqués dans la technique des enduits superficiels.

En application en SAMI, les prescriptions habituelles pour ce type d' utilisation sont observées.

3. PERFORMANCES

Ce procédé utilisé en technique S.A.M. doit répondre aux performances de la norme NF P 98-160 relative aux enduits superficiels d'usure basées sur le niveau de rugosité et l'aspect visuel.

Afin d'apprécier son comportement à la fissuration, des essais, développés dans les paragraphes suivants, ont été réalisés en laboratoire et ont permis de suivre la vitesse de remontée d'une fissure à travers le procédé.

3.1 Essai de retrait flexion

Cet essai, effectué par le Laboratoire Régional des Ponts et Chaussées d'Autun, permet de déterminer le temps de remontée d'une fissure sous l'effet de sollicitations simultanées, simulant le retrait thermique (ouverture de la fissure, continue et lente) et le trafic (flexion à la fréquence de 1 Hz) à une température de 5°C.

Le temps de remontée de la fissure du complexe à étudier est comparé à une couche témoin, cette comparaison permet de porter un jugement technique sur l'efficacité du procédé. Le complexe COLFIBRE, avec émulsion COLACID R69 + fibres (70g/m2), mis en oeuvre sous 4 cm d'enrobé RUFLEX a donné les résultats du tableau 1.

Tableau 1

Essai de retrait flexion	
Paramètres	Résultats
Temps d'amorce	120 mn
Temps totale de fissuration	> 730 mn temps maxi de l'expérimentation
Efficacité	> 1,6 valeur maximum de l'essai

L'efficacité est définie comme étant le rapport du temps de fissuration du système testé sur le temps de fissuration du témoin. Le témoin est constitué de 8 cm de béton bitumineux dont les 2 premiers cm sont en sable enrobé.

Le système est donc considéré comme très efficace.

3.2 Essai de remontée de fissure

Cet essai est réalisé par l' université de Nottingham (Royaume Uni). Il consiste en l'application d' une charge de 8,3 KN sous des sollicitations sinusoïdales à la fréquence de 5 Hz. La température d'essais est de 20°C.

Dans les conditons de l'essai 350 000 cycles suffisent à faire remonter la fissure au travers un "Rolled Asphalt" témoin sur une épaisseur de 8 cm, alors que 550 000 cycles ne parviennent pas à la faire réapparaitre lorsque ce même "Rolled Asphalt" est mis en oeuvre sur l'enduit armé aux fibres de verre.

Ce procédé atténue considérablement les efforts horizontaux, tout en transférant les charges verticales sans déformation notable.

A cet essai, le système est également considéré comme très efficace.

4. CHANTIERS INNOVANTS

Cette technique a été retenue par la Direction des Routes dans le cadre de la Charte à l'Innovation avec l'intitulé suivant: " Procédé retardant la remontée de la fissuration".

Pour évaluer l'intérêt de ce procédé, il a été décidé de le réaliser en enduit superficiel plutôt qu'en SAMI ce qui permettait d'apprécier plus rapidement son intérêt.

Trois chantiers ont été réalisés sur le réseau routier national, leurs principales caractéristiques figurent dans le tableau suivant.

Présentation des chantiers			
Chantier	**RN 6**	**RN 143**	**RN 10**
Localisation	Savoie	Indre	Charente
Trafic	> 1000 Pl/j	500 Pl/j	1600 Pl/j
Surface	22000 m²	20000 m²	20000 m²
Type d'enduit	bicouche	bicouche	bicouche
Nature du liant	Polycol X	Polycol P	Polycol X
Dosage en liant	2,5 kg/m²	2,65 kg/m²	2,35 kg/m²
Granulat	10/14 4/6	10/14 4/6	10/14 4/6
Nature du support	ancien enrobé et ancien enrobé sur grave hydraulique	ancien enrobé sur grave laitier	ancien enrobé sur grave hydraulique

La technique d'enduit retenue fut celle du bicouche en regard du trafic très lourd supportées par les chantiers de la RN 6 et de la RN 10. Afin de ne pas introduire de biais dans l'observation, la même technique fut appliquée sur la RN 143.

Pour la RN 6, 2 sections ont été traitées, l'une sur grave hydraulique et l'autre sur d' anciens enrobés.

L'importance du trafic a nécessité l'utilisation d'émulsion modifiée par des élastomères.

5. SUIVIS REALISES

Pour chaque chantier et dans la mesure du possible, des sections témoins ont été réalisés afin de pouvoir comparer le comportement de ces différentes techniques. Dans les tableaux 3 et 4 suivant, figurent quelques résultats obtenus.

Tableau 3 : Suivi réalisé RN 143

RN 143	Rive droite		Rive gauche	
	Nombre de fissures	% de fissures réapparues	Nombre de fissures	% de fissures réapparues
Section enduit superficiel armé				
- avant travaux	418		323	
- après 1 an	24	5,7 %	8	2,5 %
- après 2 ans	24	8,1 %	11	3,4 %
Section témoin				
- avant travaux	84		69	
- après 1 an	22	27 %	19	27 %
- après 2 ans	35	30 %	19	27 %

Tableau 4 : Suivi RN6 - RN 10

RN 6	Grave hydraulique		Enrobé
	Nombre de fissures	% de fissures réapparues	
Section enduit superficiel armé - avant travaux - après 1 an - après 2 ans	40 7	 18 %	Pas de réapparition de fissures de fatigue
RN 10	**Section enduit superficiel armé**		
	nombre de fissures	% de fissures réapparues	
- avant travaux	150		
- après 1 an	46	30 %	

Sur la RN 10, les fissures observées avant travaux, provenant de la fissuration de retrait thermique des graves hydrauliques inférieures, étaient pour la plupart très dégradées, ouvertes et faïencées pour certaines. Les fissures relevées au terme d'une année correspondent à ces fissures très ouvertes avant réalisation des travaux.

Il faut rappeler de plus que les trafics de la RN 10 et de la RN 6 sont très élevés. Ces sites avaient été choisies de façon à pouvoir rapidement apprécier la résistance à la fissuration de l'enduit armé : trafic fort et enduit superficiel.

Ces observations nous ont permis de préciser les domaines d'emploi et les limites du procédé et ont servi pour la rédaction du certificat.

6. CONCLUSION

La rédaction d'un certificat délivré par la Direction des Routes sur le caractère probant du comportement du procédé vis-à-vis de la fissuration est la conclusion de cette charte innovation. Son contenu, montrant l'intérêt de ce procédé, est le suivant:

Le procédé Colfibre de Colas a été expérimenté en enduit superficiel dans le cadre de la charte de l'innovation routière. Le bilan effectué à l'issue de la phase chantier de démonstration technique s'avère probant vis-à-vis de l'objectif "retardement de la remontée des fissures":

- sur les graves traitées aux liants hydrauliques pour des fissures peu dégradées,

- sur les chaussées bitumineuses pour des dégradations de type faïençage et fissuration anarchique.

Cette technique permet de s'opposer efficacement à la réapparition en surface de certaines catégories de fissures et restaure l'imperméabilité de la chaussée.

Retarding reflective cracking by use of a woven polypropylene fabric interlayer

D.J. WRIGHT and R.G. GUILD
Geotextiles Division, Don & Low Ltd., Forfar, Scotland

Abstract
Retarding reflective cracking in bituminous pavements is a problem that faces the majority of Design Engineers involved in carriageway maintenance. The use of geotextiles and related products in this end has become markedly more widespread in the United Kingdom in recent years. These products can extend the pavement lifetime to a degree that makes their inclusion justifiable or desirable. A wide range of materials has become available to offer either reinforcement or stress relief, but until recently, there were no products that have sought to solve this problem by both means together. The development of combined materials such as reinforced non-woven geotextiles, composite geogrid / non-woven geotextile materials and woven polypropylene fabric interlayers has offered the Engineer a third choice, albeit at very widely differing costs. High strength woven polypropylene fabrics have been used with great success as a part of systems to retard the reflective cracking of asphalt pavements. This paper reviews their use and the comparative performance and cost of these and other geotextile interlayers, currently widely used in the United Kingdom.
Keywords: Bituminous surfacing, geogrids, geotextiles, pavement lifetime, reinforcement, stress relief.

1 Introduction

The trend to use geotextile interlayers in the retarding of reflective cracking has, until recently centred mainly on the use of two types of material, stress relief fabrics and reinforcement grids. Their ability to bring about the requisite increase in lifetime which will justify their installed cost in the scheme varies. If the material will not give a

Reflective Cracking in Pavements. Edited by L. Francken, E. Beuving and A.A.A. Molenaar. © 1996 RILEM. Published by E & FN Spon, 2–6 Boundary Row, London, SE1 8HN. ISBN 0 419 22260 X.

lifetime extension that the Engineer can be sure will be economically viable then it is certainly, at the very least, a great act of faith by that Engineer to specify its inclusion. Reviewing current practice, how each of the types of product act to retard the onset of reflective cracking, their relative advantages and disadvantages and their relative installed cost will help the Engineer to determine which materials are viable for his scheme.

2 Current practice

There are several reasons why reflective cracking occurs and an understanding of which circumstances have contributed to the pavement damage previously may be used to decide the most appropriate action thereafter. These reasons summarised by Walsh [1] include;

1. Thermal stresses, induced in stiff bituminous bases and in cement bound macadam bases. These may be exacerbated by premature trafficking of the pavement in the case of the former or later during the pavement's life. In the case of the latter, internal thermal effects during the initial curing are noted as being a significant contributor to the later cracking of the pavement.
2. Traffic stresses, usually not the root cause of cracking in themselves albeit that the loading by traffic will undoubtedly accelerate reflective cracking.
3. Changes in construction, usually in areas where the carriageway has been widened or where service openings have been made and reinstated.
4. Settlement, in areas where the foundation has expanded, contracted or moved inducing movement in the pavement above.

Historically the Engineer had two choices how best to alleviate reflective cracking. If levels allowed then a thick overlay (or to a lesser effect, a thinner but stiff overlay) may lessen the speed with which the cracking reaches the surface and also reduce the amount of thermal or traffic stress that reaches the base of the pavement. In all other circumstances the Engineer would be driven to consider a reconstruction of the carriageway, removing the cause of the failure in the process. Two further options became available with the introduction of polymer modifying additives to the bitumen binder and the advent of geotextile interlayers. The former have been widely trialed and have unfortunately yielded results that suggest that they are not as effective as had been hoped. The latter, being the subject of this paper, have gained favour with many Engineers and continue to attract interest as the only viable alternative to the three options described above in the combating of reflective cracking damage.

The specifier currently chooses an interlayer to alleviate reflective cracking on the basis of its' ability to;

1. Reinforce by offering tensile strength to the asphalt
2. Relieve stress by transferring movement stresses into a bitumen impregnated fabric

As would be expected, there are a large number of products available to the Engineer to offer these effects. These can be divided into groupings as above. All must act by meeting the cracks that emanate through the structure of the pavement but their action in combating these cracks is subtly different. (Fig. 1)

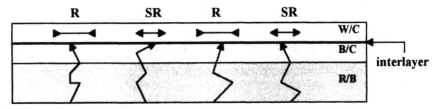

Fig. 1 Reflective cracking being controlled beneath the wearing course by reinforcing (R) and /or stress relief (SR)

2.1 Reinforcement

Asphalt has a low tensile strength. Kirschner [2] observed that this tensile strength can be exceeded by the result of strains as little as 2-3%. When this tensile strength is exceeded, cracks will form in the asphalt course, leading to its premature failure. The inclusion of a material such as a geotextile or geogrid with a high tensile strength and, often, low elongation (measured by a test method such as B.S. 6906 [3]), the crack (or potential cracking zone) is bridged by the material.

Kirschner identified the requirements of the optimum asphalt reinforcement. He determined that the tensile strength of the reinforcement should be not less than 35kN/m in both material directions and that it should have a maximum elongation at break of 15%. More specifically he concluded that the material should, at a strain of 3%, exhibit a tensile strength of at least 10kN/m in order to absorb stress even when there is little strain induced in the pavement.

These products act to stitch each side of the crack together and stop any lateral movement of the lower pavement layers. These materials are reliant upon the fact that they must have a modulus of elasticity that, by being greater than the material into which they are laid, allows them to offer that material reinforcement. To achieve this there must be an efficient bond between the reinforcing elements and the base course, a factor also identified by Kirschner. Bond between the upper course and the reinforcing elements need not be so great, albeit that it is invariably greater than that bond developed below. Theoretically, the prevention of movement within the lower course will have the effect of eliminating the continuation of cracks through the wearing course.

2.2 Stress relief

Research in the USA led to the derivation of a document referred to as the "Task-Force 25 Report" [4], the title of the committee responsible for the research leading to this guidance document. The majority of simple stress relief materials are non-woven geotextiles that comply with the requirements of this document, a minimum tensile strength of 1.2kN/m and minimum elongation of 50% at break. The crack (or potential cracking zone) is bridged by a material usually being of low tensile strength and high extension. Installation of these products is carried out by spraying a generous quantity of bitumen onto the lower pavement layer and rolling onto this binder, the stress relief fabric. By bridging the cracks with this layer of bitumen impregnated textile, movement stresses are transmitted laterally into the body of the interlayer and absorbed. Contrary to the previous case, the interlayer does not eliminate movement, but theoretically, dissipates its' effect, preventing any cracking continuing into the upper course. Bond between the interlayer and the overlying and underlying courses tends to be equal. This method has the secondary (but also significant) function of waterproofing the pavement, reducing the amount of water that may reach the lower pavement layers, possibly reducing movement.

3 Alternative practice

3.1 Reinforcing stress relief materials

Opinion is divided on which of the two methods above give best results in retarding reflective cracking. A small number of manufacturers of geotextile type products have researched this subject more closely although a definitive answer as to whether stress relief or reinforcement materials have the greater effect in retarding reflective cracking will obviously take some considerable time to discern with ongoing trials taking several years to yield results. It has been the intention of Don & Low Ltd to address this sector of the market by offering a product which fulfils the essential requirements of Kirschner in terms of tensile strength (particularly at intermediate strains) and also, by its' mode of installation, offer an impermeable, stress relieving, waterproofing layer within the pavement.

Results from projects completed will take some years to yield data which can definitively prove this stance. In some projects the design thicknesses have been varied as a result of the incorporation of a geotextile interlayer. Walsh [1], suggests that a fabric may be appraised as an equivalent to a given thickness of "x" mm of asphalt and this approach is commonplace in the USA but if this type of comparison is used to actually amend the design, the effect of the fabric cannot be accurately measured against any previously accepted construction.

3.2 Arguments for and against reinforcing and stress relief materials demonstrating the need for combination materials

Opposing proponents of each method currently use arguments as follows;

• Stress relief fabrics do not contribute significantly to the structural stability of the pavement. Their tendency to exhibit large elongation under relatively low tensile loads means that they will allow unacceptable deformation of the overlying course(s). The ability to absorb movement stress is therefore only argued to be effective as long as the stress is transferred directly to the bitumen layer, relying on the fabric only to retain the bitumen in intimate contact with the cracking zone. These fabrics tend to be lightweight non-woven materials, callendered to a varying degree. Their structure allows them to readily absorb the bitumen on which they are laid but also makes them weather sensitive after they have been laid down in so far as they will also absorb rain or surface moisture - undesirable in Northern European climates.

The contrary view;

• Reinforcement fabrics are reliant on an inherent rigidity (high tensile strength and low elongation) for their ability to effectively join and hold together the cracks beneath them. This inelasticity is incompatible with bituminous material that may expand and contract greatly in the variances of seasons and daily weather changes. The bond between reinforcing elements and the course beneath can rarely match that achieved by the stress relief fabrics, yet Kirschner suggests this is an important contributory factor to its' success in use. The refinement of some of these products to include a bitumen coating to the reinforcing elements intends to allow a bond to develop when the asphalt paved onto it heats the bitumen and this bonds with the layer below. Tests to determine the adhesion of grid to pavement are prescribed but it is argued that surety of the same degree of bond *after* paving is difficult to obtain. In damp or cold weather this bond is minimal or non-existent and the material may at the end of the paving process, be bonded efficiently only to the overlying course. The areas where two sheets lap is even more problematic as these materials are not prone to bond to themselves. Cores extracted from sites where these materials have been laid demonstrate this lack of bond and this obviously has ramifications for the lateral transmission of tensile strains across overlaps that these materials are designed to achieve. Obviously, even if they work in their own roll width, they will not perform as well (if at all) in their overlap areas. For this reason, some geogrid manufacturers recommend that adjacent widths are mechanically joined together. The reinforced wearing course may be effectively bonded to the base course at the time of installation but when expansion or contraction of the pavement occurs, albeit minimal, it may be expected that a shear plane may develop on the plane of the weakest bond since the wearing courses' expansion or contraction will be influenced by that of the reinforcement. In this event the courses above and below the reinforcement may move independently, accelerating the fatiguing of the pavement. Reinforcing grid type materials may, in some cases, have a "medium" elongation allowing more expansion and contraction but their effectiveness will still be governed by their ability to adhere to the lower course. This may have to be achieved by nailing and tensioning the product to the lower course this making installation slow and costly.

Both arguments have some substance. The position of an Engineer will be determined by considering these arguments and with local knowledge of the nature (and cause?) of the reflective cracking on his project, deciding which solution is most likely to succeed. It seems to be apparent incidentally that most Engineers do tend to strongly favour one particular argument although neither can be categorically proved a more accurate representation than the other.

Reinforcing stress relief materials have become available in response to this dilemma with the aim of combining the benefits of each method of crack retardation. Fig. 2 shows the relationship between tensile strength and elongation under load for various types of interlayer. It highlights the fact that these "combination" products perform in the middle ground between the high tensile strength/minimal extension reinforcing products and the minimal tensile strength/high extension stress relief products. Kirschner (section 2.1), in identifying the characteristics required of an asphalt reinforcement, determined that the ideal properties of such a material should be as below (i.e. lying above the "Kirschner line").

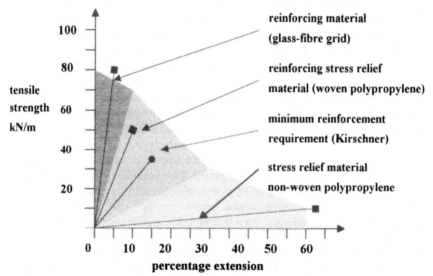

Fig. 2 Relationship between tensile strength and elongation in
pavement interlayers and the resultant effect

It may be argued that choice between stress relief and reinforcement need not therefore, be simply one or another but the Engineer who perceives benefits from both types of interlayer can now use a combination material such as that which Don & Low Ltd have sought to offer in their Lotrak Pavelay product. Several types of combination material exist. The basic principle is that the material should be a close textured fabric that can be laid on a tack-coat of bitumen in the same way as a simple stress relief fabric but that it should have reinforcing tensile strength at a lower extension. This type of material embodies the benefit of reinforcing fabrics, stitching the two sides of a crack together by virtue of the high tensile strength. Its bond to the lower course should be much more efficient due to its application onto the bitumen tack-coat and the heat transmitted to the bitumen through the fabric when paving commences ensures that the bond with the course above is efficient also. The extension of these materials tends to be greater than the plain reinforcement materials made from glass fibre or steel, minimising the risk that additional stresses are   induced in the pavement in thermal expansion or contraction. These materials also act in the same way as the simple stress relief fabrics by allowing lateral transmission of movement stresses into the fabric and the surrounding bitumen.

The question must then be posed, does this type of interlayer perform satisfactorily in reducing or eliminating reflective cracking? The most important factor in a successful installation is that the material must be laid by a competent installer. The surfacing contractor may wish to lay these materials but optimum results will be attained by using a sub-contractor who specialises in this type of work and is aware of the limitations of each material available.

Trials have been carried out by Don & Low on a variety of projects, using the Lotrak Pavelay woven polypropylene interlayer. This material acts to reinforce and to relieve stress. The initial findings of these trials are documented below.

4 Site experience

The need for a reflective cracking inhibitor was identified in the reconstruction of the high speed test track at the Motor Industry Research Association (MIRA), Nuneaton, England. The track is constructed over concrete slabs from an old airfield and in the past has seen previous overlays affected by reflective cracking from the concrete base as soon as 18 months after overlaying.

In reconstruction, the track was reduced to the concrete slab and a bituminous regulating layer laid to achieve required levels, followed by the normal base course and wearing course construction. To reduce the chance of reflective cracking recurring, MIRA specified the inclusion of a geotextile interlayer at the interface of the regulating layer and the base course. MIRA, after considering various product types together with their associated benefits, recommended the use of a reinforcing stress relief material. The woven polypropylene Lotrak Pavelay product was selected due to its ability to act both as a stress relief membrane and a reinforcing fabric. In addition, comparison of the installed product costs showed the woven polypropylene fabric to be more cost effective than other materials offering the requisite characteristics.

Lotrak Pavelay was mechanically placed on a K1-70 emulsion by a specialist sub-contractor. Prior to the over-laying of the fabric, a sudden downpour left puddles of water lying on the Pavelay product. When the weather improved, excess water was manually brushed from the surface and the fabric allowed to dry naturally for 15 to 20 minutes prior to overlaying with base course. Cores taken through the completed construction show very good adhesion between the Pavelay and the bituminous layers above and below it.

A short control section was constructed without Pavelay and the comparative performance of the two types of construction are being monitored. At the time of writing, 17 months after reconstruction, both sections are in a good condition with no sign of reflective cracking.

At the time of initial trials in 1991, the A85, Crieff Road trunk route in Perth was found to be showing unacceptably extensive levels of cracking along its length. This was judged to be due to a combination of increased traffic loadings and earlier work by public utilities. Settlement of the trench material had resulted in the cracking of the asphalt courses at the edge of the service trench and also at other areas that had been previously disturbed. Had no remedial action been taken, the surface cracking would have allowed unacceptable levels of water ingress into the foundation and this would have obviously led to the premature failure of the whole pavement. To reduce the possibility of reflective cracking recurring in these areas and to ensure that the pavement structure was effectively waterproofed, it was decided to include a geotextile interlayer.

It was recognised that the inclusion of a geotextile would not speed the maintenance scheme but if it allowed the road to remain serviceable for a longer period of time, this would have cost benefits in more ways than one. The cumulative re-laying cost of the road would be reduced but in addition, closure of this busy road in Perth would require the diversion of large traffic volumes onto more minor and less suitable carriageways, having a detrimental effect on these routes also. It was therefore decided that the use of a woven polypropylene interlayer, Lotrak Pavelay, offered the best prospect of extending the maintenance cycle at a cost that could be justified.

Tayside Contracts, the contracting arm of Tayside Regional Council, carried out the repairs to Crieff Road. The existing levels were reduced by cold-planing to a depth of 100mm, leaving a sound laying surface. A K1-70 emulsion was sprayed onto the carriageway and the Pavelay was brushed onto this tack-coat when the water had evaporated. Successive layers of 60mm of hot-rolled asphalt (HRA) base course and 40mm of HRA wearing course were laid to, or close to, the original levels. Four and a half years later, there is no sign of reflective cracking.

In both of the above cases the absence of a failure in the repair sections to date precludes any conclusions being drawn as to the lifetime extension attainable by this type of product. It is the case that all of these products have been used for too short a time in Europe for any Engineer or manufacturer to determine the average lifetime increase with any degree of accuracy. The material laid over the Pavelay in the above cases was asphalt. It is not uncommon to see very high delivery temperatures for asphalt today; improved insulation of delivery vehicles has made temperatures in excess of 175°C commonplace. Since the melting point of polypropylene is 167°C, it is occasionally perceived as a problem that asphalt temperatures may exceed this figure. On site trials by Don & Low Ltd., the manufacturer of Lotrak Pavelay have found similar results where asphalt temperatures of 175°C have led to interface temperatures of approximately 125°C. The relatively low temperature of the fabric (at or close to ambient temperature) leads to a rapid cooling of the underside of the asphalt course as is normal, hence protecting the fabric from such extremes of temperature as could cause it damage.

5 Cost benefit analysis

The benefit of these various types of interlayer may be derived by relating pavement lifetime extension to the cost per square metre of each product. However, since few installations have reached the end of their useful life, and materials have not necessarily been laid with an untreated control, this measure is difficult to determine. The only measure that can be applied universally is to relate cost per sq m installed to the cost of the overlying course per sq m. The additional cost of the interlayer is therefore measurable against the cost of asphalt giving a requisite increase in lifetime to justify its inclusion in the works. This requisite lifetime increase (RLI) should then be compared with the data available to date from ongoing trials to ascertain whether it is envisaged to be attainable. The interlayers may then be further compared by a simple relation of RLI to anticipated actual lifetime extension. Using a *notional* price per sq m of £6.00(sterling) to lay asphalt wearing course on a small to medium site and comparing to an untreated control with an estimated lifetime of, say, five years, table 1 illustrates the above analysis.

Table 1 Requsite lifetime extension in relation to notional asphalt cost of £6.00 sq m and untreated control lifetime of five years

	Product	Cost per sq.m. installed (£)	Requsite lifetime increase	Actual lifetime increase required
Stress relief materials	A	£1.50	25%	1yr 3mths
Reinforcing /	B	£2.00	33.33%	1yr 8mths
stress relief	C	£3.50	58.3%	2yrs 11mths
materials	D	£4.50	75%	3yrs 9mths
Reinforcing	E	£4.00	66.67%	3yrs 4mths
materials	F	£6.75	108.3%	5yrs 5mths

It is easy to envisage that the RLI s of 25% and 33.3% can be attained; data is readily available from trials which demonstrates this degree of pavement lifetime increase, albeit that these trials are in some cases ongoing and the actual lifetime increase may yet be considerably higher. An RLI of 75% however, is obviously more ambitious. It may be perfectly attainable but for example, a pavement where an untreated section was expected to last for say, five years, product **D** would need to prolong the lifetime of the pavement for more than two years after the cost of product **B** had been justified.

The Engineer must arguably show a certain amount of faith to decide which product is likely to justify itself in his project. The table above represents relative costs of actual materials, **A** representing a simple stress relief fabric, and **B, C** and **D** representing reinforcing stress relief materials. **E** and **F** represent high tensile strength, minimal extension reinforcement materials. All are currently available in the United Kingdom.

6 Conclusions

There is a large amount of confusion about how the retardation of reflective cracking is most effectively achieved. The two opposing views that stress relief and reinforcement are each themselves, the most effective methods, will probably remain for some time to come. A definitive answer appears to be difficult to prove.

Work to date using materials which combine the benefits of each system has offered encouraging results although this is insufficiently well developed to yield a clear proof of the superiority of these materials. However, since some of these materials have installed costs at levels close to the simple stress relief fabrics, their ability to further retard reflective cracking by their additional reinforcement function makes further trials and appraisal technically and commercially worthwhile. There have been early failures in all three categories and it must be borne in mind when designing the scheme, that the object of the exercise is to lengthen the maintenance cycle at a cost that can be justified in extra pavement lifetime. An expensive product may be expensive by virtue of the technology employed in its manufacture, not because it can necessarily offer comensurately high performance. The Engineer is almost invariably in a position where his design must be justified on technical merit and cost and an exercise in cost benefit analysis (such as in section 5) is essential in appraising the materials on offer.

Reinforcing, stress relief materials offer the Engineer an alternative to the two types of simple, one-purpose materials without compromising effectiveness in either function. Their use is forecast to increase.

7 References

1. Walsh, I. (1993) *The use of Geosynthetics in Asphalt Pavement Construction and Maintenance*, Geoconference 93, London.
2. Kirshner, R (1990) *Asphalt Reinforcing polymer grids in Airfields*, Airport Forum, Vol. No 5
3. British Standard Institution. (1987) *Methods of test for Geotextiles Part 1: Determination of the tensile properties using a wide width strip.* BSI, London. B.S.6906 Part 1
4. United States FHWA Committee Task Force 25 Report

Application of grids to prevent longitudinal cracking caused by an old concrete slab

J. VALTONEN and M. VEHMAS
Helsinki University of Technology, Espoo, Finland

Abstract
In 1930's some main roads around Helsinki were paved with concrete. The concrete pavements were quite narrow, only 6 meters wide. One of those roads was a road from Helsinki to Turku. This road was widened to 8 meters in the late 1950's and at the same time it was overlaid with asphalt concrete.

Because the edge of the road does not have concrete underneath, the heavy traffic has caused defects to the edge of the old concrete surface. The edge of concrete slab reflects to the excisting surface as longitudinal cracks.

During the last few years, many new methods to prevent cracking have been developed. Finnish National Road Administration's Uusimaa district wanted to find out the most cost effective method to prevent longitudinal cracking on a road similar to one described before. So they planned a field test together with Helsinki University of Technology.

The purpose of this study was to find out the type of grids, made out of glassfiber or steel, which would be the most suitable to prevent reflective cracking on asphalt concrete pavements. At the same time we will get knowledge of the installation of different type of grids and their effects on the evenness of the road.
Keywords: Asphalt concrete pavement, concrete pavement, defects, field test, glass fibre grid, longitudinal cracking, repairing of cracks, steel grid

1 History of the test road

In 1930's some main roads around Helsinki were paved with concrete. The concrete pavements were quite narrow, only 6 meters wide. One of those roads was a road from

Reflective Cracking in Pavements. Edited by L. Francken, E. Beuving and A.A.A. Molenaar. © 1996 RILEM. Published by E & FN Spon, 2–6 Boundary Row, London, SE1 8HN. ISBN 0 419 22260 X.

Helsinki to Turku. This road was widened to 8 meters in the late 1950's and at the same time it was overlaid with asphalt concrete.

The road from Helsinki to Turku was chosen to be a test road, because there were quite a lot of longitudinal cracks on the existing pavement. The reason for the cracking is the old concrete pavement underneath the new asphalt concrete. The edge of the old concrete reflects to the existing surface as longitudinal cracks.

One further reason for the cracking may be that the slopes of the road are quite steep and the bearing capacity of the widened part of the road has not been high enough.

2 Test procedure

Finnish National Road Administration's Uusimaa district wanted to find out the most cost effective method to prevent longitudinal cracking by carrying out a field test. The field test was planned in the summer 1995 and carried out in September 1995. The method of using glass fibre and steel grids to prevent reflective cracking on an old concrete road was new in Finland. However, geosynthetics and steel grids had been used before to prevent reflective cracking and cracking due to frost heave and poor bearing capacity. In those cases, the grids had mainly been installed lower in the unbound layers, not in the asphalt layers.

Eight test sections were chosen from the road. The lenght of the sections varied between 90 and 145 meters. According the plan, all the grids were to be placed inside the asphalt layer, not deeper in the structure. The reason for that was that construction of the test sections would be as easy as possible and construction could be done without tearing apart the existing structure of the road.

The width of the old concrete pavement and the thickness of the existing asphalt concrete were confirmed by taking samples of the road. The thickness of the asphalt overlay was found out to be 110...125 mm.

The construction of the test sections was done so, that a 40...50 mm deep and 2 m wide layer of asphalt (1 m from both sides of the crack) was scarified from the pavement. After the grid was placed correctly, a new asphalt concrete was layed over the grid. The structure is shown Fig. 1. Afterwards the whole road was repaved.

In this study, five different glass fibre grids and two different steel grids were tested. One of the steel grids was tested both in a structure described above and also without scarifying the existing asphalt.

Test sections were chosen so that they would be as homogenous as possible. Both the steepness of the slopes and the number and types of cracks were similar in each sections. The number of grids tested was limited to seven, because a larger number would have forced to shorten the sections or chosen sections with less defects. The grids used are shown in Table 1.

Glass fibre grids were chosen so, that their strength properties were close to each others and the width of the grids was 2 meters. Glasgrid 8501 is coated with bitumen and Polyfelt PGM is a combination of glass fibre grid and geotextile, other properties of grids are quite equal.

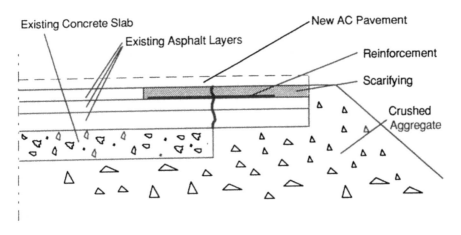

Fig. 1. Structure of the test sections.

Table 1. Used grids

Sect.	Grid type	Mesh size (mm)	Weight (g/m²)	Width (m)	Tensile strength length/width (kN/m)	Cost (FIM/m²)
1	Steel grid	50x50	4440	2.0	277 / 277	15
2	Glasfiber Geogrid	22x22	270	2.0	39 / 42	22
3	Polyfelt PGM-G 100	40x40	140	1.9	100 / 100	20
4	Glasgrid 8501	25x25	450	1.5	100 / 100	28
5	Rehau Armapal	30x30	286	2.2	50 / 50	21
6	Roadtex WG2303G1	40x26	280	2.0	35 / 56	26
7	Steel grid	50x50	4440	2.0	277 / 277	15
8	Steel grid	150x150	1480	2.0	92 / 92	5

3 Construction of the test road

Before the installation of grids took place in September, all the defects were surveyed. All test sections were badly damaged. Nearly all of the defects were reflective cracks caused by the old concrete slab. Most of the longitudinal cracks had been patched with gussasphalt during the years, in some places the patching had been necessary even once a year. Example of defect surveying is shown in Fig. 2.

The scarifying was done 0.2 meter wider than the actual width of the grids, so that the installation of the grids would be easier. It should also prevent reflective cracking occuring from the edge of the grids in the future. The longitudinal crack was always in the middle of the scarified area.

Section 6

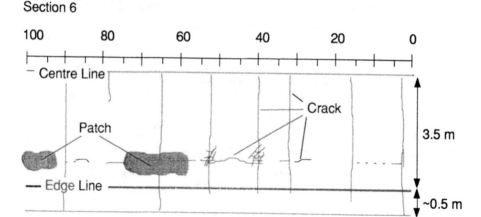

Fig. 2 Surveying of defects on section 6.

Glass fibre grids were installed first. The scarified area was treated with emulsified bitumen except while using Glasgrid 8501, which is self adhesive. The bitumen was not polymer-modified. After the bitumen had dried a short while, the grids were installed and right after the installation sections were paved. The mixture used for paving was an ordinary asphalt 0/12. Steel grids were installed straight on the scarified area and paved by using an asphalt mixture 0/18. To avoid crumpling of the grids, paver didn't push the asphalt truck in front of it. Steel grids didn't have any protection against corrosion.

During the installation and paving, traffic was using only the other side of the road so that grids wouldn't be damaged.

The installation of different grids is presented in the order of construction. During the installation work weather was fine and temperature +15 °C.

3.1 Glasfiber Geogrid, section 2
The amount of used bitumen emulsion was 0.2 kg/m^2. The grid was rolled open 10...15 meters at a time and then two men tightened it. The adhesion with the bitumen was quite good, the grid didn't come loose when trucks drove over it, instead the effect of trucks was compacting.

The whole section was successful.

3.2 Polyfelt PGM-G 100, section 3
The amount of bitumen emulsion was supposed to be 1.1 kg/m^2, but because of the spreading difficulties it was probobly sligtly smaller. The installation procedure was equal to the one in section 2, but there were some difficulties while tightening the grid. Because in this section a larger amount of bitumen was used, the process of sticking was so fast that it was impossible to avoid some folds. Paving succeeded well.

3.3 Glasgrid 8501, section 4
This was the only self adhesive grid. The bitumen was on the inner side of the grid, so the installation was done using a pipe as an axle. The grid was rolled open 15 meters at

a time and then it was layed on the surface and compacted by a rubber wheel roller. The adhesion was excellent and there were no problems with paving.

3.4 Rehau Armapal, section 5
The amount of used bitumen emulsion was 0.4 kg/m^2. The grid was rolled open 20 meters and then tightened. This emulsion broke too fast and the bitumen used seemed to be quite hard. Because of that, the adhesion was not sufficient, but it got better after the trucks had driven over the grid. The grid was a little wavy at the edges but that didn't cause any problems to paving.

3.5 Roadtex WG 2303 G1, section 6
The amount of used bitumen emulsion was 0.4...0.5 kg/m^2. The emulsion broke too fast like on section 5, but the adhesion was still slightly better. The paving succeeded quite well, some waving existed, but it was mainly because the section located in a curve and therefore the tightening of the grid was more difficult to do.

3.6 Steel grid # 50x50 mm, section 7
The grids were installed straight on a scarified surface, because of their weight the installation was done by two men. The grids had bended badly during the transportation or storing and after installation some edges were even 10 cm off the ground. While paved the grids seemed to press against the surface, but before the actual paving took place the following week the asphalt surface was broken and at some spots the grid could even be seen.

3.7 Steel grid # 150x150 mm, section 8
The installation was equal to the section 7, only it could be done by one man, because of the smaller weight of grid. These grids were in a better shape and also more flexible, so the installation work and paving succeeded well.

3.8 Steel grid # 50x50 mm, section 1
This grid was installed on an old pavement without scarifying. Like the grids used in section 7, these were also in bad shape and after installation the edges were 10...15 cm off the ground. The paving was impossible to do, because the paver got stuck to the edges of the grids and therefore pulled the grids up together with the newly layed asphalt. Because of the difficulties the grids were removed and this section was only paved with asphalt concrete.

4 Follow-up

To find out, which type of grids are the most suitable to prevent cracking the profiles of the road were measured after the final paving. The measuring was done after every 10 meters, using a 3 meter straightedge.

The follow-up measurements are planned to be done every spring and fall for a few years. On every test section there are two cross-sections from which the profiles are measured. Also the surveying of defects shall be done.

After the first winter in April 1996 no longitudinal cracks were to be seen in any section.

5 Conclusions

The installation of glass fibre grids succeeded fairly as a handwork, since the area was so small. For a larger areas some machinery is needed to avoid problems while tightening the grids.

The breaking speed of the bitumen emulsion should be slow enough, since the adhesion of the grids is better when the breaking is not completed while the grid is installed.

The paving of sections succeeded fine, except while using the steel grid # 50x50 mm, where problems were caused by the shape of the grids.

After the first winter it is not possible to say which one of the grids is the best to prevent cracking, they all seem to have worked well. After surveying the defects and measuring the profiles for some years, we'll have more knowledge of how the grids have acted. In case when cracking occures, core samples will show the reason of it: if the grid was not properly installed, or it was damaged while paved, or the quality of grid is not well enough to prevent cracking.

Apports des enrobés avec fibres dans la lutte contre les remontées de fissures

J.P. SERFASS
SCREG, St Quentin-en-Yvelines, France
B. MAHÉ DE LA VILLEGLE
SCREG Ouest, St Herblain, France

Résumé
L'observation de leur comportement en place a montré que certaines formulations d'enrobés minces avec fibres ont un effet non négligeable vis-à-vis de la réapparition de fissures.

Les sables enrobés constituent une des techniques de découplage les plus efficaces. L'ajout de fibres dans ces sables enrobés améliore leur résistance au fluage et a, en même temps, une influence positive sur la résistance à la fissuration.

La combinaison d'un sable enrobé fibré et d'une couche de roulement en béton bitumineux mince fibré donne des complexes particulièrement efficaces vis-à-vis de la remontée de fissures. Des planches comparatives ont fait l'objet d'un suivi régulier, conduisant à un bilan positif basé sur de nombreuses mesures et l'observation simultanée de témoins.

Par ailleurs, la recherche d'une macrotexture et d'une adhérence élevée a incité à concevoir des complexes associant un béton bitumineux mince avec fibres et une couche de roulement au béton bitumineux très mince. Ces complexes sont efficaces vis-à-vis de la fissuration et offrent une résistance satisfaisante à l'orniérage.

Enfin, les recherches se poursuivent avec l'expérimentation de nouvelles formules d'enrobés minces antifissures.
Mots-clés : Antifissures, béton bitumineux, découplage, fibres, orniérage, retrait-flexion, sable enrobé.

1 Introduction

L'incorporation de certains types de fibres permet, on le sait, de réaliser des enrobés bitumineux très riches en liant, ce qui est un facteur favorable pour la résistance à la fissuration. Cette technique peut être déclinée de plusieurs manières pour lutter contre la remontée des fissures. Les fibres utilisées, très fines et courtes, sont soit minérales (roche ou verre), soit organiques (cellulose).

Reflective Cracking in Pavements. Edited by L. Francken, E. Beuving and A.A.A. Molenaar. © 1996 RILEM. Published by E & FN Spon, 2–6 Boundary Row, London, SE1 8HN. ISBN 0 419 22260 X.

Elles agissent d'une part en tant que stabilisant (fixation d'une quantité importante de bitume), d'autre part en tant que micro-armatures. Ces effets se traduisent par un film de mastic très épais, donc une bonne tenue à la fatigue, au vieillissement, à l'eau et par une résistance élevée en cisaillement, traction en flexion, tout en préservant une tenue en fluage correcte.

2 Bétons bitumineux minces avec fibres

Ces enrobés sont destinés aux couches de roulement de 3 à 5 cm. Les granularités 0/14, 0/10 ou 0/6 mm sont le plus souvent discontinues. La teneur en bitume (le plus souvent 50/70) est élevée : entre 6,5 et 7,6 ppc selon les granularités et les formulations. La teneur en fines est, elle aussi, importante : le passant à 0,08 mm se situe entre 10 et 13%.

L'observation du comportement en place de nombreuses sections de ces bétons bitumineux, dont certains ont atteint jusqu'à 18 ans d'âge, a montré la durabilité de ces enrobés sur supports déformables (au sens de la déflexion) ou fatigués (début de faïençage).

Le suivi de sections de chaussées semi-rigides recouvertes d'enrobés de ce type a, par ailleurs, mis en évidence l'effet de retardement obtenu vis-à-vis de la remontée des fissures de retrait actives. De plus, les fissures réapparues restent fines et ne donnent lieu ni à dédoublement, ni à épaufrures.

3 Sables enrobés antifissures

L'idée de base consiste à réaliser une couche d'interposition très mince (1,5 à 2 cm) constituée d'un mortier bitumineux souple, à très forte teneur en liant.

La fonction première de ce sable enrobé est de dissiper les contraintes qui se concentrent au droit de fissures ou des joints et d'absorber, par plasticité interne, les déformations provoquées par le souffle de ces fissures ou joints. Cette couche d'interposition a aussi pour fonction de garantir l'imperméabilité.

Ces sables enrobés, très fermés et lisses, sont obligatoirement surmontés d'une couche d'enrobé plus ou moins épaisse.

Les granularités utilisées vont du 0/3 au 0/6 mm; elles sont généralement continues. Les pourcentages de fines (passant à 80 μm) sont élevés : 10 à 15%.

Dans la plupart des cas, le ou les sable(s) sont exclusivement issus de concassage. Un peu de sable roulé est parfois incorporé pour améliorer la compactibilité.

Le liant est, en règle générale, un bitume modifié par ajout de polymère. Cependant, pour les cas où l'orniérage n'est pas à craindre (trafic limité - au maximum T2 -, site non pénalisant, couche de roulement d'épaisseur et de résistance suffisantes), l'emploi de bitume pur est possible. Les teneurs en liant sont très fortes : de 9 à 12,5 ppc.

L'ajout de fibres permet d'augmenter l'efficacité des sables enrobés antifissures. Comme on le verra plus loin, elles ont un effet stabilisant et permettent de fixer une plus forte quantité de liant tout en évitant le fluage.

4 Complexes bicouches avec sable enrobé

Les sables enrobés peuvent être utilisés en combinaison avec divers types de couche de roulement. Les suivis de réalisations tendent à démontrer que l'épaisseur de la couche de roulement surmontant le sable enrobé est un paramètre important. En pratique, il apparaît souhaitable, pour l'épaisseur de la couche de roulement, de ne pas descendre en-dessous de 3 - 3,5 cm. Les couches de roulement considérées ci-après sont donc de deux types : béton bitumineux épais, type semi-grenu (BBSG) et béton bitumineux mince (BBM).

4.1 Complexes sable enrobé + béton bitumineux épais
Constitution des complexes
Les complexes étudiés comprennent 2 cm de sable enrobé et 6 cm de béton bitumineux semi-grenu classique au bitume pur.

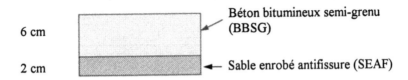

Voici, à titre d'exemple, les formulations des matériaux entrant dans certains des complexes testés :

- Béton bitumineux semi-grenu
Granulat	Microdiorite
Granularité	0/10
Passant à 6,3 mm	66%
Passant à 2 mm	37%
Passant à 0,08 mm	8 %
Bitume	5,8 ppc

- Sable enrobé - Version 1
Granulat	Microdiorite
Granularité	0/4
Liant	bitume-SBS dosé à 9,5 ppc

- Sable enrobé - Versions 2 et 3
  Idem + fibres

Essais de retrait-flexion

Les complexes ont été testés par le Laboratoire Régional des Ponts et Chaussées d'Autun. L'essai consiste à suivre la remontée d'une fissure au travers du complexe, conditionné à 5°C et soumis simultanément à :

- une traction longitudinale continue lente (0,01 mm/min), simulant l'ouverture d'une fissure par retrait thermique
- une flexion verticale cyclique, à la fréquence de 1 Hz et avec une flèche de 0,2 mm, simulant le trafic.

La progression de la fissure est suivie grâce à une série de fils conducteurs collés sur la tranche de l'éprouvette.

L'efficacité du produit testé, r, est définie comme étant le rapport du temps de fissuration du complexe testé sur le temps de fissuration du témoin bicouche (2 cm de sable enrobé 0/4 au bitume 70/100 + 6 cm de béton bitumineux 0/10 au bitume 50/70). L'échelle de jugement est la suivante :

$r = \dfrac{\text{Temps de fissuration des couches testées}}{\text{Temps de fissuration du témoin}}$	Avis sur la technique
$r < 0,7$	Inefficace
$0,7 < r < 0,9$	Moyennement efficace
$r > 0,9$	Très efficace

Plusieurs complexes ont été testés selon cette procédure. Parmi ceux-ci, les exemples suivants fournissent une comparaison entre sable enrobé sans fibres et sable enrobé avec fibres.

Les deux complexes testés étaient constitués ainsi :

Complexe 1	Complexes 2 et 3
6 cm BBSG	6 cm BBSG
2 cm Sable enrobé Composaf au liant Bitulastic E sans fibres	2 cm Sable enrobé Composaf au liant Bitulastic E avec fibres

Les figures 1 et 2 résument le déroulement des essais.

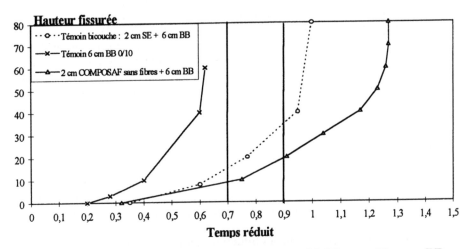

Figure 1 - Comportement d'un complexe COMPOSAF sans fibres + BB par rapport aux témoins

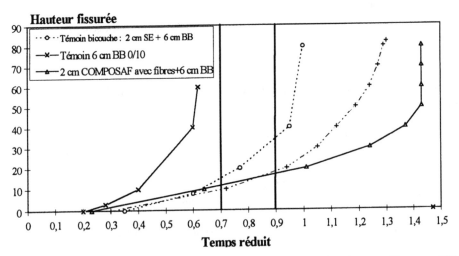

Figure 2 - Comportement de complexes COMPOSAF avec fibres + BB par rapport aux témoins

Les "efficacités" évaluées pour les complexes testés sont :
    r = 1,27 pour le complexe 1 (classé donc comme "très efficace")
    r = 1,30 pour le complexe 2 (classé donc comme "très efficace")
    r = 1,43 pour le complexe 3 (classé donc comme "très efficace")
Ces résultats confirment l'efficacité des sables enrobés antifissures généralement observée (Cf. Collège des Observateurs SETRA-LCPC). Rappelons que l'ensemble des sables enrobés testés avec la même couche de roulement (6 cm de BBSG) donne une "efficacité" moyenne de 1,15. Ils tendent, de plus, à montrer que la présence de fibres dans le sable enrobé a une influence favorable sur la tenue à la fissuration.

Essais d'orniérage
Les mêmes complexes ont été soumis à l'essai d'orniérage LCPC. Les résultats montrent que l'ajout de fibres, toutes choses égales par ailleurs, diminue l'orniérage de 14,5% à 9% (valeurs obtenues au bout de 30 000 cycles à 60°C). Le gain est très net.
    Globalement, la conclusion est que l'ajout de fibres améliore très nettement la tenue à l'orniérage, tout en conservant au moins le même niveau d'efficacité de découplage.

4.2 Complexes sable enrobé + béton bitumineux mince avec fibres

Constitution des complexes
Ils comprennent 2 cm de sable enrobé et 3 cm de béton bitumineux mince

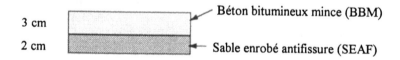

3 cm   ← Béton bitumineux mince (BBM)

2 cm   ← Sable enrobé antifissure (SEAF)

    Les matériaux entrant dans les complexes testés ci-après ont les caractéristiques suivantes :

- Béton bitumineux mince avec fibres
    Granulat   Microdiorite
    Granularité   0/10
    Passant à 0,08 mm (fines + fibres)   12,5 %
    Bitume   7,2 ppc

- Sable enrobé   Version 2 ci-dessus

Essais de retrait-flexion
La figure 3 illustre le comportement d'un tel complexe.

ESSAIS DE RETRAIT FLEXION - LRPC AUTUN

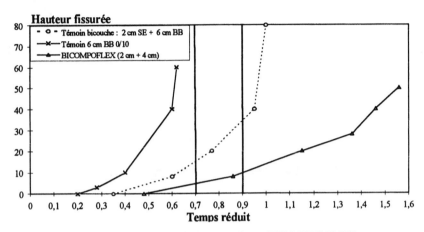

**Figure 3 - Comportement d'un complexe BICOMPOFLEX
par rapport aux témoins**

A l'ouverture maximale de fissure possible avec la machine, la fissuration n'est pas remontée jusqu'à la surface du Bicompoflex. Le complexe testé a une efficacité $\geq 1,56$ ce qui le conduit à le classer dans la catégorie **"très efficace"**.

On constate que 3 cm de béton bitumineux avec fibres sont plus efficaces que 6 cm de béton bitumineux semi-grenu classique.

Essais d'orniérage
L'essai à l'orniéreur LCPC, effectué selon le mode opératoire actuellement en vigueur, n'est pas significatif pour ce type de complexe. En effet, des phénomènes parasites (collage et arrachements en surface, glissements du sable enrobé sur la plaque d'acier) viennent perturber son déroulement.

Comportement en place
Le suivi dans le temps de plusieurs réalisations a montré l'efficacité de ces complexes vis-à-vis de la remontée des fissures (Cf. Rapport du Collège des Observateurs SETRA-LCPC - décembre 1993). Un exemple particulièrement marquant est celui des sections comparatives de la Route Nationale 23 dans la Sarthe. Sur ce chantier, suivi systématiquement depuis 7 ans, la supériorité du complexe sable enrobé + béton bitumineux mince avec fibres, par rapport aux témoins et aux autres systèmes antifissures testés, est manifeste (Figure 4).

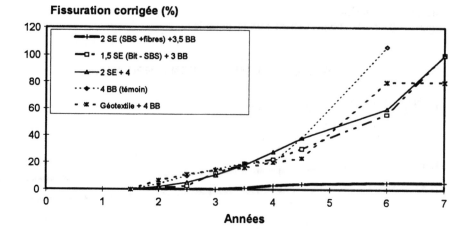

Figure 4 - RN 23 - Sections comparatives d'antifissures

D'une manière générale, il faut signaler la sensibilité des sables enrobés aux variations d'épaisseur. Un manque d'épaisseur risque de réduire l'effet de découplage. Les surépaisseurs rendent l'ensemble plus exposé au risque d'orniérage. Il faut donc veiller tout particulièrement au respect de l'épaisseur nominale de sable enrobé. Les supports déformés ne conviennent pas, ou bien ils doivent être soigneusement reprofilés au préalable.

Nota : Pour tous les complexes antifissures, l'amplitude des battements verticaux relatifs entre dalles est un facteur critique. Les valeurs limites de battements admissibles dépendent du trafic et de l'épaisseur des enrobés constituant le rechargement.

5 Complexes BB mince avec fibres + BB très mince

L'ensemble des constatations faites, tant sur les bétons bitumineux minces que sur les complexes bicouches, fournit des indications convergentes. Toutes tendent à prouver l'efficacité des enrobés structurés par des fibres, riches en mastic. Ce constat fait, du reste, partie des conclusions du Collège des Observateurs SETRA-LCPC (Rapport de décembre 1993).

Par ailleurs, les observations et mesures in situ montrent que la texture superficielle des enrobés minces riches en liant et en mastic est, fort logiquement, moyenne. Elle n'atteint pas toujours les niveaux élevés qui sont désormais exigés sur les autoroutes et routes à grande circulation.

De là est née l'idée du complexe associant un enrobé mince avec fibres en couche d'interposition et un enrobé très mince en couche de roulement, ce dernier offrant un niveau élevé de macrotexture et d'adhérence.

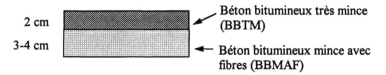

2 cm → Béton bitumineux très mince (BBTM)

3-4 cm → Béton bitumineux mince avec fibres (BBMAF)

Le béton bitumineux mince fibré joue le rôle d'antifissure; le BBTM amène un niveau élevé de macrotexture et d'adhérence.

Les deux couches contribuent à la structure et à l'amélioration de l'uni.

Pour les premières réalisations, la formulation du BBMAF était la même qu'en couche de roulement (Cf. § 3.2). Par la suite, une réflexion a été menée sur les évolutions possibles, qui a abouti à l'expérimentation d'autres formulations.

La couche de roulement est un BBTM modifié, soit avec fibres, soit avec bitume-polymère. Les grandes lignes des formulations 0/10 (les plus fréquentes) sont les suivantes :

	BBTM avec fibres	BBTM avec bitume-SBS
Passant à 0,08 mm (%)	8 - 12	7 - 9
Teneur en liant (ppc)	6,3 - 6,8	5,6 - 6,2

Essai de retrait-flexion

La figure 5 illustre le comportement d'un complexe comprenant un béton bitumineux mince avec fibres, de formulation "classique", c'est-à-dire analogue à celle d'une couche de roulement et d'un BBTM avec fibres en couche de roulement.

ESSAIS DE RETRAIT- FLEXION - LRPC AUTUN

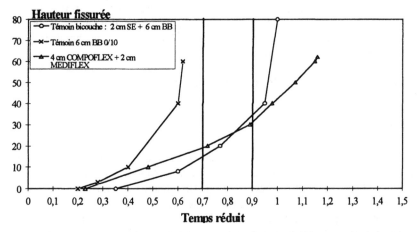

Figure 5 - Comportement d'un complexe COMPOFLEX + MEDIFLEX par rapport aux témoins

L'ensemble est moins performant qu'un complexe sable enrobé + BBM avec fibres, mais le résultat (r = 1,16) le classe largement dans la catégorie "très efficace".

Essai d'orniérage

L'essai effectué sur un complexe associant les mêmes enrobés que ci-dessus (BBM-fibres + BBTM-fibres) a donné un orniérage de 7,6% à 3 000 cycles et de 10,2% à 30 000 cycles à 60°C.

Comportement en place

Plusieurs chantiers ont été réalisés selon cette technique depuis 1993 (rechargements de chaussées semi-rigides à couche de base en grave hydraulique). Les observations à fin 95 montraient l'absence d'orniérage et de remontée de fissures.

Evolution en cours

Afin de pouvoir traiter les cas les plus difficiles (anciennes chaussées en dalles de béton ou chaussées semi-rigides à fissures très ouvertes), nous avons fait évoluer la conception et la formulation de l'enrobé mince fibré antifissures (AF). Ce nouveau type d'enrobé AF, à teneur en mastic augmentée, a été retenue par la Direction des Routes et le SETRA dans le cadre de la Charte de l'Innovation pour le thème "Lutte contre la fissuration et techniques d'entretien des chaussées béton".

Le premier chantier expérimental a eu lieu en juin 1994 sur la route Nationale 165 à Quimper (Finistère). La réalisation a été conforme au cahier des charges. Cette section fait maintenant l'objet d'un suivi régulier.

6 Conclusions

L'analyse des résultats obtenus en laboratoire et, surtout, les observations des comportements en place montrent l'intérêt des enrobés modifiés par ajout de fibres pour la lutte contre la fissuration. Ce comportement s'explique par les fortes teneurs en mastic que cette technologie rend possibles, ainsi que par le renforcement de ce mastic.

Diverses manières d'utiliser ces qualités ont conduit à la mise au point de plusieurs types de formulations et de complexes.

L'effort de recherche-développement dans ce domaine se poursuit et d'autres innovations sont en cours d'étude ou d'expérimentation.

7 Références

1 Serfass, J.P. (novembre 1991) Interposition de couches minces antifissures. Les sables enrobés. *Journées d'étude Ecole Nationale des Ponts et Chaussées*

2 Laurent, G., Serfass, J.P. (mars 1993) Un chantier comparatif de complexes antifissures. Bilan à quatre ans. *2ème Conférence RILEM*, Liège, pp. 353-359

3 Dumas, P., Giloppé, D., Laurent, G. (décembre 1993) Les systèmes retardant la remontée de fissures transversales de retrait hydraulique. *Observatoire des techniques de chaussées*. LCPC-SETRA

4 Serfass, J.P., Samanos, J. (décembre 1994) Complexes bicouches d'enrobés antifissures. Evaluation et comparaison. *Revue Générale des Routes et des Aérodromes*, N° 724, pp. 42-44

5 Serfass, J.P., Samanos, J. (janvier 1995) Bétons bitumineux avec fibres. *Revue Générale des Routes et des Aérodromes*, N° 725, pp. 52-60

6 Mahé de la Villeglé, B., Chanceaulme, M. (mars 1995) Compoflex est-il un enrobé antifissures? *Revue Générale des Routes et des Aérodromes*, N° 727, pp. 42-44

Part three

Test methods for interlayers

Laboratory testing and numerical modelling of overlay systems on cement concrete slabs

A. VANELSTRAETE and L. FRANCKEN
Belgian Road Research Centre, Brussels, Belgium

Abstract

This paper deals with the results of laboratory testing and numerical modelling of overlays on cement concrete slabs. The purpose of this study was to investigate the effect of the use of interface systems for the prevention of reflective cracking in road structures subjected to thermal and traffic loads.

Thermal cracking tests were performed to study the effect of subsequent shrinkage and expansion of cement concrete slabs as a result of temperature variations. Results are presented for several types of interface systems: SAMI, nonwoven and woven textiles, grids and steel reinforcement nettings.

Concerning the effect of traffic much effort was applied to the study of the vertical movements at the edges of cement concrete slabs. In order to simulate this effect, an experimental test device was developed. In addition to this test, 3D-computer simulations of actual road structures were carried out to study the influence of different parameters on the vertical movements at the edges of concrete slabs: type of base and subbase, stiffness and thickness of the layers, type of interface system.

Keywords: laboratory testing, numerical modelling, cement concrete slabs

1 Introduction

Reflective cracking is a major concern for engineers facing the problem of road maintenance and rehabilitation. There is evidence after several years of research and practice that overlay systems can be efficient for the prevention of reflective cracking. However it must be emphasized that there is no standard solution suited for all cases.

Reflective Cracking in Pavements. Edited by L. Francken, E. Beuving and A.A.A. Molenaar. © 1996 RILEM. Published by E & FN Spon, 2–6 Boundary Row, London, SE1 8HN. ISBN 0 419 22260 X.

The problem of reflective cracking is indeed very complex and covers a wide range of phenomena. An efficient overlay system depends on the correct choice of all its components (the interface product, the way of fixing it to the under and overlayer, the overlay), in function of the temperature and loading conditions.

This paper deals with the results of laboratory testing and numerical modelling of overlay systems on cement concrete slabs, where interface systems were used for the prevention of reflective cracking. The effect of subsequent shrinkage and expansion of cement concrete slabs as a result of temperature variations is investigated by thermal cracking tests [1, 2] performed on specimens with several types of interface systems: SAMI, non-woven and woven textiles, grids and steel reinforcement nettings. Concerning the effect of traffic it is known that, in case of cement concrete slabs, important vertical displacements can take place at the crack edges of the concrete slabs. Our evaluations in the field [3] have shown that they can give rise to cracks shortly after rehabilitation, even with the use of an interface system. In order to investigate the effectiveness of interface systems in such cases, an experimental test device has recently been developed at the Belgian Road Research Centre. In addition, 3D-computer simulations of actual road structures were carried out in order to investigate the influence of different parameters on these vertical movements: the type of base and subbase, the stiffness and thickness of the layers, the type of interface system.

2 Results of thermal cracking tests

In the thermal cracking test, schematically represented in fig. 1, specimens comprising a cracked concrete base, an interface system and a bituminous overlay are subjected to opening and closing cycles of the crack until the crack in the base becomes apparent at the surface of the overlay [1, 2].

Table 1 gives an overview of the different types of products being tested with their fixing method to the cement concrete support.

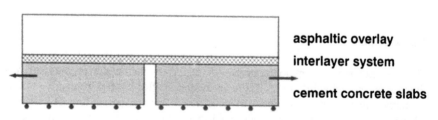

asphaltic overlay

interlayer system

cement concrete slabs

CRR-OCW 18968

Fig. 1 Schematic view of the thermal cracking test [1, 2]

Table 1: Overview of the interface systems tested in the thermal cracking test

	SAMI's	Nonwoven Textiles	Woven Textiles	Grids	Steel rein-forcement nettings	Nonwoven + grid
Base material:						
modified binder	X					
polyester		X	X .	X		X
polypropylene		X		X		
fibre glass				X		X
steel					X	
Fixing method:						
not impregnated, bonded with emulsion or binder		X	X	X		
impregnated with pure binder		X				
impregnated with modified binder		X				X
embedded in slurry					X	
nailed					X	

Fig. 2 shows the results of the thermal cracking tests performed on these types of interface systems. The tests were performed at -10 °C on samples with a 6.5 cm dense asphaltic overlay. We come to the following conclusions:

- Interface systems can be efficient to prevent reflective cracking in case of horizontal thermal movements of the underlayer. Some systems merely have an influence on the initiation phase of the crack (the crack in the overlay appears at a later time interval, see fig. 2); others lead also to changes in the propagation phase of the crack (their curves show a decreased slope).
- The results for nonwoven textiles are highly influenced by the type and quantity of the used binder, but no systematic difference in the results was found between nonwoven textiles made of polypropylene or polyester. Only full impregnation of the nonwoven textile with binder is really efficient for the prevention of cracks. This was also observed in other studies [4]. The best results are obtained if the nonwoven is fully impregnated with modified binder, allowing high deformations to take place just above the crack tip of the cement concrete base. Different curves can be obtained depending on the type and percentage of modifier. At the testing temperature of -10 °C certain modified binders can become too stiff and crack. This also holds for SAMI's.

- The results for grids depend on their type and base material. In these experiments better results are generally obtained with fibre glass than with plastic fibre grids. No systematic difference in the results could be observed between grids made of polypropylene or polyester. We also found evidence that the grid size needs to be sufficiently large to allow the stones of the overlay to penetrate easily through the meshes of the grid.
- The best results in these thermal cracking tests at -10 °C are obtained with steel reinforcement nettings ; no crack in the overlay appeared during the experiment.

- At higher testing temperatures (e.g. -5 °C) SAMI's and nonwovens impregnated with elastomeric binders often perform better than at -10 °C. Some products do not lead to cracks in the overlay, as was found for steel reinforcement nettings. We note however that the role of SAMI's and impregnated nonwovens is largely different [2]; they have a waterproofing function and mainly provide a sliding layer between the cement concrete base and the overlay, they are however not suited as reinforcement products.

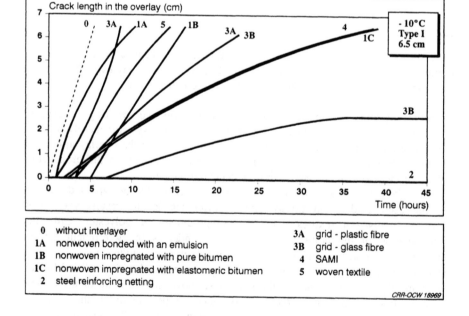

Fig. 2 Results of thermal cracking tests (-10 °C; 6.5 cm thick overlay): onset and evolution of the crack in the overlay during the experiment. For some types of interlayer systems two extreme curves are given; all tested products belonging to that category lay in between them.

3 Laboratory simulation of the efficiency of interface systems in case of severe rocking of concrete slabs

It is known from practice that rocking of concrete slabs causes extremely severe conditions for the overlay, which can lead to reflective cracking occuring very shortly after rehabilitation. In order to investigate the efficiency of different types of interface systems and to determine their limits, an experimental test device has been developed of which a schematic view is given in fig. 3. Specimens similar to these for the thermal cracking test, consisting of a cracked cement concrete base, interface system and asphaltic overlay, are placed on a flexible support on one side of the crack and on a rigid support on the other side. The specimen is then subjected to a sinusoidal compression, asymmetrically positioned with respect to the crack, causing vertical movements at the crack edges of about 1 mm if no overlay system is present. This corresponds with values measured on actual road structures consisting of cement concrete slabs without any foundation. The whole system is mounted in a thermal testing chamber, allowing to perform tests at temperatures between -20 and 35 °C. During the experiment, video recordings are taken to follow the onset and propagation of the crack in the asphaltic overlay. Measurements are performed of the vertical displacements at the crack edges in the cement concrete base and in the asphaltic overlay. The first tests are in progress now. It is the purpose to perform a systematic study of different interface systems and overlays.

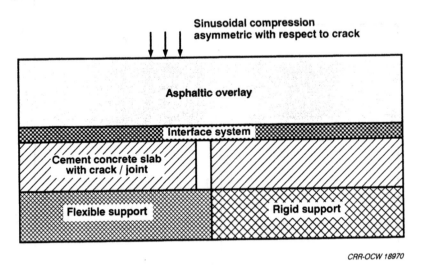

Fig. 3 Schematic view of the test device developed to study the efficiency of interface systems in case of severe rocking of concrete slabs

4 Results of numerical modelling

Three-dimensional computer simulations [5] of actual road structures were carried out in order to study the influence of different parameters on the vertical movements at the edges of cement concrete slabs: type of base and subbase, stiffness and thickness of the layers, type of interface system. The structure being modelled is represented in fig. 4 and consists of a cracked cement concrete slab with/without asphaltic overlay, with/without base and subbase, with/without interface system. The load is positioned asymmetrically with respect to the crack, in order to simulate slab rocking. The characteristics of the different layers are given in table 2. No load transfer, nor friction is present in the crack/joint.

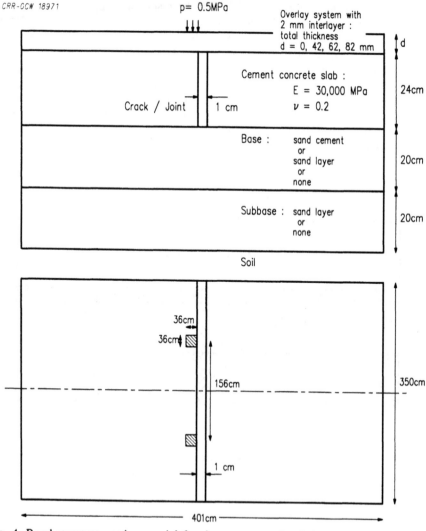

Fig. 4 Road structure used as model for the computer simulations

Table 2: Characteristics of the different layers considered in the simulations

Layer	Stiffness modulus	Poissons ratio
soil	between 5 MPa and 40 MPa	0.5
sand layer	200 MPa	0.5
sand cement layer	10000 MPa	0.3
cement concrete	30000 MPa	0.2
interface system	4 variants: no, 1000 MPa, 5000 MPa, 10000 MPa (1)	0.35
asphaltic overlay thickness: 40, 60, 80 mm	2 variants: 5000 MPa, 10000 MPa	0.35

(1) 1000 MPa : e.g. SAMI's and impregnated nonwovens at low temperatures ; 5000 MPa : e.g. some grids ; 10000 MPa : e.g. steel reinforcement nettings and some grids.

In a first step, simulations were performed on road structures consisting of cement concrete slabs without overlay system, with different types of bases and subbases. This allows us to compare the results of the calculations with measurements performed on comparable actual road structures. These were performed by the faultimeter [3] for passage of a 13 ton axle. Table 3 gives the calculated results for the slab rocking. As for the faultimeter, the slab rocking, SR, is calculated as the sum of the differences between the vertical movements at both edges of the crack/joint for passage of the 13 ton axle from one side of the crack/joint to the other side:

$$SR = \Delta u_{z,1} + \Delta u_{z,2} \qquad (1)$$

in which $\Delta u_{z,1}$ and $\Delta u_{z,2}$ are the differences in vertical displacements between both sides of the crack for the axle on one side and on the other side of the crack respectively. $\Delta u_{z,1}$ and $\Delta u_{z,2}$ are equal in the simulations, but are usually different in measurements with the faultimeter on actual road structures.

Slab rocking can also be characterized by the shear strain ε_{zx}:

$$\varepsilon_{zx} = \Delta u_z / \Delta x \qquad (2)$$

in which Δu_z is the difference between the vertical displacements at both edges of the crack/joint and Δx is the width of the crack/joint.

Table 3: Calculated slab rocking for several types of bases and subbases.

Stiffness soil	Subbase	Base	Slab rocking (mm)	ε_{zx} (x10² μstrain)
5 MPa	no	no	2.1	1050
30 MPa	no	no	0.42	206
30 MPa	no	20 cm sand layer	0.36	180
30 MPa	no	40 cm sand layer	0.12	58
30 MPa	20 cm sand layer	20 cm sand cement	0.02	10

From table 3 it is clear that the quality of the soil and the presence of a good base and subbase have a very important impact on the phenomenon of slab rocking.

This conclusion is in agreement with knowledge from measurements on site: the largest slab rocking is generally found in cases of lack of base and subbase combined with a wet season at the time of the measurements; no slab rocking is observed when base and subbase are in good state.   The calculated values are realistic and are of the same order of magnitude as those measured on site: between 0.2 and 2 mm in cases of lack of base/subbase depending on climatic conditions (temperature, humidity of the soil) and on geometric parameters (width of the cracks/joints, length of the slabs, presence of local holes under the slabs,....).

In a second step, simulations were performed on concrete slabs which are overlaid with asphalt (see fig. 4). Fig. 5 shows the slab rocking, represented by the shear strain ε_{zx}, at various depths in the asphalt overlay for different types of bases/subbases. ε_{zx} is largest in the direct vicinity of the crack/joint, but its initial value (see table 3) has decreased with a factor of about 20, as a result of overlaying. Fig. 5 demonstrates also the importance of a good base/subbase: the largest slab rocking occurs in cases of cement concrete slabs without any base/subbase.

Fig. 6 gives the shear strain ε_{zx} at various depths in the asphalt overlay for an overlay thickness of 40, 60 and 80 mm and for several interface systems: with stiffness of 1000 MPa, 5000 MPa and 10000 MPa and in case no interface is present. As can be seen, interface systems can decrease the shear strain in the vicinity of the crack tip, however the overlay thickness is more important.

Making use of the fatigue law, $\varepsilon_{zx} = K\ N^{-0.21}$, in which N represents the lifetime and K is a constant, an estimation is made in table 4 of the difference in lifetime before crack initiation between the different structures, e.g. in case of severe rocking of the slabs if no base and subbase are present. As can be seen in table 4, an increase of the overlay thickness by two cm increases the lifetime before crack initiation by a factor of 2.2; a 2 mm interface system gives an increase of a factor of about 1.8.   As can be seen, interface sytems with a small stiffness give an increase of the lifetime near the crack, but have a negative effect at distances further above the crack tip.

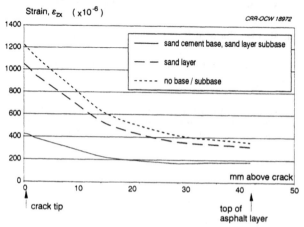

Fig. 5   Shear strain ε_{zx} (representative for the slab rocking) at different depths in the overlay, for several types of bases/subbases (stiffness soil : 30 MPa).

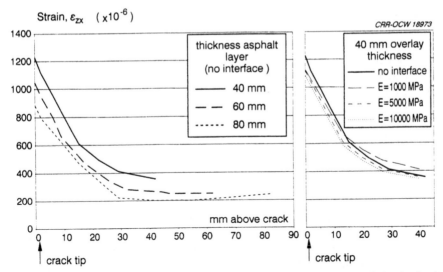

Fig. 6 Shear strain ε_{zx} (representative for the slab rocking) at several depths in the overlay, for several overlay thicknesses and interface systems. The interface systems are represented by their stiffness modulus (see table 2).

Table 4: Comparison of lifetime before crack initiation in the overlay for the road structure of fig. 4 for different overlay thicknesses and interface systems. The interface systems are represented by their stiffness modulus (see table 2).

Overlay thickness (mm)	Interface type	ε_{zx}(μstrain) in overlay, just above crack tip	lifetime compared to 40 mm overlay without interlayer system
40	no	1230	1
	E = 1000 MPa	1125	1.5 (*)
	E = 5000 MPa	1115	1.6
	E = 10000 MPa	1080	1.8
60	no	1040	2.2
	E = 1000 MPa	940	3.6 (*)
	E = 5000 MPa	950	3.4
	E = 10000 MPa	925	3.9
80	no	880	4.8
	E = 1000 MPa	770	9.3 (*)
	E = 5000 MPa	795	8.0
	E = 10000 MPa	780	8.7

(*) Further away from the crack tip, the positive effect decreases and becomes even negative at distances of more than 20 mm above the crack tip (fig. 6)

5 Conclusions

This paper deals with the results of laboratory testing and numerical modelling of overlay systems on cement concrete slabs, where interface systems were used for the prevention of reflective cracking. Thermal cracking tests were performed to study the effect of subsequent shrinkage and expansion of the slabs as a result of temperature variations. Fig. 2 gives results for several types of interface systems: SAMI, nonwoven and woven textiles, grids and steel reinforcement nettings. It can be concluded that reflection cracking in this case is highly influenced by the type of interface system and the method of bonding to the concrete base.

Concerning the effect of traffic, much effort was applied to the study of the vertical movements at the edges of cement concrete slabs (slab rocking). In order to investigate the efficiency of different types of interface systems in cases of severe slab rocking and to determine their limits, an experimental test device has been developed. In addition, 3D-computer simulations of slab rocking on road structures were carried out to study the influence of the type of base/subbase, the stiffness and thickness of the layers, the type of interface system. It can be concluded that:

- rocking of concrete slabs is largely determined by the type and state of the soil and the bearing capacity of the base/subbase. If no base/subbase are present, combined with a wet soil, severe rocking appears. This corresponds with field observations. Comparable values are found by calculations as obtained by measurements on site. In cases of severe slab rocking, it is recommended to crack and seat the slabs before overlaying.
- interface systems can decrease slab rocking, mainly close to the crack tip. Their effect is however limited compared to that of the overlay thickness.

Acknowledgments

The authors gratefully thank Mr. J.-P. Cornet and Mr. P. Vanelven (Belgian Road Research Centre) for their help. The Belgian Road Research Centre thanks the IRSIA ("Institut pour l'Encouragement de la Recherche dans l'Industrie et l'Agriculture") for their financial support in this research project.

References

1. C. Clauwaert and L. Francken, "Etude et observation de la fissuration reflective au Centre de Recherches Routières belge", Proc. 1st RILEM-Conference on Reflective Cracking in Pavements, p. 170 - 181, 1989.
2. L. Francken and A. Vanelstraete, "On the thermorheological properties of interface systems", Proceedings of the 2nd International Conference on Reflective Cracking in Pavements, pp. 206 - 219, 1993.
3. A. Vanelstraete and L. Francken, "On site behaviour of overlay systems using interface systems for the prevention of reflective cracking", Proc. of the 3rd RILEM-Conference on Reflective Cracking in Pavements, Maastricht, 1996.
4. J.-M. Rigo et al., "Laboratory testing and design method for reflective cracking interlayers", Proc. of the 1st RILEM-Conference on Reflective Cracking in Pavements, pp. 79-87, 1989.
5. SYSTUS, finite element software of Framatome SA (France).

Effect of reinforcement on crack response

A.H. DE BONDT, A. SCARPAS and M.P. STEENVOORDEN
Road and Railroad Research Laboratory, Faculty of Civil Engineering, Delft University of Technology, Netherlands

Abstract
The mechanism of asphaltic reinforcement has been causing a lot of debate for quite a number of years now. In the early stages it has been agreed that overlays in which grids, nets or wovens are applied, perform better when subjected to bending. However, the question remained as to whether a reinforcing product has a positive effect in enhancing the shear stiffness characteristics of a crack. From shear tests on plain as well as reinforced cracks, using the TU-Delft set-up, it has become clear that also in this situation the presence of reinforcement implies that the movement of the crack faces is restrained. This is possible, because cracks in asphaltic mixes exhibit a typical saw-teeth profile.
Keywords: Crack Response, Grid, Laboratory Testing, Net, Pullout Mechanism, Reinforcement, Shear, Woven.

1 Introduction

Reflective cracking is a worldwide pavement problem which has generated a lot of questions [1]. The following discontinuities in an existing pavement can be sources of this type of cracking: surface cracks, full depth cracks, construction joints and alligator cracking; the discrete sources can occur in the transverse as well as in the longitudinal direction of a road. The initiation of the previously described pavement discontinuties into the overlay, the propagation through the overlay and finally the degradation of the overlay surface can be caused by movements due to: traffic, temperature changes (between day and night and/or summer and winter) and uneven subsoil movements (not only widenings, but also effects of trees). Often a combination of these factors can be expected to occur in the field.

Reflective Cracking in Pavements. Edited by L. Francken, E. Beuving and A.A.A. Molenaar. © 1996 RILEM. Published by E & FN Spon, 2–6 Boundary Row, London, SE1 8HN. ISBN 0 419 22260 X.

If reflective cracking is to be expected several solutions are available, such as a) the application of asphaltic mixes having modified binder characteristics (e.g. polymers), b) the addition of fibres into the overlay mix, c) the construction of a so-called "stress-relieving system" (a layer of bitumen with or without a textile) which enables sliding between overlay and old surface much more easily to occur and d) last but not least the construction of reinforced overlays.

Several years ago it was impossible to assess the effectiveness of the solutions described above. To obtain a) more knowledge in the field of reflective cracking and b) to understand the mechanisms of stress-relieving and reinforcing systems, a research project at Delft University was started in 1989. The final goal of this project was to come up with a methodology and tools for the design of an asphaltic overlay against reflective cracking. In this paper a short overview is given of the experimental results of this project in the area of asphaltic reinforcement. First of all, the interaction between the reinforcement and the crack geometry is discussed.

2.   Mechanism of Asphaltic Reinforcement

Figure 1 presents a sketch of a reinforced crack which is loaded by a shear force Q as well as a normal force N; shear + tension occurs.

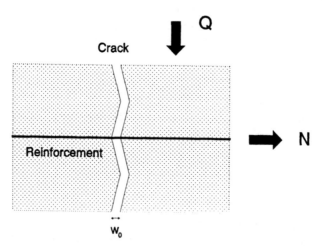

Figure 1   External Forces acting on a Reinforced Crack

From the figure it is clear that due to the presence of the normal force the rigid "blocks" at both edges of the crack move with respect to each other. If it is assumed that the reinforcement behaves as an infinite long bar, then the increase of the crack width (Δw) can be determined from the following formula [2]:

$$\Delta w = 2 \cdot \frac{N}{\sqrt{(c \cdot EA)}}$$

in which c is the equivalent shear stiffness of bond between reinforcement and surrounding material (unit: (N/mm)/mm) and EA represents the axial stiffness of the reinforcement (unit: N). The latter parameter is equal to the Young's modulus E of the reinforcing material multiplied by the sum of the cross-sectional areas of the ribs/strands A for a specific width of the reinforcing product. From the equation shown the following two extremes can be discerned:

infinite soft reinforcing system ($c \cdot EA \to 0$)   :   unlimited crack opening
infinite stiff reinforcing system ($c \cdot EA \to \infty$) :   no crack opening

It is clear that the asphaltic reinforcements (grids, nets or wovens) utilized in practice, act within the range given above. The way in which a reinforcing products resistance (= stiffness) to pullout develops in the road depends on the characteristics of the junctions between the ribs/strands of the product. In case of a grid the longitudinal (= in the direction of the applied force) and transverse ribs are connected in such a way, that the junctions have rotational stiffness; resistance to pullout is mainly generated by material located in the apertures (see figure 2). In case of a woven the strands cannot transfer forces in mutual perpendicular directions (in fact no junctions are present); resistance to pullout is generated by means of adhesion along the longitudinal strand (see figure 3). It is important to realize that this does not have to imply that wovens perform worse than grids; they operate in a different way! A net is a product which lies in between a grid and a woven with respect to the way in which force transfer to surrounding media takes place. In case of a net the ribs/strands are tightly connected in the junctions, but because these junctions have no rotational stiffness, the amount of force transfer in the transverse direction depends on the angle between the longitudinal and the transverse rib/strand [3].

In the foregoing a classification of reinforcing products has been made based on their way of transferring forces. However, this way of classifying is unfortunately not (yet) generally accepted. Especially manufacturers do not comply; they hardly give information on the properties of the connections (junctions) between the longitudinal and transverse ribs/strands of their products [3].

From figure 1 it is clear that even if the crack would initially be more or less closed ($w_0 \approx 0$), the presence of the normal force N implies that the crack will open; the crack width increases from w_0 to $w_0^{\bullet}$ (= $w_0 + \Delta w \approx \Delta w$). In case of the presence of traffic, the shear load then causes that "free" slip in the crack occurs; that is, slip without creating friction along the contact planes of the crack.

Figure 4 shows a schematic sketch of the displacements during a specific stage of the "free" slip process. The value of the "free" slip which occurs in a crack, just before the contact planes touch ($s = s_0^{\bullet}$), is equal to $w_0^{\bullet}/\tan(\alpha)$, where α represents the angle of the crack. It is clear that the possibility of "free" slip increases with decreasing α ($\to$ a less rough crack). Extensive research on aggregate interlock provided for Dutch dense asphalt concrete with an aggregate size ranging between 0 and 16 mm a value for the angle α of 15° [4].

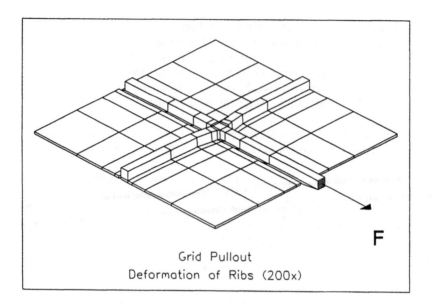

Grid Pullout
Deformation of Ribs (200x)

Figure 2   Grid Deformation in Asphaltic Overlay [2]

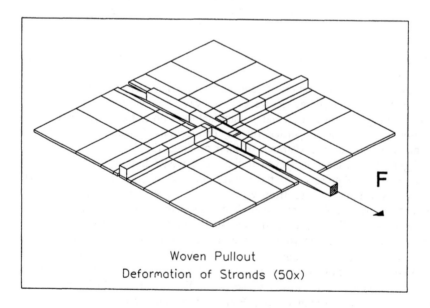

Woven Pullout
Deformation of Strands (50x)

Figure 3   Woven Deformation in Asphaltic Overlay [2]

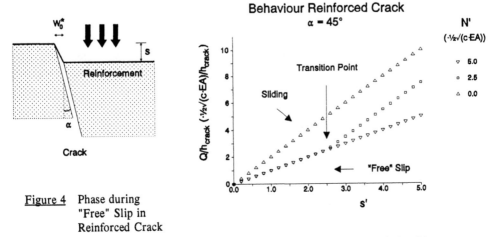

Figure 4 Phase during "Free" Slip in Reinforced Crack

Figure 5   Shear Stress - Slip Relationships

During the process of "free" slip the reinforcement is pulled out from the asphaltic mix a length equal to $\frac{1}{2}\Delta l^*$ at both edges of the crack. The moment the contact planes of the crack just touch, the extra elongation of the reinforcement, additional to the value w_0^* ($\rightarrow$ due to N!), is equal to:

$$\Delta l^* = w_0^* \left(\frac{1}{\sin(\alpha)} - 1 \right)$$

It can be seen that with decreasing angle α the reinforcement is more activated (the elongation of it increases). In the latter process an extra force in the reinforcement equal to $\frac{1}{2}\Delta l^* \sqrt{(c \cdot EA)}$ is developed; this on top of the value N which was already present. It can be derived that in the phase of "free" slip the shear stiffness of the crack has the value $\frac{1}{2}\sqrt{(c \cdot EA)}/h_{crack}$, in which h_{crack} represents the crack length.

If during the process of "free" slip no vertical equilibrium can be achieved, then the contact planes of the crack, after having touched, slide with respect to each other. From the moment that contact is being made, the externally applied forces Q and N are resisted by a) pressure Σn ($= Q\sin\alpha - N\cos\alpha$) perpendicular to the contact plane of the crack and b) the reinforcement axial force F_{rf} and the sum of the generated frictional forces Σf, which both act along this plane; the latter two have to provide a force of the magnitude $Q\cos\alpha + N\sin\alpha$. Eventually, a situation exists in which the internal forces counterbalance the externally applied shear and normal force. Figure 5 shows the normalized slip s' of a reinforced crack having an angle α, which is loaded by an increasing shear force Q, for several normalized forces N'. In this figure the contribution of the frictional component Σf has been ignored. It is clear that in case of larger tensile action the crack response is softer.

From this section it can be concluded that reinforcing systems with a high pullout stiffness (c · EA large!), not only imply a limited opening of the crack (w_0^*) due to tensile actions, but also a stiff response in case of shearing actions ($\rightarrow$ large shear stress Q/h_{crack} required to produce a specific vertical slip s).

In the next section a short overview is given of the shear tests on reinforced cracks which were carried out by Delft University of Technology in the past years. These tests represented the unique case that only a shear force Q is active (N = 0!). This means that from the previously described phases (crack opening, "free" slip and sliding along the contact planes) only the latter one took place; the contact planes of the crack immediately touch after application of the shear load.

3.   Monotonic Shear Tests Reinforced Cracks

The shear test set-up as developed and built within the framework of the Delft Research Project (see figure 6) has been used for the testing of plain as well as reinforced cracks.

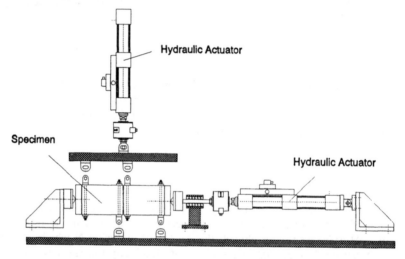

Figure 6   Shear Test Delft University of Technology [5,6]

A proper shear test for cracks (interfaces) satisfies to the following criteria [5,6]:

- provide shear without bending moment along the crack surface (interface)
- allow the application of force perpendicular to the crack surface (interface)
- enable the use of representative specimens

Within the Delft Project the set-up shown was chosen, because of the reasons mentioned above. The principle of this set-up is that by means of a careful selection of the position of the four supports, in the middle of the specimen (length 450 mm, width 110 or 250 mm and height 125 mm) a vertical plane (interface) is generated in which a high concentrated shear force is present without bending moment. This shear force is 11/15 times the force P in the vertical actuator. The set-up shows apart from edge effects, which cannot be prevented, a uniform distribution of normal and shear stresses along the crack surface (interface).

First of all, tests were performed on plain cracks which were confined by means of forces at both ends of the specimen (width 110 mm). The information which could be obtained in this way was needed for the analysis of the results of the reinforced cracks tests to be performed later. In the latter case a specimen width of 250 mm was utilized. All specimens were taken from a specially constructed road [7]. In case of the reinforced cracks no external normal force was applied (see figure 7). After an extensive procedure, using the resuls of the plain as well as the reinforced cracks, it was possible to determine the magnitudes of the contributions of the components which are active in resisting slip along the contact planes of the crack; these components are friction Σf and the reinforcement force F_{rf} (see figure 8).

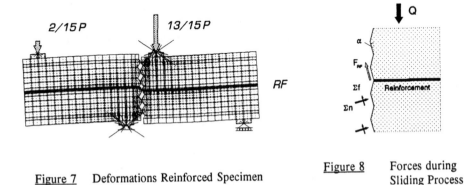

Figure 7   Deformations Reinforced Specimen

Figure 8   Forces during Sliding Process

Figure 9 presents, for a specific reinforcing system, the forces which are active along the contact planes of the crack. The curves shown are mean values from 4 individual specimens.

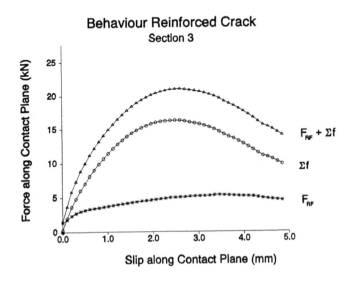

Figure 9   Test Results Specific Situation

It is interesting that the contribution of friction is even larger than the one generated by the reinforcement directly. This finding could lead to the idea that reinforcement is not very effective in this situation. However, it must be kept in mind that only by means of the presence of reinforcement, friction along the contact planes can be developed; under these conditions (no external pressure) without reinforcement no equilibrium can be achieved.

Using the monotonic displacement controlled tests performed on a total of 40 specimens (4 per situation), it was possible to determine the equivalent bond stiffnesses c from the linear phase of the derived pullout curves (see the bottom curve of figure 9 for a typical example). The parameter c describes the bond (via bearing and/or adhesion) between reinforcement and surrounding media in the generalized 2-D model [2]; its value is needed for an overlay design [8]. Figure 10 shows the values found for the tested reinforcing systems (grids, nets and wovens); according to the definition introduced in this paper only one grid was tested.

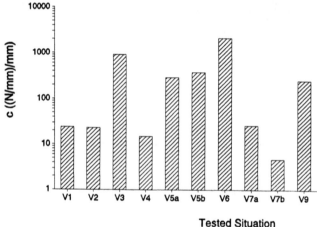

Figure 10   Equivalent Bond Stiffness Values Found for Tested Situations

It can be concluded that quite a large variation seems to occur. This is due to differences in the reinforcing products itself (surface area and roughness of the ribs/strands, junction characteristics, rib/strand spacing, axial stiffness ribs/strands) and differences in the amount and type of bitumen which was applied before or after putting the product on the road.

It is clear that different values for the equivalent bond stifness c will be found if the geometry of a reinforcing product, the amount and type of bitumen which is sprayed before or after putting it on the pavement, or even the material of which the product is composed, is changed. Fortunately, bond data for a given reinforcing product can also be obtained directly from much easier to perform pullout tests [2]; no shear tests on reinforced cracks are then necessary.

In the next section results of reversed cyclic shear tests will be discussed.

4.   Reversed Cyclic Shear Tests Reinforced Cracks

It is clear that during the passage of a crack by a wheel the shearing actions occur twice; the reinforcing product is thus twice pulled out of the surrounding media. To get insight in the cyclic behaviour of reinforced cracks, load controlled reversed cyclic tests were performed for several load/strength ratios (τ/τ_f) at a frequency of 8 Hz. Figure 11 presents typical shear stress - slip (τ-s) and dilatancy - slip (w-s) relationships recorded during the testing of a specific specimen ($\tau/\tau_f = 0.25$).

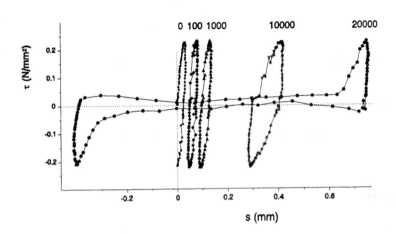

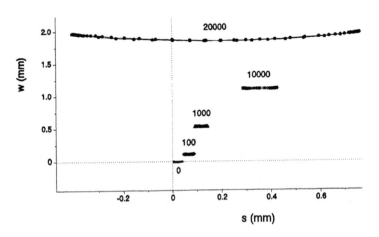

Figure 11   Typical Reversed Cyclic Shear Test Results

From the τ-s relationships it is clear that a) the shear stiffness of reinforced cracks decreases with increasing number of load repetitions (of course the degradation rate depends on the load level) and b) irrecoverable slip occurs. From the w-s relationships it can be concluded that the continuous wedging effect which takes place in the crack, implies that the width of crack increases in time. It is clear that during the process described above both degradation of the frictional response of the crack as well as of the bond capabilities along the reinforcement occurs.

5. Conclusion

It can be concluded that although asphaltic reinforcement (grids, nets or wovens) have none or hardly any dowel stiffness, a reinforced crack is still capable of transferring shear forces. This is, because adequately anchored reinforcement is capable of reducing crack opening, which implies that pressure perpendicular to and friction along the contact planes of the crack can be generated.

Acknowledgement

Both the Delft Research Project and this publication have been made possible by the financial support from the Netherlands Technology Foundation (STW).

References

1. *RILEM Conference on Reflective Cracking*, Liège, 1989 - 1993.
2. de Bondt, A.H. (1995) *Theoretical Analysis of Reinforcement Pullout*. Report 7-95-203-16, Road and Railroad Research Laboratory, Delft University of Technology, Delft, the Netherlands.
3. de Bondt, A.H. (1995) *Properties of Anti-Reflective Cracking Systems*. Report 7-95-203-22, Road and Railroad Research Laboratory, Delft University of Technology, Delft, the Netherlands.
4. de Bondt, A.H., Scarpas, A. and Steenvoorden, M.P. (1994) *Aggregate Interlock in Asphaltic Mixes (in Dutch)*. Workshop CROW, Ede, the Netherlands.
5. de Bondt, A.H. and Scarpas, A. (1993) *Shear Interface Test Set-Ups*. Report 7-93-203-12, Road and Railroad Research Laboratory, Delft University of Technology, Delft, the Netherlands.
6. de Bondt, A.H. and Scarpas, A. (1994) *Theoretical Analysis of Shear Interface Test Set-Ups*. Report 7-94-203-15, Road and Railroad Research Laboratory, Delft University of Technology, Delft, the Netherlands.
7. de Bondt, A.H. and Versluys, J.C. (1994) *Construction of Reinforced Road Sections (in Dutch)*. Asfalt - Journal of Dutch Asphaltic Contractors - Issue 2.
8. Scarpas, A., de Bondt, A.H., Molenaar, A.A.A. and Blaauwendraad, J. (1993) *CAPA: a Modern Tool for the Analysis and Design of Pavements*. RILEM Conference on Reflective Cracking, Liège.

Laboratory test on bituminous mixes reinforced by geosynthetics

G. DONDI
D.I.S.T.A.R.T. Department, University of Bologna, Italy

Abstract
The reinforcement of bituminous pavements is quite widespread especially under severe loading conditions. Designers tend to use a wide range of synthetic interlayers, not always in consideration of their technical properties. Applications vary from nonwoven low-modulus geotextiles to high strength polyester geogrids. In the paper we describe the results so far obtained in an experimental investigation carried out by our Department laboratories in co-operation with TENAX S.p.A. After a preliminary study, based upon static tests, which allowed to acknowledge the ductility increase obtained with some synthetic interlayers, a series of dynamic tests was planned. Full-scale square samples of bituminous mixes, some of which containing different geosynthetics, were dynamically loaded up to failure with the aim of improving the understanding of samples behaviour. During the tests, at intermediate stages, static tests were carried out. Experimentation is not complete yet; however here the first results regarding cracking reflection are presented. Failure cracking patterns, that are very different in the case of overlays standing upon pre-fissured bases, demonstrate the benefits of enclosing geosynthetics in pavements.
Keywords: Dynamic tests, fatigue, flexible pavements, geosynthetics, interlayers, macroreinforcement.

1 Introduction

The experimentation described in the paper takes into consideration various aspects of bituminous mixes layers with synthetic interlayers. This subject has been already

Reflective Cracking in Pavements. Edited by L. Francken, E. Beuving and A.A.A. Molenaar. © 1996 RILEM. Published by E & FN Spon, 2–6 Boundary Row, London, SE1 8HN. ISBN 0 419 22260 X.

studied in the past, by numerical methods in our Department [1] and [2]; now we are going to complete a stage of the research by a large number of experimental tests. In a first stage, for a preliminary static evaluation, we carried out some "three point bending" tests on different bituminous mixes beams: without interlayers (UR), with nonwoven geotextiles (GX) and with polyester (PET) geogrids (GG).

Then we realized a larger number of big samples with many kinds of interlayers and degrees of disturbance (Fig. 1).

Fig. 1. Preparation of full scale samples

It may be interesting to describe also the early stage of the experimentation taking into account that, due to testing characteristics, it was impossible to consider a wider selection of interlayers. In this article we will consider only macro-reinforcement; i.e. the interlayers elements are well defined in the bituminous mix. The reinforcement of asphalt, as well known, requires high temperature resistant polymers: indeed with SBS modified bitumen, during compaction, the mix reaches high temperatures (130-150

°C). For this reason it is necessary to use at least polypropylene (PP) or, better, PET geotextiles. The insertion of nonwoven geotextiles in a flexible pavement generally cause a strength decrease despite a better overall behaviour. For this reason someone prefers stiffer geogrids or composites; furthermore the latter seem to provide a better solution against reflective cracking.

2 Laboratory tests

In the first stage of the research we have verified, according to Judycki [3], that the best loading methodology was, for many reasons, the three point bending (Fig. 2): stresses and strains are more realistic. The rate of loading was approximately 50 mm/min', as suggested also by Kunst e Kirschner [4] and all the tests have been carried out using a bituminous mix, with a 5% of 80/100 bituminous binder.

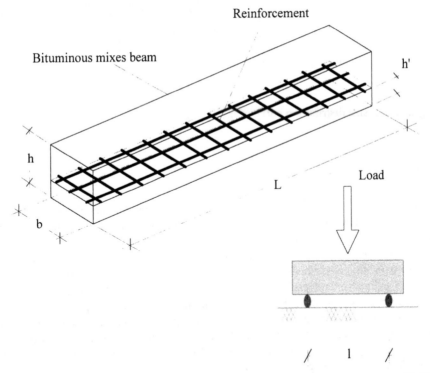

Fig. 2: Sample of bituminous mix with interlayer; b = 100 mm, h=85÷100 mm, h'=35 mm, L=600 mm, l=450 mm.

The grain size distribution curve of the aggregate is represented in Fig. 3, with the binder fuse of Italian National Roads Administration.

As interlayer we employed two geosynthetics, currently used for road pavements: a nonwoven polypropylene geotextile (Grab Test, ASTM D-4632: 450 N/25 mm, ε_f=55%); and a polyester woven geogrid (Tensile Test: 60 kN/m). The tack coat was realised with an acid emulsion containing 70% of 80/100 bitumen, modified with 5 % of SBS-R.

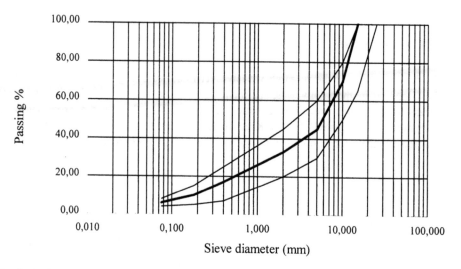

Fig. 3: Grain size distribution curve of the aggregate.

3 Results interpretation

Typical results obtained in a three point loading tests on various samples show (Fig. 4) that with geogrids there is a small increase of resistance while with nonwoven geotextiles the strength slightly decreases. But in both cases, even for large settlements, cracking never reach the upper surface of the sample. The failure of specimens with geosynthetics, at displacements 2-3 times greater than the value recorded without interlayers, is due to sliding of the geosynthetic along the interface with bituminous concrete. Furthermore the most important aspect of synthetic interlayers is the increased capability of bearing high loads also after failure, that is an higher ductility (see Fig. 4). Employing the energetic criteria to interpret the static tests we obtained [5], for the materials reinforced by geogrids, an increase ($\Delta\sigma_y$) of the equivalent ultimate strength, higher than 60%.

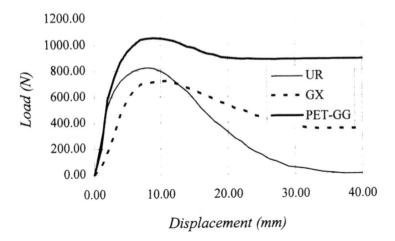

Fig. 4: Comparison between "three point loading" tests on various samples.

This last value of $\Delta\sigma_y$ appears to be more representative of the improved mechanical properties than the simple peak stress ratio.

4 Full scale load tests

After a F.E.M. study of the model (Fig. 5) we found out that the minimum dimensions of squares samples were 1.5x1.5 m.

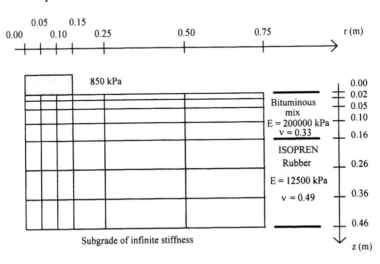

Fig. 5: F.E.M. model of the sample equivalent to the real pavement.

So we realized a strip, approximately 36 m long, in two stages: after completion of the first one we inserted various kinds of interlayers. In some cases an artificial cutting of subgrade was also realised to represent the rehabilitation of a fissured pavement. Then we built a steel box to contain the asphalt samples and the underlying layers (Fig. 6). The load was applied by mean of a circular steel plate, with a diameter of 0.3 m standing on a rubber layer with the function of minimize stress concentration related to plate stiffness.

To enhance only the behaviour of asphalt, and with the aim of reduce the uncertainties as much as possible, we decided to realise the foundation bed with a rubber characterised by a simpler constitutive law than a granular material (Fig. 5). Therefore the asphalt samples were made in situ and then cut away and disposed directly on the rubber settled down in the steel box.

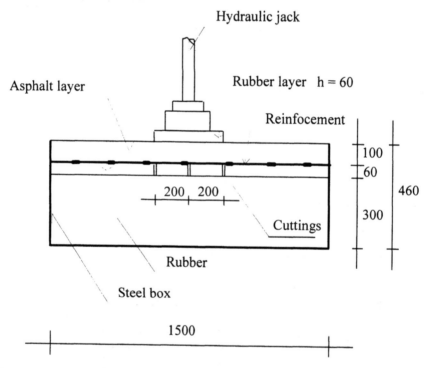

Fig. 6. Layout of a sample in the steel box.

Hence the rubber represents, as far as stiffness is concerned, both the foundation and the subgrade layers. The asphalt specimens have a total thickness of 160 mm arranged as follows (Fig. 7):

1. a first 60 mm-thick asphalt layer, in some cases fissured (indicated in the text as "LE", whereas the other non-fissured samples are referred with "NL",) by cutting it with a steel tool for a depth of 50 mm;
2. the interlayer, when present, fixed with a suitable bituminous emulsion tack coat;
3. a second 100 mm thick asphalt layer.

The load was applied by a hydraulic jack, controlled by the data acquisition system, with a frequency of 5 Hz. The shape of the loading wave is approximately sinusoidal and has an amplitude of 55 kN, in the range 5-60 kN.
Two reference grids, 100x100 mm, and 50x50 mm in the central portion, were sketched on the surface of the samples previously covered with white paint, to allow reporting failure pattern vs. n° of cycles.

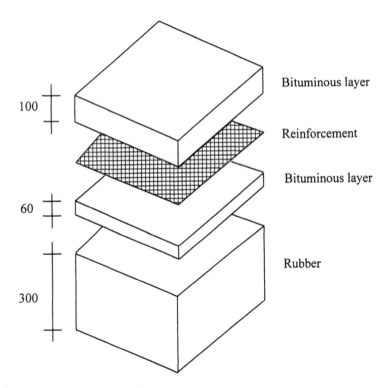

Fig. 7. Elements of the full scale sample.

Surface displacements, at different distances from the loading plate, were monitored with inductive transducers connected to an Instron data acquisition system (Fig. 8).

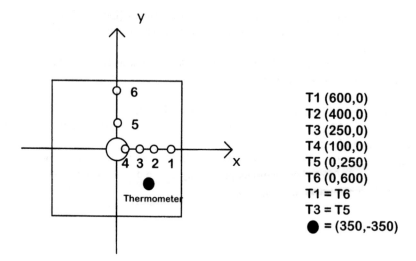

Fig. 8. Settlement transducers

5 Results interpretation

At present (but the tests are still in progress) only four samples were tested up to failure: so we can report only partial data also because the postprocessing requires a lot of time.

These samples are:

1) Undisturbed, with a nonwoven geotextile as interlayer;
2) Undisturbed, without interlayers;
3) Pre-fissured, with a nonwoven geotextile as interlayer (FIG: 9(A)).
4) Pre-fissured, without interlayers(Fig. 9(B));

Some interesting results are already available: first of all we must point out that the failure behaviour of the various types of samples is quite different.

In undisturbed samples (NL) cracking start in radial direction on the free surface: this behaviour was previously observed by other authors [6] so we could assume that this is normal for pavements whose unique factor of degradation is repetitive loading. In pre-fissured samples (LE) cracking pattern, as shown in Fig. 9 for 5×10^5 cycles, follows alignment of artificial cuttings.

Permanent deformation are surprisingly similar for samples with (GX) and without interlayers (UR) while dynamic pseudo-elastic deformations follow a different behaviour: after 4×10^5 cycles instantaneous settlements significantly increase in UR-LE sample while tend to be constant in the GX-LE specimen (see Fig. 10) Furthermore in specimens cracks appear later and their extension is much smaller than in samples without geosynthetics.

Analysing the total viscoplastic settlement of fissured samples (see Fig. 10) the two curves for UR and GX samples are so close that no differences can be pointed out.

This is consistent because we didn't expect any stiffness increase consequent to the introduction of a nonwoven geotextile. Taking into consideration the instantaneous

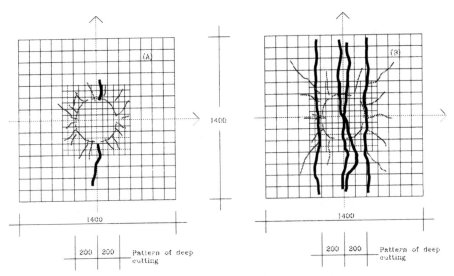

Fig. 9. Cracking pattern in GX-LE (left) and UR-LE (right) samples (5 x 10^5 cycles).

fraction of the total displacement we can see that, before a certain number of cycles (i.e. approximately 4×10^5, fig. 10) the curves are quite similar. After this limit there is a rapid increase of settlements in UR samples. But the behaviour of GX samples is better also near failure because, due to the lower degree of surface breaking, downward water seepage is prevented. This certainly plays an important role in delaying the pavement degradation.

6 Conclusions

Macroreinforcement of bituminous mixes with geosynthetics, in particular with geogrids and composites, appears to be a powerful tool first of all in critical conditions such as heavy loads and for overlays of intensely fractured pavements. It has the advantage of allowing the recycling and, as pointed out in the first part of the article, there is a significant improvement of the ductility of bituminous layers. Actually, with polyester geogrids, there is also a slightly increase in ultimate strengths, without interlayer failure.

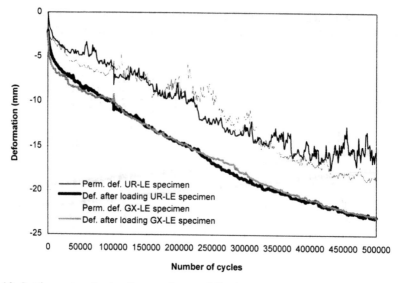

Fig. 10. Settlement under loading surface vs. N° of cycles.

Failure mechanism is also different and, in any case, the presence of a interlayer delays the cracking reflection: this was also demonstrated by the experimentation carried out on full scale samples subjected to dynamic loading in order to better represent in situ conditions. These tests, initially carried out with samples containing nonwovens, have substantially confirmed the results of static tests: the overall stiffness of the pavement is not increased but the degree of cracking is much lower. This confirms once more that nonwovens insertion in bituminous pavement rehabilitation works, with a good tack coat, is a guarantee of durability for the overlays.

Acknowledgements: we want to thank Mr. Montanelli, Mr. Rimoldi and Mr. Zinesi of TENAX S.p.A. and Mr. Donatini for his help during laboratory tests.

References

[1]   Dondi G., Righi P.V, 1990. *Rinforzo delle sovrastrutture stradali,* Quaderno AIPCR: "L' Impiego dei geotessili e prodotti affini nelle opere stradali", XXI Convegno Nazionale Stradale, 11-15 giugno, Trieste, pagg, 9-18.

[2]   Dondi G., 1994. *Three dimensional finite element analysis of a reinforced paved road,* Proceedings of the 5th International Conference on Geotextiles, Geomembranes and Related Products, Sept., Singapore.

[3]   Judycki J., 1990. *Bending test of asphaltic mixture under statical loading.* Proceedings of the fourth Intern. RILEM Symposium. October, Budapest.

[4]   Kunst P.A.J.C., Kirschner R. (1993): *Comparative laboratory investigations on polymer asphalt inlays.* Geosynthetics '93 Conference Proceedings., March, Vancouver, British Columbia.

[5]   Dondi G., 1995. *Le pavimentazioni flessibili rinforzate mediante geosintetici,* Estratto dagli Atti del Convegno Naz.:"La ricerca nel settore delle infrastrutture interportuali e aeroportuali", 28-29 giugno, Trieste.

[6]   Kief O., Livneh M., Ishai I. and Altus E., 1994 Proceedings of the 5th International Conference on Geotextiles, Geomembranes and Related Products, Sept., Singapore.

The laboratory evaluation of geogrid, APP and special steel grid for asphalt overlays on existing concrete pavements

FUJIE ZHOU and LIJUN SUN

Dept. of Road & Traffic Eng., Tongji University, Shanghai, P.R. China

Abstract

Asphalt overlay is often one of the effective measures of improving the performance of the existing concrete pavements, but a problem commonly encountered with this measure is the "reflective cracking" in the asphalt overlays. In China several measures, such as SAMI, geotextile etc., have been tried in the past, but the effect to retard reflective cracking was not significant. Therefore, it is necessary to find new measures and give them evaluation in the effectiveness to eliminate reflective cracking. Based on the environmental and traffic conditions in Shanghai, three measures(Geogrid, APP and Special Steel Grid)were selected to delay reflective cracking mainly induced by traffic loads and evaluated in laboratory at three temperatures(-3 ℃, 15 ℃, 60 ℃). The result shows that in general three measures can all effectively retard reflective cracking caused by loads; that in particular the special steel grid performed the best effects as compared to geogrid and APP in consideration of the comprehensive effectiveness at three temperatures.

Keywords: Geogrid, APP, special steel grid, reflective cracking, asphalt overlay, effectiveness.

1 Introduction

Since early 1980's, more and more concrete pavements have been constructed in the urban pavements of Shanghai. With the rapid development of the economy in Shanghai, most pavements have been carrying out the heavy loads which are much more than the design loads. As a consequence, the pavements become deteriorated seriously day and day, and more and more existing concrete pavements are facing to be rehabilitated. Compared to the asphalt pavement, it is difficult to rehabilitate the

Reflective Cracking in Pavements. Edited by L. Francken, E. Beuving and A.A.A. Molenaar. © 1996 RILEM. Published by E & FN Spon, 2–6 Boundary Row, London, SE1 8HN. ISBN 0 419 22260 X.

old concrete pavement.   Asphalt overlay is a simple and effective way of rehabilitating the deteriorated and fractured concrete pavements, providing a partial solution based mainly on economic considerations.   Nevertheless, this solution is restrained by the premature emergence of cracks on the new surface, as a result of reflective cracking from the underlying pavement.   Reflective cracking does not in itself reduce the life of the overlay, however, once reflective cracks propagate to the surface of the overlay, the pavement will become more susceptible to adverse environmental factors and traffic loads.   These factors are predominantly water infiltration and oxidation, which, combined with repeated loads, can ultimately lead to the premature failure of the pavement.

Reflective cracking in asphaltic pavements has attracted a lot of attention from researchers and road builders alike since the early 1930's[1].   Around the world effort to delay the emergence of the cracks has also been made both in the area of field trials and in the laboratory mainly by modifying of the overlay material properties, by incorporating interlayer between the old concrete pavement and the new overlay, or by treating the existing concrete pavement[2,3,4,5].   In China several measures concentrated in the use of interlayer treatments, such as SAMI and geotextile, have been tried in some small scale field trials[6,7,8].   Unfortunately the different trials have produced conflicting results on the performance of materials used to prevent reflective cracking.   These discrepancies are largely due to the lack of systematic laboratory testing, a detailed site investigation prior to the construction of corrective overlays and the different installation techniques adopted at the trials.

On the basis of the conditions of existing concrete pavements, traffic and environment in Shanghai, this paper discusses the effectiveness of Geogrid, APP and Special Steel Grid on retarding reflective cracking mainly induced by loads.

2 Measures selection

A composite pavement structures with asphalt overlays on existing concrete pavements, involve both flexible and rigid materials whose properties are very different.   Furthermore, there are many joints and cracks on the old concrete pavements, often accompanied by faults and voids under slabs, which will make the stress concentration in the asphalt overlays.   As a result, compared to asphalt overlays on the old asphalt pavements, reflective cracks in the asphalt overlays on the existing concrete pavements initiate and propagate more quickly and extensively, which affect seriously the performance of the asphalt overlays.

In China, two kinds of measures, geogrid and APP, have recently been trying in the rehabilitation of existing concrete pavements.   APP is a prefabricated asphalt sheet consisting of upper, lower layers of polymer-modified asphalt cement mixed with a filler, and an interlayer of non-woven geotextile fabric.   It has been applied in the overlay project of concrete runway in Xinmen airport.   But up to now, the quantitative and consistent performance of both measures has not been obtained.

In light of the real conditions in Shanghai, authors selected two measures: polyethylene Geogrid (GG) and APP, both of which have low moduli.   APP has good deformability and adherence to the old concrete pavement and the new overlay.   In addition, authors designed a new type of measure—Special Steel Grid(SSG): SSG1, SSG2 and SSG3. Table 1 contains a summary of the basic properties of three kinds of measures.

Table 1. Basic properties of GG, APP and SSG

Measure	Property
GG	grid size: 27mm X 27mm
	tensile strength: 7.2 KN/m
	weight: 0.66 Kg/m
APP	tensile force: 748 N/5cm
	elongation: 36%
	thickness: 4 mm
SSG1, SSG2, SSG3	height of grid: 1.5 cm
	rank of rigidity: SSG3–>SSG1>SSG2

3 Static tests

3.1 Material selection

The rehabilitation of urban road is often limited in the level, and the thickness of overlay ordinarily does not surpass 5 cm. Therefore, LH-15-I, a fine asphalt mixture was used in this test, and the gradation of aggregate is shown in Table 2. Table 3 describes the essential properties of asphalt cement utilized in this test. The optimum asphalt cement content determined by Marshall test was 5.3 percent by weight of aggregate.

Table 2. Aggregate gradation

Sieve size (mm)	15	5	2.5	1.2	0.6	0.3	0.15	0.074
Percent passing by weight	98	67	42	33	23	16.5	12.5	7

Table 3. Asphalt cement property

Penetration (1/10 mm)	Softening point ($^\circ$C)	Ductility (cm)
P(25°C, 100 g, 5 s)	T(R&B)	(15°C, 5 cm/min.)
62	48.3	120.7

3.2 Specimen fabrication

The slab specimens, instead of beam specimens, were chosen in the investigation to evaluate accurately the effects of measures on retardation of reflective cracking. The fabricated slabs have a size of 30 cm x 30 cm x 5 cm (length x width x height). Both specimens with measures and control specimens(pure Asphalt Concrete) were produced by the roller compactor according to the specifications[9]. All measures were placed at the bottom of the specimens. For each kind of measure three specimens were fabricated at a fixed temperature.

3.3 Test conditions

The variation of temperature affects not only the contraction and expansion of the old concrete slab but also the moduli of asphalt concrete and some materials used for retarding reflective cracks. As a result, based on the environmental conditions in Shanghai, three test temperatures, -3°C (representative of low temperature), 15°C (representative of normal temperature), 60°C (representative of high temperature), were considered so that the performance of three measures on delaying reflective cracking could be assessed objectively and correctly. The specimens were tested under static, four_point_bending loads, and loaded continuously. The layout of the bending device used in tests is shown in Figure 1 .

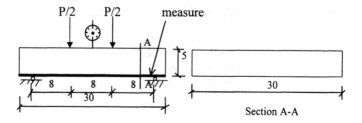

Fig. 1. Layout of the bending device, unit: cm

4 Analysis of test results

4.1 Analysis of results under low temperature(-3℃)

For all specimens figure 2 shows the relationships between the load applied and displacement at low temperature(-3℃), and Table 4 gives the quantitative information of measures about the potential effect on retarding reflective cracking. Both figure 2 and Table 4 suggest that the strengths of specimens with used measures increase very fewer than that of control specimen, that is, under low temperature the selected measures can not reinforce effectively the asphalt concrete. The reason is that the modulus of asphalt concrete is higher than or almost equal to those of the selected measures. In view of the state of failure, only the specimen with SSG1 could keep the whole, and suffer very big displacement , the others were broken into pieces. The phenomena is owing to the height of the special steel grid, which can band the asphalt concrete together. This can prevent the crack from propagating extensively.

Table 5 presents the comprehensive comparison between specimens with measures and control specimen. The results of rank to all measures are as follows:

SSG1->AC->APP->GG

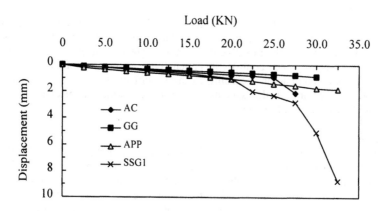

Fig. 2. Relationship between load and displacement at -3℃

Table 4. Comparison and analysis of the effectiveness for all specimens at -3 $\,^\circ\!C$

Specimen		AC	GG	APP	SSG1
Failure strength (MPa)	S: strength	9.07	9.64	10.00	9.06
	S_i/S_{AC}	1.00	1.06	1.10	1.00
	Rank	3	2	1	4
Failure deformation (mm)	D: deformation	2.13	0.87	1.83	8.76
	D_i/D_{AC}	1.00	0.41	0.86	4.12
	Rank	2	4	3	1
Deformation and	d: deformation	2.13	0.77	1.55	2.83
deformation reserve* at the	d_i/d_{AC}	1.00	0.36	0.73	1.33
failing load of AC (mm)	Rank	2	4	3	1
	D_i/d_i	1.00	1.12	1.18	3.09
	Rank	4	3	2	1
Stress and stress reserve#	σ: stress	9.07	-	-	6.66
at the failing deformation	S_i/σ_i	1.00	-	-	1.36
of AC (MPa)	Rank	2	-	-	1
Dissipated energy at the	W: energy	43.11	12.82	29.51	232.87
failure of specimen (N•m)	W_i/W_{AC}	1.00	0.30	0.68	5.40
	Rank	2	4	3	1
State of failure		pieces	pieces	pieces	whole

•deformation reserve:ratio of failure deformation to the deformation corresponded to failing load of AC
#stress reserve: ratio of failure strength to the stress corresponded to failing deformation of AC

Table 5. Comprehensive rank of the potential effectiveness for all specimens at -3 $\,^\circ\!C$

Specimen	Strength	Deformation at the failing load of AC	Deformation reserve at the failing load of AC	Failure deformation	Stress reserve at the failing deformation of AC	Dissipated energy	State of failure	Rank ♥
SSG1	4	1	1	1	1	1	whole	1
AC	3	2	4	2	2	2	pieces	2
APP	1	3	2	3	-	3	pieces	3
GG	2	4	3	4	-	4	pieces	4

♥ Deformation is more important here

4.2 Analysis of results under normal temperature(15 $\,^\circ\!C$)

For all specimens figure 3 gives the plots of load applied versus displacement at normal temperature(15 $\,^\circ\!C$). On the basis of six criteria shown above, Table 6 presents the quantitative analysis about the effect of specimen on delaying reflective cracking under normal temperature. It is found from figure 3 and Table 6 that better deformability can be obtained by means of placing APP at the bottom of the specimen. However, SSG3 is just the opposite. The specimen with SSG1 has both higher strength and better deformability, and its energy dissipated in test is 5 times as many as that of control specimen. Except the control specimen, all specimens with measures maintained the whole state at failure loads. It suggests that three measures can slow down the crack to progress.

Table 7 shows the comprehensive comparison between specimens with measures and control specimen under normal temperature. The results of rank to all measures are as follows:

SSG1->APP->SSG3->GG->SSG2->AC

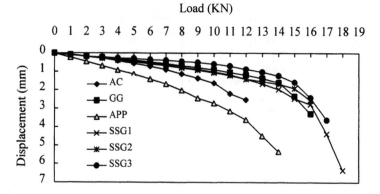

Fig. 3. Relationship between load and displacement at 15 ℃

Table 6. Comparison and analysis of the effectiveness for all specimens at 15 ℃

Specimen		AC	GG	APP	SSG1	SSG2	SSG3
Failure strength(MPa)	S: strength	3.60	4.22	3.90	5.64	4.92	5.00
	S_i/S_{AC}	1.00	1.17	1.08	1.57	1.37	1.39
	Rank	6	4	5	1	3	2
Failure deformation(mm)	D: deformation	2.52	3.28	5.34	6.35	2.56	2.61
	D_i/D_{AC}	1.00	1.30	2.12	2.52	1.01	1.04
	Rank	6	3	2	1	5	4
Deformation and	d: deformation	2.52	1.19	3.60	1.39	1.40	0.87
deformation reserve	d_i/d_{AC}	1.00	0.47	1.43	0.55	0.56	0.35
at the failing load	Rank	2	5	1	4	3	6
of AC(mm)	D_i/d_i	1.00	2.77	1.48	4.55	1.83	3.00
	Rank	6	3	5	1	4	2
Stress and stress reserve	σ: stress	3.60	4.01	2.59	4.77	4.91	4.73
at the failing deformation	S_i/σ_i	1.00	1.05	1.51	1.18	1.00	1.06
of AC (MPa)	Rank	6	4	1	2	5	3
Dissipated energy at the	W: energy	19.61	38.77	47.20	90.14	26.45	48.45
failure of specimen(N•m)	W_i/W_{AC}	1.00	1.98	2.41	5.00	1.35	2.47
	Rank	6	4	3	1	5	2
State of failure		pieces	whole	whole	whole	whole	whole

Table 7. Comprehensive rank of the potential effectiveness for all specimens at 15 ℃

Specimen	Strength	Deformation at the failing load of AC	Deformation reserve at the failing load of AC	Failure deformation	Stress reserve at the failing deformation of AC	Dissipated energy	State of failure	Rank
SSG1	1	4	1	1	2	1	whole	1
APP	5	1	5	2	1	3	whole	2
SSG3	2	6	2	4	3	2	whole	3
GG	4	5	3	3	4	4	whole	4
SSG2	3	3	4	5	5	5	whole	5
AC	6	2	6	6	6	6	pieces	6

4.3 Analysis of results under high temperature(60 ℃)

For all specimens figure 4 shows the curves of load applied versus displacement at high temperature(60 ℃). Similarly, the quantitative comparisons are shown in Table 8. It can be seen from figure 4 and Table 8 that all three measures can improve the strength and deformability of asphalt concrete. In particular, the potential effect of the Special Steel Grid is the most significant. That is owing to the fact that the asphalt concrete is temperature susceptible, and it has low modulus at high temperature. All specimens fractured completely except those with SSG1. This validates the "band" function of special steel grid again.

Table 9 describes the comprehensive comparison between specimens with measures and control specimen. The results of rank to all measures are as follows:

SSG1->APP->GG->AC

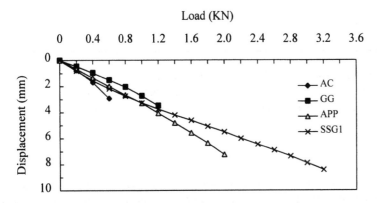

Fig. 4. Relationship between load and displacement at 60 ℃

Table 8. Comparison and analysis of the effectiveness for all specimens at 60 ℃

Specimen		AC	GG	APP	SSG1
Failure strength (MPa)	S: strength	0.19	0.51	0.96	2.59
	S_i/S_{AC}	1.0	2.6	4.9	13.4
	Rank	4	3	2	1
Failure deformation (mm)	D: deformation	2.91	3.47	7.32	8.40
	D_i/D_{AC}	1.00	1.19	2.52	2.89
	Rank	4	3	2	1
Deformation and deformation reserve at the failing load of AC (mm)	d: deformation	2.91	1.51	2.00	2.17
	d_i/d_{AC}	1.00	0.52	0.69	0.74
	Rank	1	4	3	2
	D_i/d_i	1.00	2.29	3.66	3.88
	Rank	4	3	2	1
Stress and stress reserve at the failing deformation of AC (MPa)	σ: stress	0.19	0.45	0.42	0.71
	S_i/σ_i	1.00	1.14	2.26	3.67
	Rank	4	3	2	1
Dissipated energy at the failure of specimen (N•m)	W: energy	0.97	2.28	7.60	12.41
	W_i/W_{AC}	1.00	2.35	7.84	12.79
	Rank	4	3	2	1
State of failure		pieces	pieces	pieces	whole

Table 9. Comprehensive rank of the potential effectiveness for all specimens at 60℃

Specimen	Strength	Deformation at the failing load of AC	Deformation reserve at the failing load of AC	Failure deformation	Stress reserve at the failing deformation of AC	Dissipated energy	State of failure	Rank ♠
SSG1	1	2	1	1	1	1	whole	1
APP	2	3	2	2	2	2	pieces	2
GG	3	4	3	3	3	3	pieces	3
AC	4	1	4	4	4	4	pieces	4

♠ Strength is more important here

Table 10. Comprehensive rank of the effectiveness at three temperatures

Specimen	Low temperature (-3℃)	Normal temperature (15℃)	High temperature (60℃)	Comprehensive rank
SSG1	1	1	1	1
APP	3	2	2	2
GG	4	4	3	3
AC	2	6	4	4

4.4 Comprehensive analysis under three temperatures

Table 10 presents the comprehensive comparison between specimens with measures and control specimen. The results of rank to all measures are as follows:

SSG1->APP -> GG->AC

5 Conclusions

The measures evaluated in this paper show different effects on retarding reflective cracking at different temperatures. The measures can improve the strength of asphalt concrete, increase its deformability, or both of that. In general, three measures have potential effectiveness to retard reflective cracking predominately caused by loads, and the special steel grid performed the best effects as compared to geogrid and APP in consideration of the comprehensive effectiveness under three temperatures. The results of tests show that the shape and size of special steel grid have a significant effect on the reflective cracking.

6 References

1   Von Quintus, H.L. et al.(1979) Reflection Cracking Analysis for Asphaltic Concrete Overlays, Proceedings of Association of Asphaltic Paving Technologists, Vol.48, pp.447-506.
2   Monismith, C.L. et al.(1980) Reflection Cracking:Analyses, Laboratory Studies, and Design Considerations, Proceedings of Association of Asphaltic Paving Technologists, Vol.49, pp.268-313.
3   Brown, S.F., Brunton, J.M., Hughes, D.A.B., and Brodrick, B.V.(1985) Polymer Grid Reinforcement of Asphalt, Proceeding of Association of Asphaltic Paving Technologists, Vol.54, pp.18-41.

4 Majidzadeh, K., Ilves, G.J. and Kumar, V.R. (1985) Improved Methods to Eliminate Reflection Cracking, FHWA/RD-86/075, Federal Highway administration, Washington, D.C..

5 Botton, J.W., and Lytton, R.L.(1987) Evaluation of fabric, fibers and grids in Overlays, Proceedings, sixth International Conference Structural Design of Asphalt Pavements, Vol.1, pp.925-934.

6 Song, S.G., and Zhou, Q.(1992) Study of S.B.R. Bitumen Overlay on the old Cement Concrete Pavement, east China highway, No. 4(Total No. 77), pp.55-77.

7 Wan, G.Z. and Ling, X.C. (1990) Study on Application Semi-rigid Based with Bituminous Decking, east China highway, No. 1(Total No. 62), pp.1-8.

8 Yu, B.M.(1992) Study of Reflective Cracking and Design of Asphalt Overlays on Existing Concrete Pavement, Ph.D. Dissertation, Tongji University.

9 MOC, P.R. China (1994) Specifications and test methods of bitumen and bituminous mixtures for highway engineering(JTJ 052-93), people's Communication Press.

Reflective cracking: a new direct tensile test and using a computerized programme of finite elements with a model of continuous damage

F. PEREZ-JIMENEZ and R. MIRO-RECASENS
Universidad Politécnica de Cataluña, Barcelona, Spain
C.H. FONSECA-RODRIGUEZ
Instituto Tecnológico de Monterrey, Mexico
J.M. CANCER
Courtaulds España, S.A., Barcelona, Spain

Abstract
Being aware of the development of Reflective Cracking in the diferent types of pavements, we have tried as a target in this research work to find materials or asphalt composites that, when used as wearing courses, can release this phenomenon, and whose main feature are: high toughness, ductile fracture, flexibility, high deformation up to fracture and their capacity to develop a high fracture energy.

The performance of single layers used as the upper layer of semirigid pavement built with tenacious materials, asphalt composites enhanced by adding long synthetic fibres (> 10 mm) or the use of modified binders which, with an inferior thickness, can be more effective than the conventional technique of antifissure layer plus an asphalt overlay, is compared.

Using a computerized programme of finite elements, know as OMEGA, and using a model of continuous damage in traction, an elastic nonlinear analysis is made simulating the two fracture mechanisms in the pavement's upper layer, the one produced by horizontal stress, thermal gradient; and the other due a shearing stress, traffic load.

Likewise, a laboratory experimental study is carried out for wich a new direct tensile test is proposed, BTD test (*Ensayo Barcelona de Tracción Directa*), to evaluate tenacity of different asphalt composites.
Keywords: Direct Tensile Test, Flexibility, Fracture Energy, Long Synthetics Fibres, Overlays, Reflective Cracking, Semirigid and Rigid Pavement, Toughness.

Reflective Cracking in Pavements. Edited by L. Francken, E. Beuving and A.A.A. Molenaar. © 1996 RILEM. Published by E & FN Spon, 2–6 Boundary Row, London, SE1 8HN. ISBN 0 419 22260 X.

1 Introduction

For many years, some researchers have been trying to find a solution to Reflective Cracking, one of the process of road deterioration which has excessively worried road maintenance and restoration technicians. When the wearing course of a flexible pavement is cracked, a possible restoration technique is to place a new overlay so as to recover, up to a great extent, its functional capability and thus prevent deterioration from expanding. This technique is also applicable to rigid pavements, where concrete slabs may be cracked and their joints in a damaged condition. On the other hand, in a semirigid pavement, the base course stabilised with a hydraulic conglomerate is liable to early cracking owing to thermal retraction, which may increase due to summer thermal changes and condition the wearing course to settle on a cracked layer. In the three cases, the wearing courses or overlays, as a result of the horizontal and vertical movements of joint and/or crack edges, will experiment the reflection of their patterns advancing through the upper layers until being noticed on the surface. Since the phenomenon of Reflective Cracking is known and has been identified, the aim of research has been that of finding techniques that, although they do not eliminate it completely, at least slow down this phenomenon increasing pavement durability and, in many cases, perhaps reducing maintenance and restoration costs and encouraging the task of keeping the road network of a given country in a good condition.

2 Analytical study

2.1 Application of a continuous damage model in the analytical study

The starting point of the research work presented below is an analytical study where a finite elements program was used, known as OMEGA (Organised Moduli for Engineering General Analysis) and created by the Structures and Materials Department of the Polytechnic University of Catalonia [2]. In such analytical study a continuous damage model was applied [3], with a non linear performance determined by a scalar variable known as damage or degradation variable "d". The damage variable adopts values ranging from 0 to 1, 0 being the material in a perfect condition and 1 the completely damaged material. The damage variable is used to simulate the loss of the material's secant rigidity.

The general aim of the analytical study was to study the effect of the mechanical properties of the material on the pavement performance when it is used to build a single anticracking wearing course of a semirigid pavement (100, 160 and 200 mm thick), and compare it with the performance the same pavement shows when built with a SAMI special anticracking interlayer (30 mm thick) plus a standard wearing course (150 mm thick) [1]. This led to the carrying out of an analytical study were the geometrical features of the special anticracking wearing course of the semirigid pavement and their mechanical properties were changed: modulus of elasticity, E, peak tensile strength, Ft and fracture energy by area unit, Gf. In all the analysed cases, the values corresponding to the mechanical properties and the thickness of the base course and subgrade remained unchanged. Figure 1 is a schematic representation of the studied cases, which were analysed under the conditions of horizontal displacement, dH, and vertical displacement, dV, simulating the Mode I and Mode II fracture respectively.

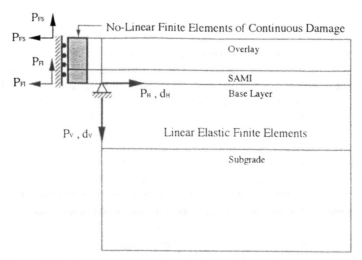

Fig. 1. Schematic Representation of the Analysed Cases.

Results as those shown in Figure 2, were obtained in the analysis of all the cases, for Mode I fracture as well as for Mode II, where it is worth noting the overlay load capacity in relation to the gradual displacement applied. This fact stresses the influence changes on the values of mechanical properties have, conditions that are not considered by the elastic and viscoelastic models.

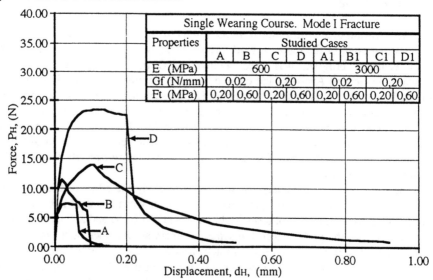

Single Wearing Course. Mode I Fracture								
Properties	Studied Cases							
	A	B	C	D	A1	B1	C1	D1
E (MPa)	600				3000			
Gf (N/mm)	0,02		0,20		0,02		0,20	
Ft (MPa)	0,20	0,60	0,20	0,60	0,20	0,60	0,20	0,60

Fig. 2. Force versus Displacement in an Analysed Case.

Horizontal displacements, dH, necessary in each case to achieve the breaking of the pavement upper fibre when it seats directly on the base course are grouped together in Figure 3. We can come to the conclusion that more tenacious materials and with a higher fracture energy need the same displacement, with an inferior thickness, than a fragile material with little fracture energy and thicker. In other words, the fracture of a pavement upper fibre is achieved at the same displacement with a high energy 100 mm thick material than with a low fracture energy 200 mm thick material. This reveals the need of searching for more flexible and tenacious materials, able to develop a high fracture energy, to use for building semirigid pavement wearing courses or for restoring rigid pavements. Such an achievement would significantly reduce the thickness of such overlays and its economic consequence would mean an important saving in the building of new pavements or in the restoration of those already in existence.

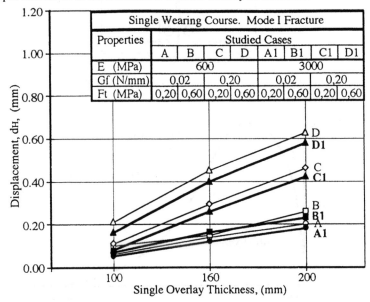

Fig. 3. Horizontal Displacements. Upper Fibre Fracture, PFS.

When comparing the stress absorbing membrane interlayer SAMI (30 mm thick) plus a standard wearing course (150 mm thick) with the single anticracking wearing course (100 mm thick), both placed on the cracked cement-bound granular layer, the special single wearing course showed a similar performance to the bilayer case. In such cases it is worth mentioning the difference in thickness, 100 mm versus 180 mm.

3 Experimental Study

3.1 Direct tensile test used in the experimental study

The searching for new materials to disperse the Reflective Cracking phenomenon and their simple and quick characterisation was carried out using a new laboratory testing method developed at the Civil Engineering Laboratory of the UPC and known as BTD® Test (*Ensayo Barcelona de Tracción Directa*) [1] [7]. By means of this test, Figure 4 , the mixture resistance to cracking propagation is determined, measuring its flexibility, deformation capacity and fracture energy by area unit as it is shown in Figure 5. The effect fibres have on the material mechanical properties is revealed when comparing the results obtained in the materials test without fibres and with acrylic fibres. When changing the mixture grading and establishing a relation between fibres length and aggregate maximum size, at affears as a key of fibre anchorage performance, the results of the test show the change in performance of the asphalt mixture. The effect of the type and content of asphalt binder can also be observed using this test.

Notched Cylindrical Specimen

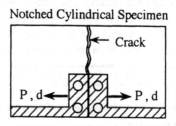

Fig. 4. Notched Cylindrical Specimen and Specimen Anchoring Device.

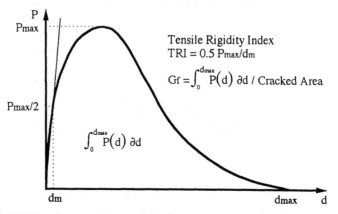

Tensile Rigidity Index
$$TRI = 0.5\, P_{max}/dm$$

$$Gf = \int_0^{dmax} P(d)\, \partial d \ / \ \text{Cracked Area}$$

$$\int_0^{dmax} P(d)\, \partial d$$

Fig. 5. BTD Test Strength-Deformation Graph.

Three grading have been used in the tests carried out: a dense mixture D8 type of the Spanish Instruction, a stone mastic asphalt with a 0/10 mm calcareous material and another stone mastic asphalt with a 0/5 mm calcareous material, hereinafter called D8, GI and MI respectively. The D8 mixture, generally used in wearing courses of semirigid pavements is quite rigid and shows a very fragile fracture, it also shows little resistance to fracture propagation through it. This mixture was considered as a reference mixture to compare mechanical properties and performance versus the other two mixtures, GI and MI. These two mixtures are richer in binder, being more flexible and able to support a greater crack opening without cracking completely. The GI mixture keeps a relation fibre length and maximum size of aggregate inferior to that of the MI mixture.

Different types of normal binders were used, with 20/30, 40/50, 60/70, 80/100, 150/200 penetrations and a modified binder with SBS, with 4.5%, 5.5% and 6.5% contents, to which 2 types of acrylic fibres with a different thickness and length (16 and 24 mm) were added. Those fibres were supplied by the company *Courtaulds España S.A.* and their mechanical properties are shown in references [5] [6]. The results of composites produced with both fibres are compared with the results obtained when using short mineral fibres, such as 3 mm long Rock Wool .

The BTD test results at 25°C are presented showing the enhancement of mechanical properties in GI and MI mixtures when acrylic fibres are used and they are compared with the properties of the dense D8 mixture. Figure 6 shows the results of a dense D8 fibre-free mixture (SF) and of a GI fibre-free mixture (SF) and with 0.4% of 900/16 mm Sekril fibres (CF), all of them produced with 5.5% of B-60/70 binder. In the case of the results of tests with fibre-free mixtures it is noticeable that the GI mixture is more flexible than the D8 mixture, which means that it can support greater displacements before cracking completely. Now, if acrylic fibres are added to this GI mixture, a slight increase of its maximum load can be observed (this phenomenon is greater as the binder hard increases), and tenacity also increases significantly adopting a higher fracture energy ranging from 75 to 100% more than the GI fibre-free mixture. With this energy increase the GI mixture with fibres supports a load close to 50% of the maximum load of the fibre-free mixture with a displacement for which the latter has zero load.

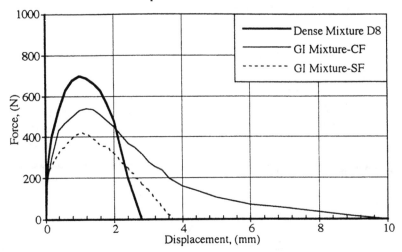

Fig. 6. Comparison of Force-Displacement Curves of D8 and GI Mixtures, B-60/70 Binder (5.5%)

The same applies to the MI mixture, Figure 7, however, when compared to the GI mixture, Figure 6, it is observed that the granulometry with less thick aggregate and more mastic asphalt creates a denser medium with a greater cohesion, which allows for a greater fibre anchorage and work, allowing it to break and avoiding the pull out. This increases the fracture energy of the material, which becomes more tenacious.

The effect of binder hard for the GI mixture is shown in Figure 8. Similar results were obtained for the MI mixture, where strong increases in fracture energy by area unit were obtained due to the anchorage provided to the fibre. In view of the results it is worth mentioning the reduction of the stiffness index when fibres are used, greater when the fibre has a good anchorage.

A good anchorage is achieved increasing the relation between fibre length, LF, and maximum aggregate size, TA, (LF/TA), Figure 9. If hard binders, with a low penetration, are used the fibre anchorage in the mastic asphalt increases, working more on the fibre, condition that appears with less hard binders at low temperatures. The cracking process appears in these extreme conditions and it is when asphalt mixtures with long synthetic fibres show a better performance.

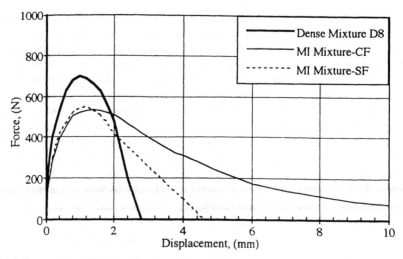

Fig. 7. Comparison of Force-Displacement Curves of D8 and MI Mixtures, B-60/70 Binder (5.5%)

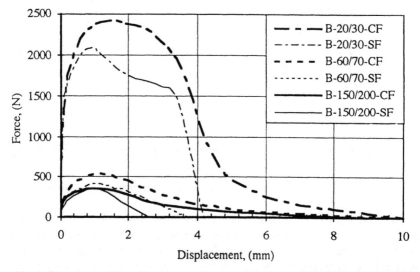

Fig. 8. Comparison of Force-Displacement Curves of GI Mixture with Different Binders (5.5%)

This is due to the fact that the fibre manages to distribute the efforts in a greater material volume working on a higher proportion in a solidary way. When work is redistributed, the efforts or loads do not concentrate on a single breaking plan but there is cracking in various plans of the cracked area. This has been observed in the type of fracture of specimens produced with MI mixture, and a strangulation of the cracked area was also observed. Consequently, the increase in the fracture energy of the MI mixture in comparison to the GI mixture can be noticed.

When using a modified binder with SBS, the first aspect to be noticed was that the modified binder gives more flexibility to the mixture, obtaining less rigid mixtures than those obtained with a standard binder. The second aspect is that the modified binder increases the tensile strength of the mixture, obtaining peak load values higher than those obtained with a standard binder with similar characteristics. Such an increase in tensile strength depends not only on the modifier used and its contents but also on the penetration of the standard base binder used.

It was also noted that deformation values for total fracture for mixtures with both binders were very similar, under 4 mm. Therefore, modified binders do not contribute to the mixture tenacity increase. Nevertheless, when adding long acrylic fibres to the mixtures with modified binder, increases in the fracture energy or tenacity of the already mentioned nature were obtained, the values for total fracture displacement being much higher, over 8 mm.

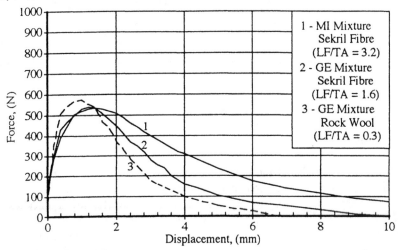

Fig. 9. Force-Displacement curves for different fibre length to maximum aggregate size (LF/TA) relations.

All the results shown up to now have been obtained using the BTD test method applying a displacement of 0.5 mm/minute. When speed was increased in the last tests, 10 mm/minute, sensitivity to fibre content raised. As expected, the composite response is more rigid, higher load, condition that improved the fibre anchorage in the mastic asphalt, producing a higher post-peak work, that is a higher fracture energy, Figure 10. The observation of these results shows that 940/24 mm Sekril fibre had a better performance.

The changes in mechanical properties of asphalt mixtures brought about by Sekril fibres are a greater tenacity and flexibility [5] [6]. They also increase the mixture cohesion, making it more resistant to impact and traffic abrasive effects, a fact that has been revealed by the UCL® Method (Universal Method for Binders Characterisation) [4] [8]. On the other hand, synthetic fibres have an effect on the asphalt mixture thermal susceptibility because they are less fragile at low temperatures and they increase the fatigue and cracking strength, properties that raise the durability of the asphalt mixture [5]. In short, acrylic fibres of the 900 and 940 Sekril type have an effect on rate of spread as another component with which to obtain a high quality asphalt mixture, giving at the same time good adherence to the binder and inalterability as far as mixing temperature and water are concerned.

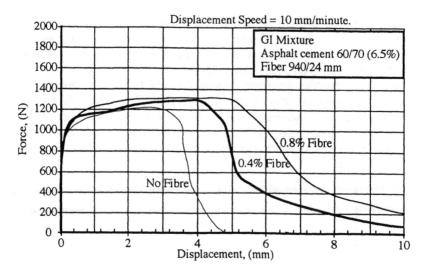

Fig. 10. Force-Displacement Curves at Different Speeds.

4 Conclusions

According to the results of the numerical analysis, trying to use tenacious materials, with a high fracture energy by area unit and a low modulus of elasticity, is important. It was proven that when using these materials, a single overlay lower thicknesses are equivalent to a membrane interlayer (SAMI) plus a standard layer as wearing course. Therefore, the tendency must be finding materials that can be used as multifunctional single wearing courses, i.e., having the functional and structural characteristics of a good wearing course and being, at the same time, resistant to reflective cracking, slowing down the appearance of cracks on the surface.

It is clear that the use of acrylic fibres 900/16 and 940/24 types helps to increase the material tensile strength, they make it more flexible, increasing its deformation capability without cracking as well as its tenacity and fracture energy by area unit. This can lead to the use of much more inferior thicknesses that those recommended in most specifications,   200 mm or over, in wearing courses placed on bases treated with cement or concrete slabs able to resist reflective cracking effectively.

It is important to bear in mind the fibre length and aggregate maximum size relation so as to obtain a better fibre anchorage and, therefore, a better mixture performance with regard to reflective cracking.

On the other hand, the use of this kind of fibres, which prevents the fracture pattern of the inferior layer from being transmitted with the same opening to the overlay, allows the latter to increase its self-restoration capacity during the hot seasons. In the track tests applied to these mixtures, when acrylic fibres are used, the performance with respect to plastic deformations proved to be highly satisfactory.

5 References

1. Fonseca-Rodríguez, C.H. (1995) *Estudio de Capas Antifisuras para Retardar el Inicio y Propagación de Grietas en Pavimentos Flexibles y Semirrígidos*, Tesis Doctoral, ETS de Ingenieros de Caminos, Canales y Puertos, Universidad Politécnica de Cataluña, Barcelona, España.
2. Fusco, A. y Cervera, M. (1991) OMEGA *User Manual*, ETS de Ingenieros de Caminos, Canales y Puertos, Universidad Politécnica de Cataluña, Barcelona, España.
3. Galindo, M. (1993) *Una Metodologia para el Análisis Numérico del Comportamiento No lineal de Presas de Hormigón bajo Cargas Estáticas y Dinámicas*, Tesis Doctoral, ETS de Ingenieros de Caminos, Canales y Puertos, Universidad Politécnica de Cataluña, Barcelona, España.
4. Miró, R. (1994) *Metodología para la Caracterización de Ligantes Asfálticos Mediante el Empleo del Ensayo Cántabro*, Tesis Doctoral, ETS de Ingenieros de Caminos, Canales y Puertos, Universidad Politécnica de Cataluña, Barcelona, España.
5. Pérez-Jiménez, F.E., Miró, R., Cancer, J. M., Alba, A. y Sainton, A. (1992) *Capas de Rodadura Bituminosas: Mejoras Obtenidas en el Comportamiento de las Mezclas Porosas Mediante la Incorporación de Fibras Acrílicas*, I Congreso Nacional de Firmes, Valladolid, España.
6. Pérez-Jiménez, F. E., Miró, R. y Cancer, J. M. (1993) *Uso de Fibras Sintéticas en la Construcción de Carreteras*, XII Congreso Mundial IRF, Tomo II, Madrid, España.
7. Pérez-Jiménez, F. E., Miró, R. y Fonseca-Rodríguez, C. F. (1995) *Aplicación de Modelos No Lineales de Daño Continuo y de un Ensayo Nuevo a Tracción Directa para el Estudio de la Propagación de Grietas en Pavimentos*, Revista RUTAS, Número 50, XX° Congreso Mundial de Carreteras, pp.13-19, Madrid, España.
8. Pérez-Jiménez, F. E., Miró, R., Cancer, J. M. et Fonseca-Rodríguez, C.F. (1996) *Enrobes Bitumineaux Tenaces, Tres Resistants au Fissurage, grace a l'Incorporation deFibres Acryliques Longues*, Eurasphalt and Eurobitume Congress, Strasbourg, France.

Geogrid reinforcement of asphalt overlays on Australian airport pavements

W.S. ALEXANDER
Geofabrics Australasia Pty Ltd, Victoria, Australia

Abstract
Full scale field trials of pavement overlays are reported for Airports at three Australian locations, Tullamarine and East Sale in Victoria and Edinburgh in South Australia.   Alternative overlay treatments are discussed, with the performance results to date, after up to four years service.   Treatments used included geogrids, differing A.C. mixes and concrete pavers. Final results are yet to be determined since any reflective cracking takes several years to show through thick A.C. layers. Pavement cores taken at Tullamarine show no crack initiation. This may be due to insulation of the concrete pavement by the A.C. overlay from temperature changes or the relative newness and flexibility of the A.C. overlay and ability to accommodate stresses from movements in the underlying concrete pavement.   The importance of proper installation, to manufacturers specification is observed.
Keywords: Airport pavements, asphalt, geogrid, geotextile, Reflective cracking.

1 Introduction

Overlaying rigid concrete pavements with asphalt can lead to the problem of reflective cracking of the overlay. In southern Australia, heavy duty airport taxiways and hardstand areas have traditionally been rigid pavements constructed from thick reinforced and unreinforced concrete. In Australia mean ambient temperatures vary in the range of 6°C to 13°C in winter and 14°C to 26° in summer, although extreme temperatures of - 2.7°C in winter and 45.6°C in summer have been recorded.

After 30 years many of these pavements are nearing the end of their service

Reflective Cracking in Pavements. Edited by L. Francken, E. Beuving and A.A.A. Molenaar. © 1996 RILEM. Published by E & FN Spon, 2–6 Boundary Row, London, SE1 8HN. ISBN 0 419 22260 X.

lives and geosynthetic treatments are being used within A.C. overlays to help rehabilitate the aged concrete pavements and resist reflective cracking.

Flexible A.C. overlays are considered more appropriate than rigid overlays since they are significantly less expensive and can be constructed in less time, resulting in less runway down-time, which is valuable.

Three projects are discussed, Melbourne Airport at Tullamarine, Victoria, East Sale RAAF Airbase in Victoria and Edinburgh RAAF Airbase in South Australia.

2 Melbourne Airport project

Monteith (Ref 1.) has documented in detail the A.C. overlay trials that were carried out at Tullamarine Airport in April 1992. An F.A.C. (Federal Airports Commission) commissioned report by Dickie predicted thermally inducted cracking of the untreated concrete surface of the order of 1.0 mm.

Overlay thickness was predominantly 90 mm A.C. and interlocking concrete block pavers were also trialled. The concrete pavers have cracked at the panel joints of the concrete pavement. They are therefore deemed to have failed and will be discussed no further.

Two types of asphalt were used in the trials:

- conventional Class 320 bitumen asphalt
- Shelphalt TRX Styrene - Butadiene - Styrene polymer modified binder asphalt

The A.C. was placed in two layers. First layer was a 30 mm thick size 10 mm intermediate course followed by a 60 mm thick, 14 mm wearing course. The dual layer overlay allowed the Hatelit and Glassgrid geogrids to be installed within the AC mat rather than at the bottom, as per the manufacturers recommendation.

2.1 Trial section treatments
These included the following:

1. Three trial areas using geogrid reinforced asphalt overlays. Products trialled included Hatelit 30/13 geogrid, Glassgrid and Tensar AR1 geogrid. See Fig. 1.
2. 80 mm thick interlocking concrete blocks, laid on a 15 mm layer of bedding sand. These cracked very soon after installation.
3. Control Section.

2.2 Preliminary surface, joint and crack treatment
As reported by Monteith, existing concrete pavement transverse and longitudinal concrete joints were ground clean, cleared of all debris, then prepared in one of three ways:

- left open
- tightly filled with ordinary hemp rope
- sealed with a foam backing verd/poured silicon sealant type joint treatment system (Dow Corning 888).

Fig. 1. Tensar AR1 fixed with Hilti nails to A.C. shoulder

2.3 Results

Objections were raised to the use of chip sealing on airport runways as required with the method of fixing Tensar AR1 geogrid. Loose stones entering aircraft engines could cause major damage. See Fig. 2.

The interlocking concrete pavers cracked soon after installation at all the panel joints. They are therefore deemed to be a failure. As of May 1996 all the trial sections, including the control section all appear in excellent condition, with no cracking apparent. The only cracking has occurred in the run-on ramp areas, where the A.C. layer is thinner, to a depth 5 - 35 mm. Fine transverse cracks have appeared above the slab joints in both end ramps, in those areas with conventional 320 A.C. but not in those areas with polymer modified A.C.

2.4 Pavement Coring

Pavement cores were taken in April 1996 in 10 locations corresponding to the position of the underlying concrete panel joints for each of the areas treated with geogrids and control sections. See Fig. 3.

Interestingly, no evidence was found of cracking throughout the cores, and no cracking had even been initiated in the A.C. overlay above the existing joints in the concrete pavement. It is not possible to differentiate the performance of the conventional Class 320 bitumen asphalt and the polymer modified binder asphalt.

In the core taken over the Tensar AR1 it was found that the Tensar delaminated from the core. This may indicate that the chip seal used over the grid to fix the Tensar to the concrete surface acts as a debonding layer, preventing adhesion of the A.C. overlay to the concrete surface.

2.5 Discussion of Results

The fact of no initiation of any cracking in the A.C. overlay may be explained by the following:

1. The 90 mm thick A.C. overlay acts to insulate the concrete pavement from temperature change and hence reduce thermal movement in the concrete pavement. This possibility could be tested by installing thermocouples in the pavement and measuring the changes in temperature.

2. After four years both the conventional and polymer - modified A.C. overlays retain sufficient flexibility to accommodate small thermally induced strains. This implies that long term monitoring, over many years, would be required to observe pavement performance as the A.C. overlay becomes more brittle and less able to accommodate stresses.

Fig. 2. Chip seal of AR1 grid

Fig. 3. Pavement core showing Tensar AR1 grid

3. East Sale Airbase, Victoria

The existing rigid concrete airport pavement had reached the end of its service life. Gaps at the concrete panel joints were up to 40 mm wide. Some two years ago the concrete pavement was overlaid with 75 mm of A.C. (See Fig. 4).

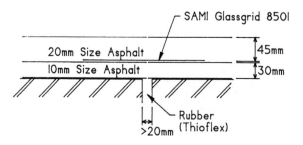

Fig. 4. Jointing detail

The overlay procedure was as follows:

1. The joints with gaps up to 20 mm were filled with Thioflex 600 (polysulphide joint sealer). On some of the wider joints, those with gaps up

to 40 mm, 75 mm wide galvanised sheet metal strips were placed as well as the Thioflex.

2. An initial asphalt layer was applied consisting of 10 mm size by 30 mm thick A.C.
3. SAMI Glassgrid 8501 was then placed on this A.C. layer. Glassgrid is a woven fibreglass, bitumen impregnated self-adhesive reinforcing fabric.
4. The second layer of A.C. was then applied, which consisted of 20 mm size by 45 mm thick A.C.

3.1 Performance
Problems encountered included the following:

1. The initial A.C. overlay cracked during constructions at the joints, prior to applying the geogrid. Some joints in the second layer of A.C. cracked during construction. This occurred at those joints 20 mm wide and greater. The joints 40 mm wide cracked severely.
2. It was noticed that the Thioflex, being like rubber compound, caused the A.C. to rebound and crack on the surface.
3. In those areas where the galvanised steel strips were placed, the Thioflex underneath was proud of the concrete and in some places caused the strips to buckle.
4. After two years reflective cracking has occurred at the pavement surface at the pavement surface at virtually all the joints.

4. Edinburgh RAAF Airbase, South Australia

In March 1994 A.C. overlay work was carried out at Edinburgh Airbase. Within the overall existing A.C. surface there were a number of thick concrete pads, some 60 m x 60 m which were used as hardstand areas. These pads were some 30 years old and there was a concern that any A.C. overlay would be subject to reflective cracking.

4.1 Geogrid treatment
Tensar AR1 geogrid was applied to the concrete surface according to the following specification:

"The Contractor shall treat the entire area of concrete pavement to be covered by bituminous concrete overlay, as follows:

1. Lay "Tensar AR1" grid directly on existing concrete pavement such that there is a minimum 100 mm overlap at edges of rolls and that any longitudinal joint is a minimum of 500 mm from any wheel track, or existing construction joint. Where cutting of the grid is required any part filaments shall be removed.
2. Fix leading edge of grid to existing pavement using proprietary fasteners. Average spacing of fasteners; 15 per 3.8 m roll width for "Hilti Hammer screws" (drill) and 35 per 3.8 m roll width for nails (gun type using "Hilti

DX450").

3. Roll out length. Connect tensioning bar (provided by supplier) to free end of full grid length and tension to 0.5% strain (i.e. 500 mm per 100 m length at 20°C).

4. Fix tail end to existing pavement as per (2) above. Overlap adjacent rolls by 2 grid spacings. Fasten grid at any localised points as required to remove ripples or seat grid on pavement surface.

5. Apply tack coat at 0.35l/m², and protective seal of bituminous concrete mix, of 10 mm nominal size, to a compacted thickness of 25 mm. Bituminous concrete temperature shall not exceed 145°C.

6. Construct bituminous concrete overlay using mix of 20 mm nominal size, as specified, to the grades and levels shown in the drawings."

Note: Drawings not shown in this paper.

4.2 Problems encountered

Difficulty was first encountered fixing the geogrid to the hard brittle concrete surface with "Hilti" nails. Once the grid was fixed with the Hilti nails it was expected to adhere the geogrid to the concrete surface with a light cold emulsion spray only. However when the grid was trafficked by the mechanical paving machine the emulsion caused the grid to adhere to the wheels of the paver and floated into the A.C. overlay.

High ambient temperature of 24°C together with the high pressure from the paver caused the emulsion to enliven and adhere the geogrid to the pavers wheels.

These initial stages had to be milled off and relaid using a chip seal to adhere the grid to the surface.

5 Conclusion

1. Use of geosynthetic treatment should not be considered a cure-all for all pavement problems. Advice should be sought from the manufacturer as to the suitability of proposed products.

2. On going monitoring of controlled trials is necessary, often for many years after the trial is initially installed. This is often difficult to achieve due to a change of personnel, changes of budget priorities and loss of interest.

3. Advice should be sought from the manufacturers of the geogrid products as to how these products should best be installed. Failure to follow these recommendations can lead to installation problems and poor longterm performance.

4. It is important that any geosynthetic treatment be installed with great care. It is recommended that specialist contractors install these products; i.e. people trained and approved by the manufacturer. A general civil contractor does not have the skilled personnel who understand or can carry our the tasks required.

6 References

1. Monteith, R. & Bendon, P (1994) Paving the way rigid pavement overlay trials. Airport Technology International 1994/95.

7 Acknowledgments

The author would like to gratefully acknowledge the assistance of the following:

1. Mr. Ross Monteith, Senior Pavements Engineer, AirPlan Pty Ltd
2. Mr. Richard Knott, Pavements Engineer, Works Australia

Asphalt reinforcing using glass fibre grid "Glasphalt"

F.P. JAECKLIN
Geotechnical Engineers, Ennetbaden, Switzerland
J. SCHERER
Scherer & Partner Bausysteme, Brunnen, Switzerland

Abstract
A new type of asphalt reinforcing has been developed using a strong glass fibre geogrid mounted on Polypropylene nonwoven geofabric. This 'Glasphalt' combines the features of veritable reinforcing and of water proofing of base layers beneath and easy installation with the nonwoven geofabric attached to. This paper describes the results of test performed at three leading European laboratories for proof of performance effects under severe and controlled conditions.

Figure 1 : Glass grid and overlay being placed

The glass fibre geo-grid is directly placed and nailed onto the asphalt sprayed with bitumen binder. Then asphalt is spread and compacted. In case of geogrid with nonwoven geofabric, heavy spraying provides water proofing and eliminates nailing.

1   Introduction

Past experience with Polymer geogrids were positive for increased longevity under adverse conditions even though the module of polymers do not permit true reinforcing like steel in concrete, except for severe cracking from deformations. For this reason the

Reflective Cracking in Pavements. Edited by L. Francken, E. Beuving and A.A.A. Molenaar. © 1996 RILEM. Published by E & FN Spon, 2–6 Boundary Row, London, SE1 8HN. ISBN 0 419 22260 X.

high module of glass fibre geogrids offer an immensely increased potential for substantive reinforcing asphalt overlays: The glass absorbs forces much earlier than any Polymer can do. This study concentrates on the reinforcing effect in comparison with previously available data using Polyester grids. Three series of testing were performed: one for dynamic performance, one for pullout, and one for reinforcing performance under cyclic temperature deformations.

a) Dynamic Testing at the Netherlands Pavement Consultants Laboratory. They used the Dutch testing procedure previously known for dynamic testing [1]. The comparison of glass fibre reinforcing compares very favorably with other samples using Polyester grids of the same tear strength or non woven geofabrics. Additional testings determine dynamic performance under varying temperature ranges. This is interesting, because fissuring may occur under cold winter conditions, when the asphalt is hard and the elastic module is high.

b) Pullout Testing at Scherer & Partner Laboratory. Specimens were prepared on an actual asphalting site, with the glass fibre geogrid [2]. The actual tests were performed similar to the testing procedures at Delft University, measuring pullout forces for various glass fibre grids with and without nonwoven geofabric, using meshes 30/30 mm and 10/10 mm.

c) Cyclic Temperature Loading Tests at The Belgium Road Research Centre in Brussels. The glass fibre geogrid was tested simulating slow cyclic temperature loading and a static load. This previously known testing procedure investigates the glass fibre reinforcing effect in comparison with the other materials such as Polyester geogrids and Polypropylene nonwoven fabrics [3].

2   Limits of State of the Art Reflective Cracking Prevention

Current state of the art for effective reduction of reflective cracking uses geotextiles and geogrids. The procedures have some limits however: Geotextiles (nonwoven geofabrics) are embedded in a thick layer of bitumen sprayed onto the existing asphalt and act as an impervious layer that allows for SAMI, (stresses absorbing membrane interlayer). SAMI performs fine generally, except for very limited deformation of non modified binders during very cold winter nights, because the bitumen binder might be 'frozen' when the maximum temperature shrinking occurs and thus the maximum of stress absorbing effect is not available. Contrasting geogrids reduce reflective cracking by absorbing some of the stress that would otherwise affect the asphalt. This reduces stress peaks and asphalt deformations which in turn reduces crack potential. The low module of elasticity limits the effect of polymer geogrids however requires substantial deformation and sometimes first fissures to pick up such deformation stresses. This limits the effect on longevity substantially.

The new generation of geogrids use coarse mesh glass fibre grids with a deformation module about ten times higher. The physical potential of high module grid material affects longevity and performance. This had to be verified quantitatively by systematic testing which is presented here. This new generation of glass fibre geogrids is manufactured with a nonwoven Polypropylene geotextile for ease of installation, for waterproofing, and for the SAMI effect.

3   Dynamic Testing

3.1   Testing Concepts - Netherlands Pavement Consultants performed special testing using the dynamic testing machine with asphalt specimens in which the Glasphalt glassfibre geogrid was embedded and tested after a 4-week curing period. The results are described in the NPC report 95'155 of September 1995 by T. Coppens W. v. Brand and A. Versluis [1].

Figure 2: Asphalt specimen and configuration

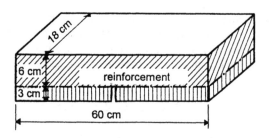

All specimens are 600 x 180 x 90 mm. The glass fibre grids are embedded at 30 mm and specimens were 'precracked' by sawing a 25mm indent at the center.

All tests have been carried out in a four point loading test on specimens 600 x 180 x 90mm. The force controlled mode uses a maximum of 4.5 kN applied both ways at a frequency of 29.3 Hz at a temperature of 5°C. The grids are embedded at 30 mm and specimens were 'precracked' by sawing a 25 mm indent at the center. The following Geogrids and geotextiles were studied:

a)	**Glass fibre geogrid with Polyprop. nonwoven**	NPC code Di 2318
b)	**Polyester geogrid**	NPC code Di 2319
c)	**Polypropylene nonwoven geotextile**	NPC code Di 2320

Figure 3 :   Scheme of Test and Dynamic Force Application

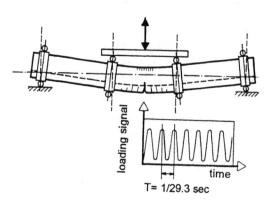

All tests use four point loading. The force is controlled for a maximum of 4.5 kN applied both ways at 29.3 Hz frequency and at 5°C Temperature.

3.2   Test Method - Figure 3 shows the mechanical scheme of the test procedures. Each test specimen is secured in four frames: Two end frames sitting on the base, the cyclic loading force acting on the two inner frames. For each reinforcement sample two specimen beams were tested counting the loading cycles until failure, measuring the force, the deformation at the beam center, and the crack propagation if possible. These readings allow for Fourier transformation of displacement amplitude and average also phase angle between force and displacement. Then dissipated energy is calculated from force, displacement and phase angle (W diss), finally add cumulated energy dissipation.

The force controlled determination of failure reflects the crack growth within the beam. Failure of the geogrid reinforced beam and crack growth through the beam may not coincide. Yet when the crack has grown completely through the beam, the force drops down and the phase angle is no longer measurable. So the number of load repetitions until failure (Nf) is clearly determined in each test. All these specimens were tested at 5°C temperature with loading cycles reaching 4.5 kN at a frequency of 29.3 Hz.

3.3   Calculation Method - The artificially induced force is kept constant throughout each test as is the constant temperature level. The dimensions of each specimen may slightly vary for which compensating structural calculations are applied considering the normal bending stress sigma max.

3.4   Testing Procedures - The testing procedures consisted of 9 tests with each type of geotextile or geogrid reinforcing. The specimen were cured for 4 weeks, then cut from the asphalt slab and tested in the dynamic climate chamber.

3.5   Test Results of Non Reinforced Specimens - For the reference beam tests were made with no reinforcing on various force levels to find the fatigue line for this material and for this test pattern. Figure 4.1 shows the test results for beams tested at 5°C temperature and at 4.5 kN force level with a load frequency of 29.3 Hz. The tests at other temperatures such as -10° and - 20° C are less reliable, because one specimen was tested only.

Table 1 : Number of Load Cycles until Failure

Tests performed at NPC 95-155		Reference no reinforcing	Polypropylen nonwoven	Polyester grid 60 kN/m	Glasphalt 60 kN/m
average displacement	μm		24'254	15'992	26'900
at rupture (last value)	μm	21'400	21'444	15'380	25'913
Load Cycles until rupture at 5°C		13'700 28'000 29'900	71'243 74214	84'757 95'123	181'128 187'163
Average		23'870	72'728	89'940	184'151
Average in %		100%	304%	370%	772%
Life Expectancy under dynamic Loading at 5°C		1-time	3-times	3-4-times	7-8-times

Figure 8 : Load Cycles until Rupture for Various Reinforcements - Glass fibre
geogrids substantially increase longevity by true reinforcing effect.

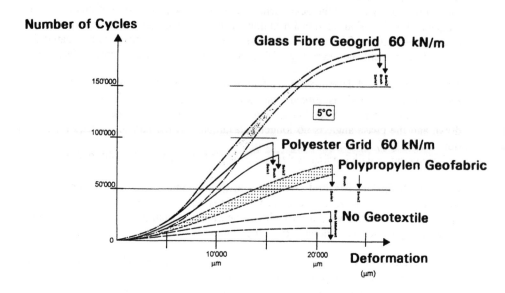

Number of Cycles

Figure 4 : Life Expectancy under Dynamic Loading at 5°C

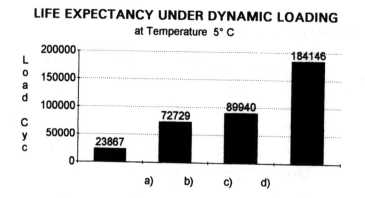

The table and the
graphics demon-
strate Glashalt rein-
forcing (glass fibre
geogrid with non-
woven Polypropy-
lene geotextile),
performing for a
life expectancy
about 7 to 8 times
greater than a non-
reinforced asphalt
slab.

Essentially the specimens with no reinforcing a) took 24'000 load cycles, the nonwoven
Polypropylene geofabric b) took 73'000 and the Polyester grid d) approximately 90'000
load cycles, whereas the glass fibre grid reinforced specimens d) took about 200'000
cycles.

5   Pullout Tests

Figure 5 : Pull out Test Configuration

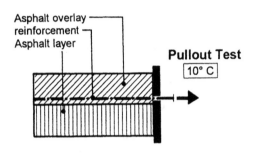

The actual tests were performed similarly to the testing procedures at Delft University, measuring pullout forces for various glass fibre grids with and without nonwoven geofabric, using meshes 30/30 mm and 10/10 mm.

5.1 Scope of Testing - Testing was performed at the Scherer & Partner Laboratory to check pullout and adhesion between the two asphalt layers containing the reinforcement [2]. The following reinforcements were tested:

a) **Glass fibre geogrid** 30 x 30mm, without nonwoven
b) **Glass fibre geogrid** 30 x 30mm, **with** nonwoven Polypropylene
c) **Glass fibre geogrid** 10 x 10mm **(fine mesh)**
d) **Glass fibre geogrid** 30 x 30mm and **crushed rock aggregate**

Table 2 : Results of Pullout Tests

Sample	Pullout Force kN			Average Force		
	1.	2.	3.	kN	%	%
a)	2960	2780	2810	2850	102%	+2%
b)	2730	2650	3010	2797	100%	0
c)	1870	1950	2015	1945	70%	- 30 %
d)	3210	3530	3400	3380	121%	+ 21 %

Figure 6 : Results of Pullout Tests

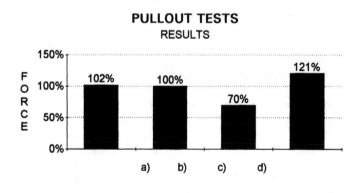

PULLOUT TESTS
RESULTS

1. identical pullout forces are found with or without nonwoven geotextile
2. the fine mesh results in 30% less pullout force
3. crushed rock aggregate placed above the geogrid increases pullout forces about 20%.

5.2   Interpretation - Adhesion and pullout force depend largely on the grain size and material type directly next to the geogrid. The asphalt roller compaction presses the grains into the geogrid meshes and result in an additional bonding and interlocking effect. A nonwoven geotextile does not affect this bonding and interlocking, yet it guarantees for an even thickness of the bonding layer. Fine geogrid meshes reduce interlocking and adhesion and are therefore problematic. - The S&P Glasphalt geogrid with 30x30 mm meshes allows for maximum bonding and interlocking.

6.   Cyclic Temperature Loading Simulation Tests

6.1 Scope of testing - The Belgium Road Research Centre (BRRC) in Bruxelles developed procedures for testing the adverse effect of low temperatures [3]. The temperature or road salt induced asphalt deformations on the side of a crack are simulated in a climate chamber.

6.2   Sample Preparation - The samples with dimensions 600 x 150 x 150 mm were prepared at the BRRC, composed as shown in figure 7.

Figure   7 :   Test Specimen for Temperature Testing

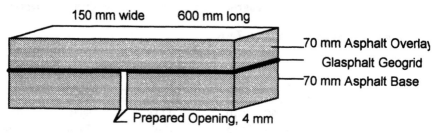

- a base layer of two cement blocks separated by a 4 mm indent wide as a discontinuity to simulate a crack beneath the overlay.
- an interface consisting of glass fibre geogrid combined with the nonwoven geotextile
  - impregnation with modified binder, 80/100 penetration, SBS, sprayed 0.8 kg/m² before placement of geotextile/geogrid and 1 kg/m² after placement of the interface product for good adherence between lower layer, geotextile / geogrid, and asphaltic overlay; and binder quantity as set by US task force 25.
- the asphalt overlay was 70 mm Belgian type dense wearing course.

Figure 8 : Test Conditions Simulating Temperature Loads

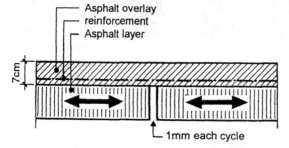

The static load is varied slowly producing load cycles of 3 hours and crack opening of 1 mm. Load cycles are counted, measured until failure.

6.3 Test Conditions - Two test specimens are compared, one without geogrid interface for reference and one with Glasphalt. Each specimen was tested at temperature of -10°C until failure. The temperature load on the asphalt specimen was simulated by applying cyclic variations of crack opening by 1 mm. The cycle length was selected at 3 hours to simulate slow changing temperature effects.

6.4 Test results - The test performance is shown in figure 2 : The specimen without interface survived only for 2 cycles, developing a crack immediately after the first loading, developing to 65 mm length during the second cycle displacement of 0.46mm was measured in the asphalt layer close to the crack tip during the first cycle, that grew to 1.0 mm in the second and last cycle before failure.

It took 5 h 20 minutes or about two cycles to break the reference sample with no reinforcing. Contrasting the glass fibre reinforced sample took 56 cycles or 168 hours. No complete failure was observed. A 30 to 40 mm long crack was formed during the first 20 cycles of the test. This crack did not change during the rest of the test, neither in length nor in width.

The specimen with the glass fibre grid and geotextile developed 0.21 mm displacement in the asphalt layer closed to the crack tip after the first cycle or less than 50% of what is found without interface and 0.37 mm with the last cycle (56th) or 37% only.

6.5 Conclusions - The glass fibre grid reinforcing consists of a glass fibre geogrid and a nonwoven geotextile. It performed well in thermal cracking test, even at the low temperature of -10°C.
1. The test lasted for 56 cycles during 168 hours without failure observed, whereas the non-reinforced reference specimen failed after two cycles and only about five hours.
2. Thus the reinforced specimen accepts 56/2 = about 28 times more cycles and deforms sensibly less than the non reinforced specimen.
3. The tests further suggest the strong bond obtained between the concrete base the Glasphalt-grid interface and the overlay.
This really means that the reinforcing effect is substantial and even dramatic.

Figure 9 : Longevity of Nonwoven Geofabric in Overlay [4]

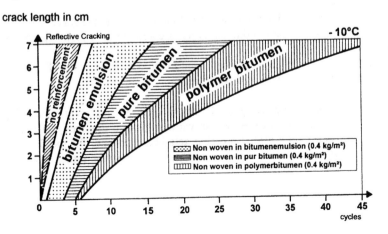

Figure 9, was compiled from other test data of BRRC on textiles [4]. It indicates that longevity largely depends on the type of tack coat selected :

1. Non-modified bitumen emulsion without real impregnation of the textile results in a small improvement compared to the case with no reinforcement or no geofabric at all.
2. Impregnation of the textile with pure bitumen results in much better performance and distinctly capable many more load cycles than emulsion.
3. Full impregnation with polymer modified bitumen results in significantly better performance.

In conclusion only polymer modified bitumen should be used in conjunction with nonwoven geofabric overlays, because the modified version only allows for the crucial stress relieving effect at very low temperatures.

Figure 10 : Longevity of Various Types of Grid Reinforcing

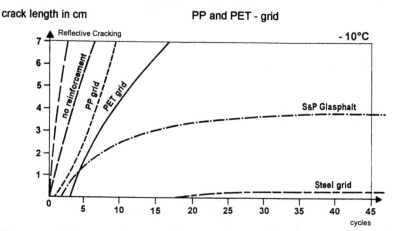

The tests performed at BRRC finally extended the existing data available on asphalt overlay reinforcement. Figure 10 demonstrates :

1. Poor performance of non reinforced asphalt under the described temperature test, capable of taking a few load cycles only.
2. Polypropylen PP grid and even more so Polyester PET grids significantly increase the load cycle capacity.
3. The glass fibre grid may start deforming similarly to polymers, yet the curve then turns to horizontal and many additional cycles are acceptable without significant additional cracking.
4. The steel grid reinforcing acts as a significant reinforcing with none or very little deformation - similar to reinforced concrete.

In conclusion the Glasphalt polymer geogrid demonstrates an important reinforcing effect with much higher load cycle capacity than without geogrid interface, yet tied to notable deformations. Glass fibre grid reinforcing acts like a polymer for the first few load cycles, then however turns around and changes to a behavior rather like steel reinforcing.

7.   Final Conclusions

This paper summarizes the performance of reinforcing and their proportional effect. Bitumen impregnated nonwoven geotextiles proofed to be successful in reducing reflective cracking. Obviously such stress relieving effect depends on the type of bitumen sprayed: use polymer modified quality only.

The glass fibre grid uses such a geotextile and an additional large mesh glass fibre geogrid of 60 kN/m breaking stress at about 3% elongation only.

The glass fibre geogrid absorbs substantial forces because of its very high stiffness. Even daily temperature changes are clearly sensed and thus protect the area above an existing crack from repeated deformations and fatigue cracking. This reduces the crack propagation and thus increases life expectancy of the asphalt overlay. This effect is known from use of Polypropylene or Polyester geogrids. However the effect is in multiple proportion because of this capability of glass fibre grids to absorb stresses at much lower deformation levels.

The tests performed have shown that glass fibre geogrids improve the longevity of asphalt overlays by immense factors : about twice as long life expectancy compared with Polyester grids or nonwoven geotextiles and about 7 to 8 times longer life expectancy than no reinforcing. This factor greatly offsets the extra cost for glass grids and installation by much better performance. Parallel pilot projects and field applications are under way for compiling data and experiences. -   16.05.96 MAAST4.WPS

*   *

*

The authors wish to thank Mr. T. Coppens, W. v. d. Brand, and A. Versluis at the NPC and Dr. A. Vanelstraete at the BRRC for the excellent cooperation, the professional testing and reporting finding new avenues to explore.

10.   Literature

[1]   T. Coppens, W. v. d. Brand, and A. Versluis, *' Reinforcement Research Test Report'*, Netherlands Pavement Consultants Laboratory, NPC report 95'155, NL-3871 KM Hoevlaken, September 1995.

[2]   Joseph Scherer at Scherer & Partner Bausysteme and Laboratory, Brunnen, Switzerland, *'Ausreissversuche'*, Brunnen 26.9.1995

[3]   Dr. A. Vanelstraete : *'Thermal Cracking Tests on S&P Glasphalt'*, Centre de Recherches Routiers, Boulevard de la Wolvuwe 42, 1200 Bruxelles, Belgique, Department Research and Development, BRRC Report Number EP 3765/3847, 5. September 1995.

[4]   A. Vanelstraete and L. Francken, 'Anticracking Underlayers', BRRC - Bulletin, Nr. 3, (1995).

Addresses :
Dr. Felix P. Jaecklin, Geotechnical Engineers, Geissbergstrasse 46,
    CH-5408 Ennetbaden, Switzerland, Tel. 0041 56 222 0722, Fax 0041 56 221 1344
Joseph Scherer, Scherer & Partner Bausysteme, alte Kantonsstrasse 16,
    CH-6440 Brunnen, Switzerland, Tel. 0041 41 820 5633, Fax 0041 41 820 4525

Design and performance of overlay combined with SAMI for concrete pavement

NAGATO ABE and YOSHIO SAIKA
Toa Doro Kogyo Co., Ltd., Research Laboratory, Yokohama, Japan
MASAKI KAMIURA
Japan Freight Railway Co., Ltd., Tokyo, Japan
TERUHIKO MARUYAMA
Nagaoka University of Technology, Nagaoka, Japan

Abstract

The overlay method has frequently been adopted as a method for repairing concrete pavement, but there are some problems about this method. One of the problems is that the concrete slab causes reflective cracking to the asphalt overlay. We have prepared a test sample of an overlay with a stress absorbing membrane interlayer (SAMI) which utilized bituminous sheets or highly elastic asphalt to reduce reflective cracking, and carried out a repeated loading test in the laboratory. As a result of this test, it was found that SAMI with utilized highly elastic asphalt is highly effective for the reflective cracking.

In May, 1989, we applied the overlay method, utilizing this material, to the concrete slabs in container platform of which cracking ratio was 75%, and have made an investigation of the road surface condition and its secular change by means of FWD. No reflection crack has been caused so far. Obviously, SAMI has prevented the transmission of the movement of the damaged concrete to the overlay. It is already more than five years since the construction of the overlay with SAMI in this site; however, no crack has been found on the surface of the pavement and its high performance has been maintained. This proves that this method is effective for repairing the pavement having joints or reflective cracking on its surface.

Keywords: SAMI, concrete slab, reflective cracking, RC sheets, FWD.

1. Introduction

In recent years, there has been serious damage to pavement as traffic volume grows and vehicles become increasingly larger. Breaks in corner areas, as well

Reflective Cracking in Pavements. Edited by L. Francken, E. Beuving and A.A.A. Molenaar. © 1996 RILEM. Published by E & FN Spon, 2–6 Boundary Row, London, SE1 8HN. ISBN 0 419 22260 X.

as cracks due to bending force, are being observed both in asphalt and concrete pavement.   Reflective cracking is also taking place in the asphalt mixture overlay.   Replacement and overlay are known as common methods to be employed for repairing concrete pavement.   The former, however, entails the problem of waste disposal because concrete slabs must be removed, while the method of overlaying pavement with asphalt mixture causes reflective cracks in concrete slabs to extend to the layer of the asphalt mixture.

In search of a way to reduce reflective cracks, laboratory testing was conducted on the method of overlaying pavement applying reflective crack prevention sheets (RC prevention sheets) or stress absorbing membrane interlayers (SAMI).   On the basis of the outcome of the testing, on-site overlaying of old concrete slab was carried out by using SAMI in addition to asphalt mixture.   The site selected for overlaying work was the container yard of Funamachi Station that belonged to Japan Freight Railway Company, and yearly changes in performance were subsequently monitored there by using FWD and through a survey of road surface conditions.

2. Laboratory Testing

In a laboratory, tension and compression tests were conducted to confirm the mechanical properties of the materials which were considered to be effective in absorbing stress.   Table 1 shows the property values of the RC prevention sheet, which is characterized by fiber meshes that make the sheet resistant to tensile stress.   In general, transverse tension is more difficult to counter than longitudinal tension, but the RC prevention sheet has been proven to withstand the transverse tension of 366 N (37.3 kgf).

Fig.1 indicates the aggregate grading of the asphalt mixture used for SAMI. This asphalt mixture was prepared as an open-graded one with target void rates of 20% to 25%, for the purposes of securing strength through the engagement of aggregate grains and alleviating stress and strain by the voids and the viscosity of asphalt.   In producing two types of highly viscous asphalt binder, the aggregate grading in Fig. 1 was adopted and straight asphalt and SBS polymer were added respectively for modification.   These two types of modified asphalt were used in the tests in order to determine the viscoelastic characteristics of the asphalt mixture.   Table 2 introduces the values indicating the mechanical properties of the two types of asphalt binder used for the tests, and the results of

Table 1. Values for mechanical properties of RC sheet

Test item		Test result
Tensile strength (N)	Longitudinal	432
	Transversal	373
Elongation percentage (%)	Longitudinal	58.6
	Transversal	63.8
Tearing strength (N)	Longitudinal	95
	Transversal	86

Grain diameter (mm)	19.0	13.2	4.75	2.36	0.6	0.075
Range of grading (%)	100	97	31	13.2	8.4	2.5

Marshal property value	Void rate (%)	Stability (N)	Flow value (1/100 cm)
Open Grade Mix	22	4350	32

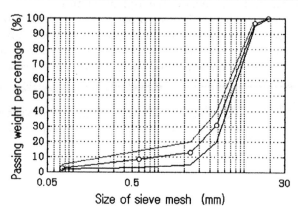

Fig. 1.   Ranges of the aggregate grading and property values of
asphalt mixture used for SAMI

toughness-tenacity comparisons are shown in Fig. 2.   On comparing the load
and displacement curves of the two types of asphalt binder, the one modified by
SBS polymer was found to excel in both "grip" and adhesiveness, exhibiting better
cohesion between aggregate grains.

Fig. 3 presents the relationship between measurement temperatures and
viscoelastic moduli of the two types of open-graded asphalt mixture in which the
above-mentioned materials were used as binder.   In the tests concerning stress
absorption, which employed a bar 50 mm long, 50 mm wide, and 100 mm high,
the stress was made σ_1 when compression strain stood at 0.2% (0.2 mm
compression), and the time that elapsed before σ_1 attenuated to 1/e (36.8%)
was counted.   The time thus obtained corresponds to the viscoelastic modulus of
the asphalt mixture.   As the figure demonstrates, the smaller the viscoelastic
modulus, the greater stress absorption becomes.

The results of the viscoelastic modulus test accord with those obtained from
the toughness-tenacity test, verifying that the asphalt binder modified by SBS
polymer is more capable of absorbing inner stress, as compared with the one in
which straight asphalt was used for modification.

With a view to verifying the aforesaid test results, test specimens were
produced by overlaying concrete plates in four different ways: simple overlay with
no crack-preventive materials, overlay to which RC prevention sheet was applied,
overlay combined with SAMI consisting of asphalt mixture, and overlay combined
with SAMI in which aggregates and an emulsion were contained.   The concrete

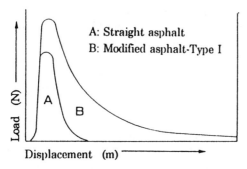

Fig. 2.   Results of toughness-tenacity test of asphalt binder

Table 2. Property values of asphalt binder

Test item	Asphalt modified with SBS polymer	Straight asphalt
Penetration (25°C,1/10mm)	59	67
Softing point (°C)	55.5	48.0
Ignition point (°C)	320	316
Residual penetration after thin-film oven test (%)	78.2	61.8
Toughness (25°C,N·m)	9.8	5.2
Tenacity (25°C,N·m)	4.1	1.2

Conditions for stress absorption test,   Predetermined strain: 0.2 %
Loading speed: 50 mm/min,   Specimen: 50 × 50 × 100 mm

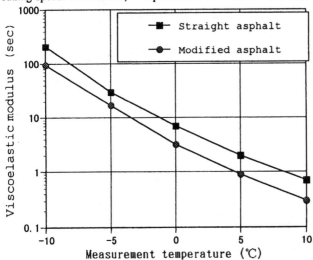

Fig. 3.   Difference in viscoelastic modulus between two types of asphalt binder

slabs to be overlayed for the testing had side joints of 5 mm wide, as shown in Fig. 4. The asphalt mixture for overlay, the concrete block, and the rubber mat with a hardness of 30 were of the same thickness of 50 mm. While the thickness of the RC prevention sheet was adjusted to 1.4 mm, the single layer of the SAMI comprising asphalt mixture was 30 mm in thickness, and another type SAMI had a thickness of 25 mm. Each specimen was fitted with a strain gauge on the side 35 mm away from the paved surface. Then, a wheel tracking tester was employed to apply a wheel load amounting to 720 kPa contact pressure (equivalent to a live load of 90 kg), thereby examining the relationship between the number of wheel passings as well as the strain generated[1]. The results are clear in Fig. 5. During the testing, temperature was maintained 30 ℃ .

In the case of the overlay with no crack-preventive material, damage began to appear after 2,100 wheel passings of the wheel. Overlay combined with either type of SAMI did not see any cracks until the number of wheel passings reached 8,000. In other words, overlay combined with stress absorbing membrane interlayer can withstand four times more wheel passings compared with simple overlay.

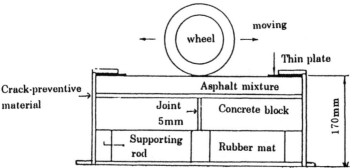

Fig. 4.   Configuration of specimen used for evaluation test

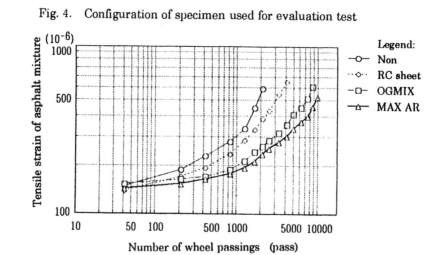

Fig. 5.   Relationship between the number of wheel passings and
          tensile strain

3. Conditions Before Repair

As the laboratory testing led us to the conclusion that the overlay method of adding SAMI to the ordinary layer of asphalt mixture was most effective in retarding the occurrence of cracks caused by joints, actual overlaying work with SAMI was implemented on concrete slabs that had been damaged.

The site designated for the overlaying work was the container platform for Lines 8 and 9 at Funamachi Station of Japan Freight Railway Company. After the concrete slabs of the platform were laid in 1968, cracks as serious as 80 cm/m ^{2}, which corresponds to a cracking ratio of 75%, with a maximum of 35 mm difference in height appeared before a survey was made in February 1989. Fig. 6 outlines the pavement structure of the platform. As Fig. 7 shows, the container platform was 180 m long and 20 m wide on average, approximating 3600 m ^{2} in area. For the purpose of comparing with the SAMI-added overlay, two sections were selected in the passage of container-loaded trucks, whose area totaled 300 m ^{2}; one with RC prevention sheet and the other with a simple overlay. In each of these

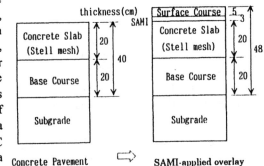

Concrete Pavement to Reflective Cracking ⟹ SAMI-applied overlay

Fig. 6.   Pavement structure

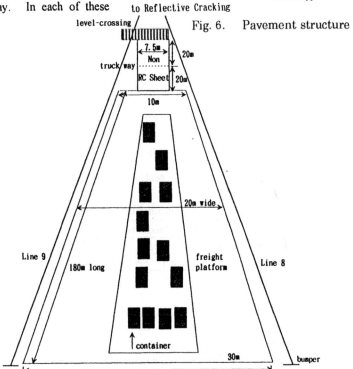

Fig. 7   The container freight platform at Funamachi Station of JFR

overlays, the thickness of the surface and the binder was 4 cm respectively.   The damage caused to the concrete plates is seen in Fig. 8.   Most striking are the break of corner areas and innumerable cracks due to the bending fatigue of concrete slabs.

Fig. 9 illustrates the longitudinal cross section of the deflections which were measured by FWD before the overlaying work was done with SAMI.   Noting the fact that the degree of deflection is larger in an area with more cracks, an analytical study was conducted by dividing the section in question into two sections. The analyses enabled us to estimate elastic moduli, as indicated in Table 3. In calculating the apparent bearing coefficient K_{30} of base, the method of using deflection area, which was developed by Darter et al[2], was employed. The calculation brought us the results shown in Fig. 10.

The estimated bearing coefficient K_{30} of the base, which was obtained from the load intensity

Fig. 8.   Damage caused to concrete slabs

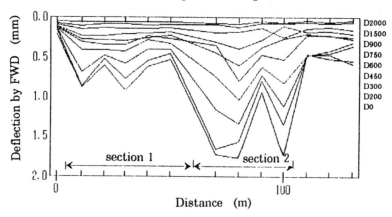

Fig. 9.   Longitudinal cross section of deflection before repair

Table 3.   Estimated elastic modulus of different layers

Pavement material	Elastic modulus (MPa)			
	Before repair		After repair	
	Section 1	Section 2	Section 1	Section 2
Surface layer	—	—	8000	8000
SAMI	—	—	1000	1000
Concrete slab	3900	800	4500	1500
Crushed stone	110	80	130	85
Subgrade	60	45	65	50

* Values of surface layer and SAMI were corrected with temperature.

when the base subsided by 1.25 mm due to the plate whose diameter was 30 cm, ranged from 20 MPa/m to 40 MPa/m (2 kgf/cm^3 - 4 kgf/cm^3). These values were considerably smaller than the design value of 69 MPa/m (7 kgf/cm^3), presumably due to voids generated between the concrete slab and the base course.

4. Design of Overlay Combined with SAMI

The thickness of the overlay combined with SAMI was investigated by using data gathered in Section 2, where deflections were more pronounced. The thickness of the overlay to which SAMI was applied affected the deflection directly under the surface layer, as shown in Fig. 11, while the thickness of such overlay related to the tensile stress under the surface layer, as Fig. 12 illustrates. An increase in the thickness of the SAMI-added overlay contributes to the reduction of deflection and tensile stress. In designing the thickness of SAMI-added overlay, consideration was given to a requirement that no damage should be caused to the surface layer for five years under a wheel load of 104 kN applied by a 12 ft (3.6 m) container fork lift. This meant that design deflection directly under a loading point should be 0.9 mm and design stress under the surface layer be 0.5 MPa or 50% of the stress level which causes damage to the layer[3]. The pavement structure thus determined comprised a 30 mm thick SAMI, which was made of asphalt binder modified by SBS polymer, and a 50 mm thick surface layer of asphalt mixture.

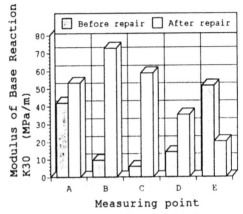

Fig. 10.   Bearing coefficients of the base before and after repair

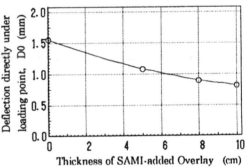

Fig. 11.   Relationship between the thickness of SAMI-added overlay and deflection

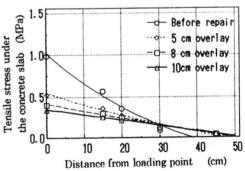

Fig. 12.   Relationship between the thickness of SAMI-added overlay and tensile stress

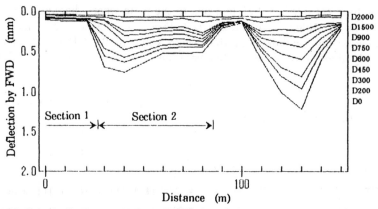

Fig.13. Longitudinal cross section of deflection after overlaying work using SAMI

5. Performance After Repair Work

As of September 1995, no reflective cracks attributable to the movement of concrete slabs were seen on the site where repair work was carried out using SAMI-added overlay in June 1989. Fig. 13 shows a longitudinal cross section of deflection as measured by FWD in October 1994. In section 2, deflection directly under the loading point averaged 0.65 mm, which was notably smaller than the value

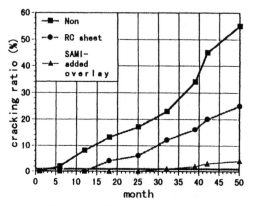

Fig. 14. Cracking ratio monitored after the repair work

predicted at the design stage. The elastic moduli of the concrete slabs and the crushed stone were found to have increased after the repair work, probably because the overlay combined with SAMI helped suppress the movement of the pavement which had been damaged.

Fig. 14 indicates the yearly changes in cracking ratio that have been monitored after the repair work. In the section with an 8 cm overlay, the cracking ratio reached 8% in 12 months and then 17% in 25 months after the repair work. After the lapse of 42 months, the cracking ratio in the overlayed section amounted to 45%, while the section for which RC preventive sheet was applied indicated a cracking ratio of 20%. In view of these results, those sections previously selected for comparison were also repaired again March 1994.

Compared with a section repaired with a simple overlay, it took twice as long for the section overlaid with RC preventive sheet to see a cracking ratio of 20%. In the case of the section with SAMI-added overlay, the length of time before the appearance of such a cracking ratio was longer by a factor of 4 or 5.

6. Conclusion

In the past six years following the repair work, no cracks have been produced on the surface layer to which SAMI was applied. This demonstrates the high performance of the SAMI-applied overlay, along with the effectiveness of the method of using stress absorbing materials to repair pavement with joints where cracks have appeared.

References

1. Takuya, I.(1988) A Method for Evaluating Crack-Preventive Materials Through Laboratory Testing, Japan Road Construction, No.487, pp.61-67.
2. Darter, M. I., Smith, K. D. and Hall, K. T. (1992) Concrete Pavement Backcalculation Results from Field Studies, Transportation Research Record 1377, pp. 7-16.
3. Francken, L. and Vanelstraete, A.(1993) Interface System to Prevent Reflective Cracking, The 7th International Conference on Asphalt Pavements, Vol.1, pp. 45-60.

The development of a grid/geotextile composite for bituminous pavement reinforcement

A.J.T. GILCHRIST and R.A. AUSTIN
Netlon Limited, Blackburn, United Kingdom
B.V. BRODRICK
University of Nottingham, United Kingdom

Abstract

Polypropylene geogrids have been extensively used to reinforce bituminous pavements and have been shown to provide considerable benefits to the long-term pavement performance with regard to the reduction of rutting, improving the fatigue life of the pavement and in the control of reflective cracking.

The paper describes the development and testing of a composite combining a stiff polypropylene geogrid with a geotextile thus producing a material with the handling and installation benefits of a geotextile combined with the performance benefits of a stiff geogrid.

Keywords

Geogrids, geotextile, composite, pavement trafficking, weak foundation testing reflective cracking, slab testing, overlap testing, core testing,

1 Introduction

Stiff biaxial geogrids were first used for the reinforcement of asphalt in 1982 at Canvey Island, near London, England, where 10 000m^2 of grid were used to control reflective cracking over a cracked concrete pavement.

Numerous geogrid installations have been carried out world-wide since this initial project and the use of geosynthetics for pavement applications is becoming more common. More recently geocomposites have been developed for pavement applications, combining the performance of a geogrid with the ease of installation associated with a geotextile.

Reflective Cracking in Pavements. Edited by L. Francken, E. Beuving and A.A.A. Molenaar. © 1996 RILEM. Published by E & FN Spon, 2–6 Boundary Row, London, SE1 8HN. ISBN 0 419 22260 X.

2 Initial research

Prior to the introduction of stiff polymer grids for pavement applications, in 1981 a four year programme of research was started to investigate the benefits of using such materials in bituminous materials. Laboratory test work investigated the benefits of the grid with respect to the control of surface deformation, the control of reflective cracking and the improvement of the fatigue life of the pavement. This early research work, at the University of Nottingham, UK, consisted of simulative testing with installation techniques being developed in field trials and installations. Details of the laboratory work undertaken have been published by Brown et al (1985a & 1985b).

3 Full scale trials

Many full scale trials have been carried out in different geographical locations throughout the world to confirm the results of the laboratory testing and to develop efficient methods of installing the geogrid, Nunn & Potter[1]. From the results of the laboratory and field trials it was possible to derive an analytical design method incorporating Tensar geogrids which is based on the strain capacity improvement of the pavement related to the number of load applications, Brown [2].

4 Composite Development

Although the methods developed for installing the geogrid gained rapid acceptance, methods to improve the technique were constantly being investigated. In 1990 Netlon Limited began to examine the benefits of bonding a geotextile to the asphalt reinforcing geogrid in order to provide a material which could be installed without the need to fix, tension and dress prior to paving by machine.

The development work undertaken resulted in the composite product known as Tensar AR-G. In this composite, Tensar AR1 geogrid is thermally bonded to a thin, non woven polypropylene geotextile at the geogrid nodes. Bonding the two materials in this manner provides a sufficiently bonded composite for installation purposes with no reduction in tensile performance of the oriented grid since there is no heat damage to the highly oriented geogrid ribs.

Installation of the composite requires a regulated road surface to be sprayed with either a straight 200 pen bitumen or rich bitumen emulsion at a net rate of 1.1-1.5 l/m^2. The composite is then rolled out either by hand or by machine under light tension and brushed to ensure intimate contact with the pavement. When the bitumen has fully cured and the fabric is fully stuck down, mechanical paving can commence. Due to the high bond strength between the grid and the fabric in the composite, the grid is held down sufficiently during paving by normal paving equipment.

Several trial installations of this patented use of geosynthetics have been carried out in many countries using bitumen emulsions, polymer modified emulsions and straight bitumen. In these full scale trial pavement installations, cores have been taken which have confirmed the good bond achieved between the reinforcing composite and the asphalt layers.

5 Composite research

With the development of the composite reinforcement product a further research programme commenced. Two main areas of research were re-evaluated at the University of Nottingham to assess the improved performance of the composite material:

    i.   Performance of pavements with weak foundations. (i.e. flexure of the bituminous layer).

    ii.   Control of reflective cracking.

5.1 Performance of pavements with weak foundations

5.1.1 Pavement construction

The basic philosophy was to compare the performance of a reinforced pavement under wheel loading in the pavement test facility with an unreinforced or control pavement. The Pavement Test Facility (PTF) comprises a 1.2m deep pit which is 7m long and 2.4m wide and is spanned by a frame housing a moving loaded wheel. Wheel loads up to 15 kN with a contact pressure of 760 kPa can be generated.

    For the weak foundation tests, three test sections were installed in the pit in the following configuration:

    1. Control Section - no reinforcement in a 80mm thick DBM layer

    2. Composite at the middle of a 80mm thick DBM layer

    3. Composite at the base of a 80mm thick DBM layer

The test pit is permanently filled with approximately 1.1m of a local Keuper marl subgrade. To generate a soft formation a 60mm layer of unfired clay bricks was added, with CBR being reduced by controlling the moisture content. A single pressure cell per test section was pressed into the subgrade clay centrally in the wheel track to 20mm below the surface, prior to placing of a 160mm thick sub-base layer. A 15mm thick regulating course of 6mm asphalt was spread thinly over the sub-base on all sections to provide an overall seal and a suitable surface for tack coating on the composite section.

    The tack coat of bitumen emulsion was applied at the rate of 1.5 l/m^2 and the composite brushed into the tack coat and left to cure. A 40mm layer of 14mm aggregate dense bitumen macadam (DBM) wearing course, was placed and compacted. Following compaction of the first 40mm layer, the composite in the mid asphalt position was placed following the same procedure and the second bituminous layer placed and fully compacted.

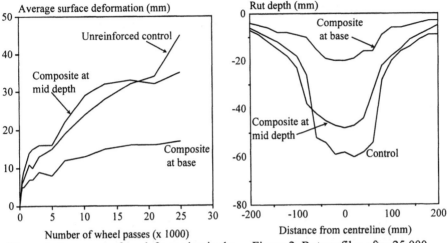

Figure 1 Average surface deformation in the wheel path vs. number of passes

Figure 2 Rut profiles after 25 000 wheel passes

5.1.2 Loading

The pavement was tested with the wheel running in single track mode, at a wheel load of 6kN and contact pressure of approximately 400 kPa, to provide a more concentrated load thus ensuring rapid deformation of the pavement. Surface profiles were taken at regular intervals. Throughout the test, subgrade stress readings were taken from the pressure cells. The test was continued for 25000 passes of the wheel load.

5.1.3 Deformation results

A fundamental indicator of the performance of each test section is the deformation in the wheel track. The results from an average of three readings down the length of each test section are shown in Figure 1. In all cases there is an immediate deformation increase at the start of the test as all layers compact under the action of the wheel which was seen to deflect the pavement when running.

Deformation then continues at a slower rate as rutting develops, i.e. material "flows" and is characterised by shoulders on either side of the wheel track. In these circumstances, where the pavement has been deliberately weakened by using a soft subgrade, failure was defined as the point at which the wheel track visibly sheared along each side. This occurred for the control section at 24 000 passes, Figure 2 shows the test section surface profiles at the end of trafficking.

The results of the rut depth measurements are summarised in Table 1, which shows that the effect of the composite at the base of the asphalt layer was to reduce the rut depth to a third of that for the control section. The test section with the composite at the bottom of the bituminous layer showed the best performance throughout and although larger deformations were observed for the composite in the middle of the layer, side wall cracking in the wheel track was considerably less than in the control.

Table 1 Summary of average rut depths for weak foundation test

Test section	Rut depth (mm)	Rut depth / control rut depth
Control	60	1.0
Composite in middle of asphalt layer	48	0.80
Composite at base of asphalt layer	20	0.33

This type of cracking is inevitable once a significant rut has developed as the asphalt surface tensile strain limits are exceeded.

5.1.4 Subgrade stresses

Subgrade stresses were measured with pressure cells of the strain gauged diaphragm type. Figure 3 shows readings initially at around 40 kPa and then they gradually increase as the subgrade becomes more stressed due to deterioration of the bituminous layer. This effect is more noticeable for the control test section, indicating the reinforcing effect of the composite.

5.1.5 Summary of weak foundation testing

1. By using a soft formation, substantial surface rutting was achieved in a relatively short time period.
2. The control section failed, with substantial cracking adjacent to the rut. The section with the composite at the bottom of the asphalt layer gave the lowest rut depth whilst the section with composite in the middle of the asphalt layer also gave better performance than the control section.
3. Stress measurements in the subgrade generally show expected trends during the test although their absolute value was higher than expected.

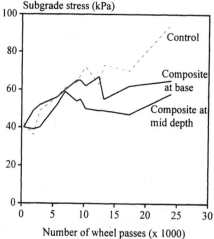

Figure 3 Subgrade stress vs number of wheel passes

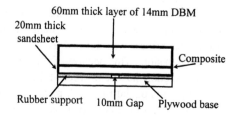

Figure 4 Slab construction

5.2 Control of reflective cracking

Improved performance against reflective cracking was investigated using a series of slabs trafficked by a moving wheel load over a rubber support. The slabs, 1.0m long x 0.2m wide by 0.08m thick were constructed as shown in Figure 4 and were tack coated to a plywood support with a 10mm gap maintained in the support to initiate cracking in the slab. In the test rig, the plywood supports rested on a rubber sheet to permit the slab to flex during trafficking. Strain measuring transducers were fixed to the edge of the slab to permit strain monitoring as the specimens were loaded.

Three slab types were tested. A control, a slab reinforced with the grid alone and a slab reinforced with the grid/geotextile composite. A sandsheet was used to simulate the actual site conditions when the composite is used, i.e. a smooth regulated surface.

The slabs were trafficked with a wheel load of 3kN with load and strain being continually monitored. The wheel loading on the slab caused a repeated opening and closing of the gap between the plywood supports which generated a cyclic maximum tensile strain in the sandsheet. This caused a crack to develop at the edge of the gap which propagated through this section.

The results of the slab testing are summarised in Table 2. The control slab became severely cracked at an early stage in the testing, with cracks propagating across the sandsheet/DBM interface by 1700 passes. Crack growth continued rapidly until the slab was completely cracked through by 3300 passes. For the grid reinforced slab the crack propagated through the sandsheet at 4500 passes. The crack then travelled to 5mm above the grid level and stopped at this position until 7600 passes. After 13000 passes the slab became cracked through due to the crack propagating from the base of the slab joining a surface crack. The surface crack had been caused by a rocking movement of the slab as the wheel moved on and off the slab at the end of its travel, which caused tension in the slab surface.

For the composite reinforced slab, the sandsheet cracked though at 4600 passes. This crack widened but did not transfer into the DBM until 12700 passes had been completed. At 18900 passes a surface crack developed which eventually met the base crack at 25000 passes. Trafficking continued to 30000 passes with still only a very fine crack visible in the DBM layer and a value of 2500 microstrain recorded.

The installation of the grid and composite into the slabs was found to improve the crack resistance of the slab under a moving wheel load with reference to the control slab which became severely cracked. This has been observed before by Brown [2]. Even when a crack propagates through a grid reinforced beam it still retains its structural integrity. The addition of the fabric to the grid to form the composite further improves reflective crack resistance.

Table 2 Summary of slab test results

Slab type	Number of wheel passes to	
	Crack through sandsheet	Crack through slab
Control	1700	3300
Grid	4500	13000
Composite	4600	25000

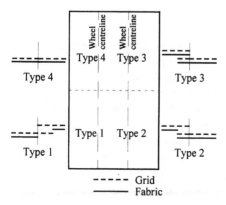

Figure 5 Plan of pavement test facility
          showing overlap configurations
          in section

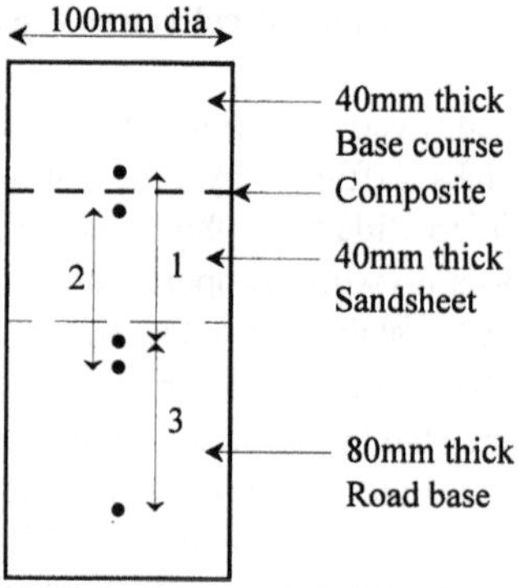

Figure 6 Instrumentation location on
          core samples

At the end of the test it was observed that the composite debonded from the sandsheet for about 150mm either side of the gap. It is thought that the composite may control cracking through this mechanism rather than strengthening the interface.

5.3 Overlap testing.

5.3.1. Pavement Construction

During installation of the composite overlaps will occasionally coincide with the wheel path. A series of tests were set up in the PTF to consider the effects, if any, on pavement performance of different overlap arrangements and to look at the effects of compaction and loading over the joint. A number of different overlap arrangements were tested.

In the PTF, 160mm of sub base was compacted over the clay formation as before. The bituminous construction was a lower DBM roadbase, sand sheet regulating course and an upper HRA basecourse. The composite was installed directly onto the sand sheet. The layout of the different overlaps tested was as per Figure 5. The composite was installed using a K1-60 bitumen emulsion and the final HRA coat installed and compacted.

5.3.2 Trafficking

The first loading phase was along the centre line of overlap types 2 and 3 at a 12kN load for 50,000 passes. This was followed by 30,000 passes at the same load directly adjacent to the previous track. The next phase was with the edge of the tyre running along the centre line of overlap types 1 and 4, again at 12kN for 50,000 passes. Finally, the wheel was run along the other side of the centre line but with the edge of the tyre offset by 50mm.

5.3.3 Visual Inspection

No cracks associated with compaction were evident at the start of the trafficking tests. During trafficking there was a build up of permanent deformation which subjected the shoulders of the rut to tension. The average rut depth was measured as 15mm but as there was no upward heaving at the shoulders, which occurs if the bituminous material flows and is displaced under loading, so deformation was by compaction under the loaded wheel. On completion of the trafficking there was no signs of any cracks associated with the location of the longitudinal joint between the grids either for direct loading or offset loading.

Transverse sections were cut out across the centre of each quadrant down to the sub-base. This exposed vertical sections through the asphalt and through the joint, and the mating faces could be examined on the excavated blocks. There was no evidence of any cracking associated with trafficking of any of the joint types. The upper base course layer was well bonded to the composite material and in turn this was well bonded to the sand sheet layer.

5.3.4 Testing of Cores

To further investigate any possibility of vertical deflection at the joint locations, cores were taken from both the trafficked and un-trafficked areas. Displacement transducers were fitted over 3 gauge lengths of 50mm, Figure 6, and the cores were tested at 1 cycle/sec in a compression testing machine. Tests were carried out at $9^{o}C$ to limit permanent deformation during loading. Each specimen was cycled about a mean load of 2.6kN over a peak to trough range of 4.5kN so that the specimen always remained in compression.

Deflection readings were taken both wholly within an asphalt layer and across the asphalt-composite interface. As all the readings were taken over a 50mm gauge length before permanent deformation could accrue it was possible to make direct comparisons using the deflection data. The results showed that no significant large deflections were associated with the joint arrangements compared to the readings away from the joints and outside the trafficked areas. Although there are some variations, they are small enough to be attributable to material stiffness variability.

5.3.7 Discussion of Results.

Wheel trafficking both along the joints and offset to them did not generate any surface cracking attributable to the discontinuity of the overlap between the composite layers. All overlap types tested exhibited similar performance under load. Cyclic compression tests on cores did not show any effect at the joint which would suggest an unsatisfactory joint type. For the materials used in the tests there was no evidence of local flexure along the line of the jointing or overlapping of the composite.

6 Case studies

6.1 Case Study 1

The B3004 road in Hampshire, England, carries heavy local traffic over soft saturated ground on low embankments. Failure of the existing surfacing had been caused by geotechnical failure of the subgrade leading to the formation of a longitudinal crack along the centre of the lane with evidence of several lateral and vertical deformation.

In 1991, a hot rolled asphalt overlay was laid on the road two sections of which were reinforced using proprietary systems leaving an unreinforced control section. The existing road surface was first regulated with fine asphalt to a thickness of 30mm. Bitumen emulsion was then sprayed at a rate of $1.51/m^2$ and Tensar AR-G composite immediately rolled out as the emulsion cured. To ensure full curing of the emulsion, vigorous brushing of the composite into the emulsion was then carried out.

When the bitumen emulsion had fully cured, 40mm of hot rolled asphalt was overlaid by conventional paver at a maximum temperature of $165^{\circ}C$. No evidence of movement of the composite was detected during paving. The site has now been open to traffic for nearly five years and there is no evidence of any cracking in the section reinforced with the geogrid composite material. The control section cracked through within four months and has since been removed and repaired several times.

6.1 Case Study 2

The A587 runs between the coastal town of Blackpool and the fishing port of Fleetwood, UK. It carries a variety of traffic which is particularly heavy during the tourist season. Sub soil conditions are predominately silty sands of marine origin and are generally soft with a high water table. Prior to overlaying the road was suffering from severe fatigue failure with extensive cracking on the road surface and localised rutting, depressions and pot holing.

A deflectograph survey carried out to assess the condition of the existing pavement determined that to give a serviceable life of 15 years, at present traffic volumes, 50mm would require to be planed off followed by a 200mm bituminous overlay. This would have caused disruption to the existing vertical geometry of the road requiring extensive kerb raising and alterations to the considerable number of access points to the road, from both commercial and private properties.

Based on the work previously described, the inclusion of the composite material can allow for an overall reduction in the thickness of the bituminous overlay. From the deflectograph data, and using an analytical design method, the re-constructed thickness, using the composite material below the overlay, was 120mm. Taking account of the regulating layer below the composite material, this represented a reduction in final pavement levels of 60mm when compared to the un-reinforced option.

In May 1995, the planed road surface was regulated with 20 mm of rolled asphalt and the composite was simply installed without the need for any fixing, tensioning or dressing. Use of a 200 pen bitumen sprayed at a rate of 1.1 l/m^2, enabled paving of the additional 120 mm of bituminous material to commence almost immediately.

7 Conclusions

1. Tensar AR-G geogrid/geotextile composite is suitable for the effective reinforcement of asphalt layers in pavements and appropriate installation techniques have been developed.
2. Pavements reinforced with the geogrid composite showed considerably improved performance with respect to permanent surface deformation.
3. When the composite is installed immediately above an existing jointed or cracked surface prior to the application of an asphaltic overlay, effective resistance to reflective cracking is provided.
4. The geogrid composite is able to resist large horizontal strains that develop in the bituminous material at the level of the grid. This is of particular value on roads with weak or unstable sub-grades.
5. For pavements over soft formations, subgrade stresses are reduced by the inclusion of the geogrid reinforcing composite in the bound layers of the pavement.

8 References

1. Nunn, M.E. & Potter, J.F. (1993), *Assessment of Methods to Prevent Reflection Cracking.* Proceedings of the Second International RILEM Conference, Liege, Belgium
2. Brown, S.F. et al, (1985), *Polymer Grid Reinforcement of Asphalt,* Annual Meeting of the Association of Asphalt Paving Technologists, San Antonio, Texas
3. Brown, S.F. et al, (1985), *The use of Polymer Grids for Improved Asphalt Performance,* Eurobitume Conference, The Hague, Netherlands.
4. Hughes, D.A.B. (1986), *Polymer Grid Reinforcement of Asphalt Pavements.* PhD Thesis, University of Nottingham, UK.

Jointless asphalt overlay on bridge decks

H. VAN DUIJN and M. NIJLAND
Latexfalt bv, Koudekerk a/d Rijn, Netherlands

Abstract

Prior to carrying out practical tests with a variety of joint constructions, a laboratory test arrangement was developed to simulate the horizontal movement of a joint construction with a continuous asphalt layer.

The operation of a number of joint systems was then tested. Tests were carried out on the SAMI, reinforced SAMI, with a highly modified bitumen emulsion binder and asphalt concrete directly on a concrete base layer. The test temperatures were 20°C and 0°C. A dense asphalt 0/16 type tarmac mixture was used. Drops in temperature were simulated by movement-controlled stretching, increases by movement-controlled compression. The tests were recorded using a video camera and analyzed in more detail by means of image analysis.

Practical tests were then carried out on 292 metre long bridge road, which is divided into sections by 6 expansion joints. At each joint, a joint construction was laid using a different system, using the modified emulsion [Latexfalt Safegrip] as a binder. The entire road was then surfaced, 2 x 5 cm dense asphalt, using a binder modified with EVA. The road was then visually checked to see if cracks appeared, and if so, when. Cyclical measurements were also taken of the expansion and shrinkage of the various joints at a range of temperatures.

Keywords: reflective cracking, asphalt-overlay, jointless, practical tests, bridge deck.

Reflective Cracking in Pavements. Edited by L. Francken, E. Beuving and A.A.A. Molenaar. © 1996 RILEM. Published by E & FN Spon, 2–6 Boundary Row, London, SE1 8HN. ISBN 0 419 22260 X.

1. Introduction

Beside the use of bitumen emulsion as a conservation coating for concrete, asphalt, etc., there is also interest in the use of bitumen emulsion as SAMI under asphalt, bridging small moving and non-moving cracks and joints. The top asphalt layer can then be applied as a continuous layer. It seems only sensible to use this option, especially for cracks and joints where large vertical movements are not expected.

The problem is how to direct the vertical and horizontal movement of the joint or crack so that stresses and distortion in the asphalt layer remain within the permissable range.

Laboratory tests were carried out before a practical test, which were exclusively to investigate the horizontal movement. The point of departure is a concrete surface on which is covered with an asphalt construction.

The aim of the construction with polymer-modified bitumen emulsion as SAMI is to reduce the tension in the asphaltic concrete caused by the horizontal movement of the joint/crack, and to minimize the variable forces in the adhesion of asphalt to concrete.

2. Description of the Laboratory Tests

2.1 Test Arrangement

The test arrangement was such that the same type of movement could be created as is found in the construction with a crack or joint.

Tests were carried out with a L6000R Lloyd tensile strength tester. This tester can be used with a climate cabinet, to control the temperature. The maximum test piece length in the climate cabinet is 550 mm.

The maximum force which can be applied is 30 kN. As we assume a relatively slow deformation control, this will be sufficient. The tensile tester is computer controlled and the information [force and deformation] are automatically recorded. The following test set up was devised as the test pieces have to be fitted vertically in a tensile tester.

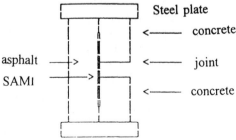

Figure 1: diagram of test set-up

The asphalt layer and the concrete test pieces were glued with their head ends to a steel plate. The steel plate is fixed to the tensile tester clamp. The clamp is constructed in such a way that the plates remain virtually parallel during the test. The joint construction is turned through 90 degrees, as it were.

A grid of lines was placed on the side of the test pieces. Any changes in these grid lines were recorded during the test using a video camera. Resolution can be made using image analysis which records the distortion very accurately to 0.01 mm.

2.2 Manufacture of Test Pieces

In order to produce the test pieces, panels of asphalt concrete were made on a concrete surface in which a 50 mm opening was made. This opening was temporarily filled with a wood mould during the asphalt application. The dimensions of the panel are 600 x 600 x 600 mm. Four test pieces 600 x 150 mm were made from each panel. The height of the test piece was 100 mm minimum [60 mm asphalt concrete, minimum 40 mm cement concrete].

Figure 2: lengthwise cross-section [dimensions not in relation]

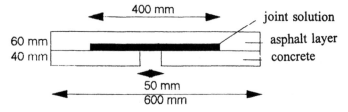

2.3 Tests with Systems

Test pieces 550 mm long, 150 mm wide and about 100 mm high were used for the tests. A length of 400 mm for the joint system was selected with a joint opening of 50 mm.

Table 1: tested variants

type no.	layers	joint construction
1	1	1,0 kg/m² Latexfalt Safegrip emulsion with Roadtexgrid
	2	1,5 kg/m² Latexfalt Safegrip emulsion scattered with 5 kg/m² Dutch chippings 2/8
	3	1,5 kg/m² Latexfalt Safegrip emulsion scattered with 8 kg/m² Dutch chippings 2/8
2	1	1,0 kg/m² Latexfalt Safegrip emulsion
	2	1,5 kg/m² Latexfalt Safegrip emulsion scattered with 5 kg/m² Dutch chippings 2/8
	3	1,5 kg/m² Latexfalt Safegrip emulsion scattered with 8 kg/m² Dutch chippings 2/8

Three tests per type of joint construction were carried out on one test piece, at the temperature as shown in the Table below.

Table 2: testresults at different temperatures

	jointconstruction					
	T = 20°C			T = 0°C		
	distortion cycle			distortion cycle		
	max. distortion [mm]	cycle duration [min.]	number of cycles	max. distortion [mm]	cycle duration [min.]	number of cycles
test 1	10	120	1	8	120	1
test 2	10	60	7	8	60	7
test 3	15	60	1	15	60	1

* test 1: an extra slow distortion test to find out if the test piece would collapse [2 hour cycle];
* test 2: a "fatigue test" of 7 cycles [cycle = 1 hour];
* test 3: after the fatigue test, an extra fast distortion test to 15 mm, to find any differences between the systems.

2.4 Results

Table 3: testresults of test 1 and test 2

Test 1

| type | 20°C | | 0°C | |
	max.tension [N]	max. pressure [N]	max.tension [N]	max. pressure [N]
1	842	180	9446	4500
2	331	190	7677	6150

Test 2: Maximum tension force [N] [first cycle] for Test 2

type	20°C	0°C
1	730	7361
2	406	12830

The same tendency as in Test 1 was observed at the second temperatures. Although the distortion control is twice as fast, the increase in force is limited. After the first cycle, the maximum force for both tension and pressure reduces a little, but otherwise remains virtually constant.

The fatigue test was carried out over 7 cycles. A typical example of the force - time diagram at 0°C is shown in Figure 5.

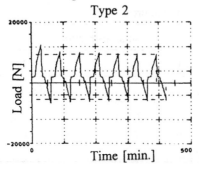

Figure 3: Example of Test 2 for Type 2 [temperature = 0°C]

Test 3

In order to find if there are any differences between the systems, a quick test up to 15 mm stretch was carried out on the test pieces which had not collapsed. A summary of the maximum forces during this test is shown in the Table below.

Table 4: maximum force [N] for Test 3

type	20°C	0°C
1	1124	15280 [collapsed]
2	817	16849 [collapsed]

A comparison of the force level shows that this relatively fast distortion-controlled test [15 mm per hour] gives considerably higher maximum forces than in Test 2.

No single test piece collapsed at 20°C before maximum stretch was reached. All test pieces collapsed at 0°C, but only after the 10 mm distortion was considerably exceeded.

2.5 Expectations

It was demonstrated that the use of a SAMI constructed with the Latexfalt Safegrip System over a specific length over a joint provides the possibility for continuous asphalting over a construction with cracks or joints, as far as horizontal distortion is concerned.

Repeated horizontal distortion [7 times, in this case] of the joint construction with SAMI did not lead to the asphalt breaking up. The joint movement was then:
- at 20°C, 10 mm over a 50 mm joint opening;
- at 0°, 8 mm over a 50 mm joint opening.

Calculated at 400 mm stretch length under the asphalt, this means in general:
-   at 20°C, about 2.5 % stretch under the asphalt;
-   at 0°C, about 2% stretch under the asphalt.
No distinction can be made from the "fatigue test" between the various joint systems, with or without reinforcement.

Pressure forces during reformation show that the joint constructions hardly recover elastically and the material has to be pressed back together.

3. Description and Results of the Practical Test

3.1 Description of the Practical Test
In order to test the expectations described above in practice, six variants were used on a bridge in Zaandam.

3.1.1 The Bridge
It was found possible to test six different joints on a bridge in Zaandam. This bridge has a total length of 295.50 metres and has expansion joints at 56.25 metres [2x], and 45.00 metres [4x]. The bridge can be represented schematically as follows:

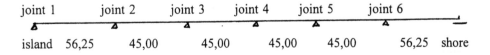

joint 1	joint 2	joint 3	joint 4	joint 5	joint 6	
island 56,25	45,00	45,00	45,00	45,00	56,25	shore

Figure 4: schematically drawing of the bridge

The bridge sections of 56.25 metres and 45.00 metres are each supported at three points, so that the angular rotation at the expansion joints can be ignored. The supports are fitted with sliding foil. The width of the concrete construction is 10.60 metres, divided into 6 metres roadway and 4.60 metres for footpath and railings.

The calculated movement in the expansion is:
- joint 1   ±   9,85 mm
- joint 2   ± 17,72 mm
- joint 3   ± 15,72 mm
- joint 4   ± 15,72 mm
- joint 5   ± 15,72 mm
- joint 6   ± 17,72 mm

3.1.2 The Systems

Latexfalt bv fitted the following constructions in the road section in all expansion joints:

         - milled slot 13 cm wide, 5 cm deep;
         - cleaned slot;
         - applied primer layer of Latexfalt JE Primer, about 150 g/m²;
         - applied backing filler;
         - applied joint filler;
         - fitted steel plate, 10 cm wide, 3 mm thick;
         - pour joint filler over steel plate.

Then the Safegrip system, consisting of 2 x 1.5 kg/m² polymer-modified bitumen emulsion, scattered per layer with 8 kg/m² and 10 kg/m² Dutch grit 2-6 mm, was applied over the whole concrete road deck [6 metres wide].

The asphalt covering consisted of 1 x 3 cm and 1 x 4 cm modified DAB 0-11 with 6% binder [Esso Flexxipave 106].

The following variants were applied to the joints:

Joint 1:   A 0.60 metre strip of Latexfalt Samtape. A 1 cm wide saw-cut in the first layer of asphalt [modified DAB 3 cm] 0.75 metres distance from the joint on the bridge side. 3 metres layer of double Safegrip on the first layer of asphalt.

Joint 2:   A 0.60 metre strip of Latexfalt Samtape. A 3.00 metre long strip of asphalt reinforcement [Glasgrid 8501] between the two asphalt layers.

Joint 3:   A 0.60 metre strip of Latexfalt Samtape.

Joint 4:   Two 1 cm wide saw-cuts in the first layer of asphalt [modified DAB 3 cm] at 0.75 metres distance from either side of the joint. A 3.00 metre long double Safegrip layer on the first layer of asphalt. A strip of asphalt reinforcement [Glasgrid 8501] on the Safegrip layer.

Joint 5:   A 3.00 metre long double Safegrip layer on the first layer of asphalt [modified DAB 3 cm].

Joint 6:   Without any extra provisions.

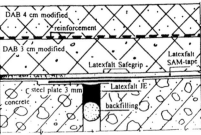

Figure 5: cross-section of joint 2

3.2 Execution

22 May 1995 Asphalting first layer of DAB [180°C]
23 May 1995 Asphalting second layer of DAB [180°C]

Joints 1, 2 and 3 gave no problems during or after asphalting. Joints 4, 5 and 6 showed marks. A day after asphalting, Joint 4 displayed a small 10 cm long crack in the middle of the road.

3.3 Measurements and Findings

1 June 1995
Measurements points 150 mm apart were made on both sides of the expansion joints, using a engraving pen on the edge beams of the concrete construction. As this edge beam does not continue through at the island side and the land side, displacement could not be measured; this is why no measurements are included below for joint 1. The measurements were taken between 16:00 and 18:00 hours, unless stated otherwise.

T°C in the Table below indicates ambient air temperature, while the numbers under Joint 1, Joint 2, etc, are the expansion joint widths in mm.

Table 5: joint width at different temperatures [per month]

date	T°C	joint 2	joint 3	joint 4	joint 5	joint 6
01-06-1995	18	25	27	28	23	23
03-07-1995	15	24	28	28	21	22

Table 6: 24-hour measurements from 18:00 to 18:00, taken every 2 hours
Date: 11-07-1995

time	T°C	joint 2	joint 3	joint 4	joint 5	joint 6
18.00	30	15	19	22	11	14
20.00	29	15	20	22	11	15
22.00	25	16	20	23	12	15
24.00	23	16	21	24	13	15

H. Van Duijn and M. Nijland

Date: 12-07-1995

time	T°C	joint 2	joint 3	joint 4	joint 5	joint 6
02.00	18	17	21	24	13	15
04.00	15	18	22	25	14	17
06.00	20	18	22	25	15	18
08.00	28	19	22	26	16	18
10.00	28	19	22	25	16	18
12.00	26	18	21	24	14	17
14.00	25	17	21	24	13	16
16.00	23	16	21	23	13	16
18.00	20	17	20	24	12	15

Table 7: joint width at different temperatures [per month]

date	T°C	joint 2	joint 3	joint 4	joint 5	joint 6
28-07-1995	22	20	23	25	16	19
30-08-1995	16	25	28	28	21	23
03-10-1995	12	26	29	29	23	24
02-11-1995	7	30	32	31	26	26
05-12-1995	-5	36	39	37	34	33
12-01-1995	9	31	33	31	28	27

Table 8: 24-hour measurements from 18:00 to 18:00, taken every 2 hours
Date: 26-01-1996

time	T°C	joint 2	joint 3	joint 4	joint 5	joint 6
18.00	-5	38	41	37	37	35
20.00	-8	38	41	38	37	35
22.00	-9	38	41	39	36	35
24.00	-10	38	42	39	37	35

Date: 27-01-1996

time	T°C	joint 2	joint 3	joint 4	joint 5	joint 6
02.00	-10	39	42	40	38	36
04.00	-11	39	43	40	38	36
06.00	-10	38	43	39	37	35
08.00	-9	38	43	39	37	35
10.00	-8	38	42	39	37	35
12.00	-8	38	43	39	37	36
14.00	-7	38	43	39	37	35
16.00	-5	38	42	39	37	35
18.00	-5	38	41	38	37	35

Largest movement at Joint 5   : -12 mm / +15 mm
Least movement at Joint 4   : -5 mm / +12 mm
Greatest crack width at Joint 6 [16mm] : movement -9 mm / +13 mm
Smallest crack width at Joint 2 [7mm]   : movement -10 mm / +14 mm

JOINT-FREE TRANSITION

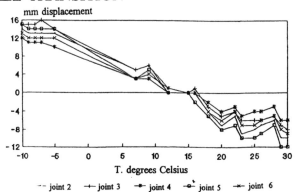

Figure 6: relation between displacement and temperature

4. Conclusion

Although the asphalt over all joints has cracked, it appears that the construction of Joint 2, with an underlayer of SAMtape, a membrane of bitumen emulsion on the concrete and Glasgrid reinforcement in the asphalt gives an excellent result. The movement at this joint is 24 mm, whereas the crack width is only 7 mm. The reinforcement will have restricted the crack width, and the Safegrip membrane has acted as a stress absorber. A further practical test, with stronger reinforcement and extra emulsion-layer is certainly justified.

Design methods:
theory and practice

Part four

Design methods:
theory and practice

Design method for plain and geogrid reinforced overlays on cracked pavements

A.A.A. MOLENAAR
Road and Railway Research Laboratory, Delft University of Technology, Netherlands
M. NODS
Huesker Synthetic GmbH & Co., Gescher, Germany

Abstract

This paper describes a simplified method to analyze the crack reflective resistance of unreinforced and geogrid reinforced overlays. The method is based on a theoretical model developed at the Texas A&M University for the analysis of stress conditions in cracked pavement, laboratory testing on reinforced and plain asphalt as performed by the Netherlands Pavement Consultants, and work done at the Delft University of Technology on crack propagation in asphalt layers and pavement evaluation. Special attention is given on the collection of the needed input data in order to be able to run the overlay design model.

Keywords: asphalt reinforcement, geogrid, modulus of elasticity, overlay design, reflective cracking, stress intensity factor.

1 Introduction

Overlays are an effective maintenance strategy for increasing the structural pavement life. Many of the overlay design procedures which are available nowadays are based on reducing the stresses and strains in the existing pavement.

Such an approach is fine as long as the existing pavement doesn't show too much cracking. If, however, the overlay has to be placed on a severely cracked pavement, these cracks will have a significant effect of the lifetime of the overlay. This is because such cracks tend to propagate from the existing pavement through the overlay quite rapidly.

It is obvious that for such conditions an overlay design method is needed that properly addresses the reflective crack phenomenon. Such an approach is even more needed to allow realistic assessments to be made of the effect of new maintenance techniques such as the application of reinforcements in overlays.

From a theoretical point of view it is quite obvious that in designing an overlay that is resistant to reflective cracking, use has to be made of finite element techniques and fracture mechanics principles. Although nowadays powerful personal computers allow finite element analyses to be made within a reasonable period of time, this technique is still considered to be a researchers type of tool given its complexity and the needed skills of the user. The same holds for the application of fracture mechanics.

For that reason a simplified, practical approach for the design of plain and geogrid reinforced overlays was developed. The procedure is based on sound engineering principles and uses techniques that can be easily verified by potential clients.

The method is based on a theoretical model developed at the Texas A&M University [1] for the analysis of stress conditions in cracked pavement, laboratory testing on geogrid reinforced and plain asphalt as performed by Netherlands Pavement Consultants [2], and work that has been done at the Delft University of Technology on crack propagation in asphalt layers [9] [10] and pavement evaluation [10] [12]. Special attention is given on the collection of the needed input data in order to be able to run the overlay design model and for that reason several examples are given.

In the chapter here-after, the basic theoretical principles of the method will be described first of all. Then attention will be paid to how the model can be used in practice and how the needed input data can be collected. Finally two overlay design examples are given.

A remark should be made on the modelling of the effect of the reinforcement. Ideally one should take into account the stiffness and area of the reinforcement and the characteristics of the bond between the reinforcement and the surrounding material (overlay and existing pavement). However such an approach is not yet ready to be used on an every days basis. Therefore it has been decided to model the effect of the reinforcement in the overlay as an increased resistance to cracking of the overlay material. As mentioned, the results presented in [2] have been used to obtain values for the increased crack resistance of geogrid reinforced materials.

2 Principles of Crack Propagation in Pavements

Assume one is dealing with the pavement structure which is shown in figure 1.

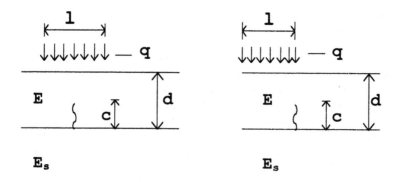

A: bending case   B: shearing case

Fig. 1   Schematized cracked pavement structure

The pavement is a two layer system with a bound layer (thickness d, elastic modulus E) on a subgrade (elastic modulus E_s). The wheel load that is placed on the pavement is characterized by means of the contact pressure q and the diameter of the contact area l. In the bound layer there is a crack growing from bottom to top and at a given moment, the length of this crack equals c. Two cases can be recognized; case A when the load is placed precisely on top of the crack and case B when the load is placed on one side of the crack. In case A the crack will propagate due to bending while in case B shear forces are the reason for crack propagation. The question now is how fast the crack will propagate through the bound layer.

In order to be able to make this analysis one has to use some basic "laws" of fracture mechanics. One of these "laws" is so called Paris law which describes the rate of crack growth as:

$$dc/dN = A * K^n \qquad (1)$$

where dc/dN =   increase in crack length per load cycle
         K =   stress intensity factor
         A,n =   material constants

In principle the lifetime of a given overlay with thickness h can be calculated using:

$$N = h / (dc/dN) \qquad (2)$$

It is a well known fact that around the tip of a crack, peak stresses will develop. The stress conditions at the tip of the crack are described by means of the stress intensity factor K. Figure 2 gives a schematic representation of the stresses in a cracked plate.

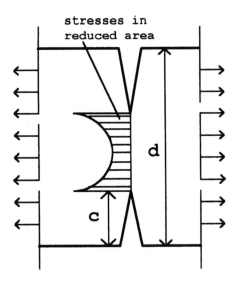

Fig. 2   Schematic representation of stresses in a cracked plate

It will be obvious that K is dependent on the overall stress σ and the length of the crack. K will increase when both the overall stress σ and the crack length c increase. In general one can write:

$$K = f(\sigma \sqrt{c}) \tag{3}$$

Given the fact that the square root of the crack length is involved gives K the somewhat strange looking dimension of N/mm$^{1.5}$.

Mathematical solutions are available to calculate K for the condition as shown in figure 2 but also for the conditions shown in figure 1 which are of special interest for the pavement engineer. These mathematical solutions have been published by researchers of the Texas A&M University [1] [3].

3 Analysis of Crack Propagation in Overlays placed on Cracked Pavements

In reality pavements are a bit more complicated than the one which is shown in figure 1. Especially this is true in case of an overlay. An example of a realistic case is given in figure 3. Here an overlay has to be placed on a pavement where a crack has propagated from the cement treated base through the asphaltic top layer, and the overlay has to increase the pavement life.

In figure 3 it is also indicated how the real pavement can be schematized by means of the simple two layer model. It means that all the bound layers are taken together as one layer. This is a realistic schematization since the crack propagation will take place in those bound layers. For the calculations it means that the elastic modulus of the overlay, the asphalt layer and the cement treated base have to be combined to one single overall effective modulus E.

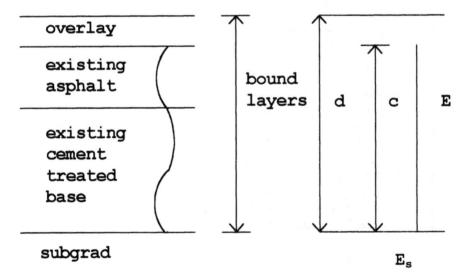

Fig. 3   Example of a cracked pavement with a cement treated base and it schematisation by the model

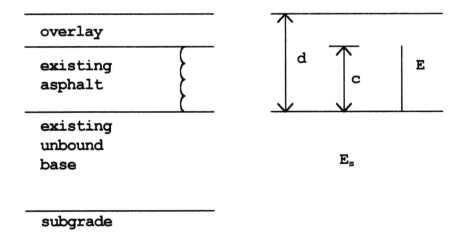

Fig. 4   Example of a cracked pavement with an unbound base and its schematization by the model

The procedure to arrive to such an effective modulus will be presented here-after.

In figure 4 it is shown how the case of an overlay on a cracked asphalt layer that is placed on top of an unbound base has to be modelled. In this case an effective modulus E has to be derived for the combined overlay, existing asphalt layer. Also an effective modulus E_s for the combined base, subgrade has to be derived.

In both cases the length of the crack c is taken as the thickness of the cracked bound layers. Since it is known how various pavements have to be modelled, attention has to be paid to how the effective moduli of the combined layers have to be determined. This will be discussed in the next chapter.

4 Assessment of the Effective Modulus of the Bound Pavement Layers and the Effective Modulus of the Unbound Supporting Layers

In the previous chapter it has been indicated that one needs to know the effective modulus E of the bound layers where the crack is propagating and the effective modulus E_s of the unbound layers which are supporting the bound layers. In order to do so basically two methods are available which are:
- back calculation of E and E_s from measured deflection profiles,
- assessing the moduli of the various layers individually using simple tests and combining them afterwards using Odemark's and Nijboer's theory [8] [13].

Both methods will be discussed in detail here-after.

4.1 Back Calculation Procedure

Layer moduli can be successfully backcalculated from measured deflectionprofiles provided the deflection survey is performed by means of a falling weight deflectometer (f.w.d.). The reason for this is that deflection profiles measured with a deflectograph or a Benkelman beam need significant corrections due to the applied low loading speed and because of the

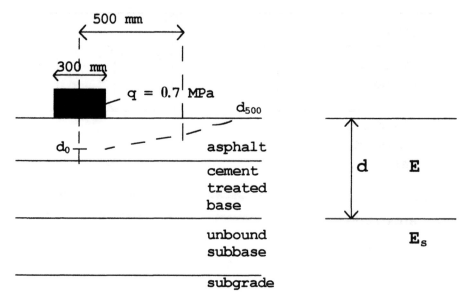

Fig. 5   Needed information from the deflection test

fact that the measured deflections might be heavily affected by the movement of the support system.

The short loading times generated in the f.w.d. test (t ≈ 0.025 s) means that most materials behave elastically even at elevated temperatures. Furthermore the f.w.d. deflections need not to be corrected for movement of the support system.

The moduli of a two layer system can easily be determined from the measured deflection profile if the thickness d of the combined bound layers is known. Figure 5 shows the needed information.

Assuming an f.w.d. load of 50 kN (q = 0.7 MPa), the maximum deflection d_0 and the deflection at 500 mm from the load centre d_{500} have been calculated for various total thicknesses of the bound layers d and various values for E and E_s. From these calculations evaluation charts have been developed, an example of which is shown in figure 6.

In figure 6 the quantity SCI is defined as:

$$SCI = d_0 - d_{500} \tag{4}$$

From figure 6 it is clear that knowing d, E and E_s can be estimated very easily if d_0 and SCI are known. For intermediate values of d, interpolation has to be used to arrive to the desired E and E_s value.

In back calculating layer moduli from deflection profiles, one has to realize that the measured deflections are influenced by temperature in case the pavement comprises bituminous layers. The commonly used approach is that the measured deflections are corrected to a reference temperature which could e.g. be the weighted mean annual asphalt temperature (WMAAT) as defined in the Shell Pavement Design Manual [4]. The reader is referred to [4] for details on the calculation of WMAAT.

The asphalt temperature during the deflection survey can be calculated using the proce-

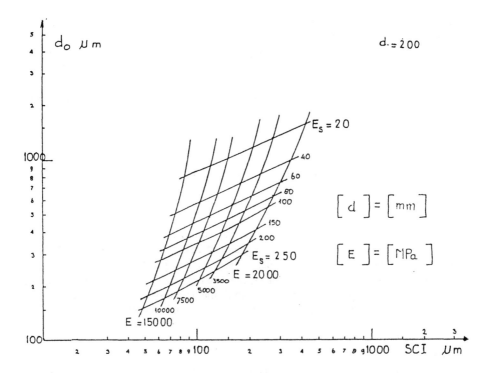

Fig. 6   Example of a modulus evaluation chart

dure that is described in reference [11]. From that procedure it is obvious that only the surface temperature has to be measured during the deflection survey.

The maximum deflection and the SCI have to be corrected to the reference temperature using:

$$d_0 = d_0\,(T)\,/\,TNF \tag{5}$$

$$SCI = SCI(T)\,/\,TNF \tag{6}$$

When d_0 and SCI are the maximum deflection and curvature index at the reference temperature. The values $d_0\,(T)$ and SCI (T) are the respective values at the measurement temperature (T).

The reader is referred to reference [12] to obtain all details on how the correction factor TNF has to be calculated.

4.2 Assessment of Layer Moduli using Simple Tests

In almost every textbook on pavement design several procedures are described to estimate the moduli of the subgrade, unbound subbase and base, bound base and asphaltic layers by means of simple tests. The reader is referred to [4], [5], [6] and [7] to obtain details on these procedures.

After the moduli of the various layers have been estimated, one has to arrive to a combined modulus for the asphaltic top layer and bound base layer as well as to a combined modulus for the unbound (sub)base and subgrade. Again various procedures are available in literature to do so. It is recommended to use the procedure developed by Nijboer [13] to obtain a combined modulus for the asphaltic top layers and the bound base layer and to the use Odemark's [8] procedure to arrive to a combined modulus for the unbound (sub)base and subgrade.

It is not considered to be necessary to repeat these procedures in detail here.

5 Estimation of the Crack Growth Characteristics A and n

As has been indicated in chapter 2, the characteristics A and n of the overlay material should be known in order to be able to make an analysis of the crack propagation through a plain or reinforced overlay. Extensive work has been done at the Delft University of Technology in the assessment of the above mentioned crack growth characteristics [9] [10]. It has been shown that the slope of the relation between the logarithm of the loading time vs the logarithm of the stiffness modulus of the mix is an extremely important parameter (see figure 7). At the loading time conditions for which the design should be made, the slope m of the curve is determined. The crack growth characteristic n is calculated from:

$$n = 2 / (m * c) \qquad\qquad\qquad (7)$$

The correction factor c is dependent on the mix stiffness S_{mix} and the bitumen stiffness S_{bit}. This relation is based on extensive work reported in [9].

The constant A is dependent on a large number of factors, extensive research however

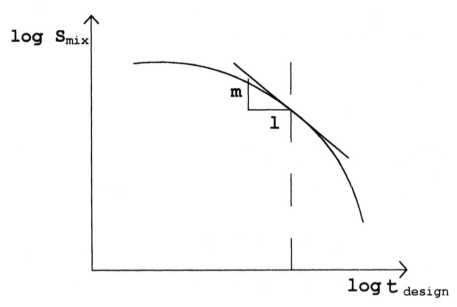

Fig. 7   Schematic representation of the log S_{mix} vs log t relation

[9] [10] has shown that for design purposes a sufficient accurate estimation can be made using:

$$\log A = a_0 + A_1 * n \tag{8}$$

The effect of a reinforcement is a quite complicated aspect to describe since it depends on a fairly large number of factors.

A study made by NPC [2] however showed that the overall effect of the reinforcement can be simulated by taking a reduced value for A. A reinforcement has no effect on the value of n. For the geogrid investigated one can take for the reinforced overlays:

$$A_{geogrid} = 1/3 \; A_{plain \; asphalt} \tag{9}$$

$$n_{geogrid} = n_{plain \; asphalt} \tag{10}$$

6 Examples

Several examples have been made using the overlay design approach which is described in the previous chapters.

The first example was dealing with the design of an overlay which is placed on top of a severely cracked pavement having a cement treated base (200 mm asphalt on top of 300 mm cement treated material). It was shown that if a 100 mm plain asphalt overlay was designed originally, a 70 mm geogrid reinforced overlay was a better solution since it would give an allowable number of load repetitions which is 1.74 times higher than the plain, unreinforced overlay.

A second example was dealing with a severely cracked pavement on an unbound base (100 mm asphalt on top of 300 mm unbound base material). Because of height restrictions, the maximum overlay thickness that can be placed is 50 mm. From the analysis it was shown that in this particular case the lifetime of the geogrid reinforced overlay is 3 times higher than the lifetime of a plain unreinforced overlay.

Although one should keep in mind that the values mentioned here are dependent on the input values that are used, it is however believed that they show clearly the effectiveness of the geogrid reinforced overlays when compared to the behaviour of plain, unreinforced overlays.

7 References

1. Lytton, R.L. (1989) Use of geotextiles for reinforcement and strain relief in asphalt concrete, *Geotextiles and Geomembranes*, Vol.. 8, No. 3.
2. Kunst, P.A.J.C.; Deenekamp, Q.; Wattimena, J.S. (1990) *Vergleichende Untersuchungen an unbewehrten und HaTelit bewehrten Asphaltprobekörpern*, Huesker Synthetic, Germany.
3. Jayawickrama, P.W.; Smith, R.E.; Lytton, R.L.; Tirado-Crovetti, M.R. (1989) Development of asphalt overlay design program for reflective cracking. *Proceedings 1st Int. Conference on Reflective Cracking in Pavements*, pp. 164-169 Liege.

4. Shell International Petroleum Company Ltd. (1989) *Shell Pavement Design Manual*, London.
5. AUSTROADS (1992) *Pavement Design - a guide to the structural design of road pavements*, Sydney.
6. Walker, R.N.; Paterson, W.D.O.; Freeme, C.R.; Marais, C.P. (1977) The South African mechanistic design procedure, *Proceedings 4th Int. Conference Structural Design of Asphalt Pavements*, Vol. II, pp. 363-415, Ann Arbor.
7. Shell International Petroleum Company Ltd (1990) *The Shell bitumen handbook*, London.
8. Odemark, N. (1949) *Investigations to the elastic properties of soils and design of pavements according to the theory of elasticity* (in Swedish), Report nr. 77, State Road Institute, Stockholm.
9. Jacobs, M.M.J. (1995) *Crack growth in asphaltic mixes*, Dissertation Faculty of Civil Engineering, Delft University of Technology, Delft.
10. Molenaar, A.A.A. (1983) *Structural performance and design of flexible pavements and asphalt concrete overlays*, Dissertation Faculty of Civil Engineering, Delft University of Technology, Delft.
11. Stubstad, R.N.; Lukanen, E.O.; Baltzer, S.; Ertman-Larsen, H.J. (1994) Prediction of AC mat temperatures for routine load/deflection measurements, *Proceedings fourth Int. Conf. on Bearing Capacity Roads and Airfields*, Vol I, pp. 401-412, Minneapolis.
12. Van Gurp, C.A.P.M.. (1995*) Characterisation of seasonal influences on asphalt pavements with the use of falling weight deflectometers*, Dissertation Faculty of Civil Engineering, Delft University of Technology, Delft.
13. Nijboer, L.W. (1955) *Dynamic investigations of road constructions*, Shell Bitumen Nomograph No. 2, Shell Int. Petroleum Company Ltd., London.

New anti-reflective cracking pavement overlay design

D.V. RAMSAMOOJ and G.S. LIN
California State University, Fullerton, USA
J. RAMADAN
University of Southern California, USA

Abstract
An analytical solution for the stresses and deflection in asphalt concrete (AC) overlays on rigid pavements is presented. The stresses and deflection are obtained from fracture mechanics using the relationship between the deflection and the stress intensity factor for a crack or joint. A new design for pavements is proposed to eliminate reflective cracking by the use of continuous steel reinforcement across the joints.
Keywords: stresses, deflection, elastic joints, rigid pavements.

1. Introduction

There has been considerable research on the problem of reflective cracking in AC overlays, mostly by trial-and-error methods. The fracture mechanics approach to the reflection cracking problem was introduced by Ramsamooj [1]. This paper presents an analytic method for determining the stress intensity factors, the deflection and stresses under vehicular or continuous steel reinforcement across the joints (Sherman, [2]) for eliminating reflective cracking.

2. The Stress Intensity Factor

Consider an asphalt concrete overlay on a rigid pavement as shown in Figure 1, with a through-crack of length $2c$ in the concrete slab. For a constant bending moment over the crack surfaces, Lin and Folias [3] obtained a theoretical solution for the stress intensity factor (SIF) given by

Reflective Cracking in Pavements. Edited by L. Francken, E. Beuving and A.A.A. Molenaar. © 1996 RILEM. Published by E & FN Spon, 2–6 Boundary Row, London, SE1 8HN. ISBN 0 419 22260 X.

$$K_I = K_{Iref} = \frac{\sigma\sqrt{c\pi}}{1 + a(\lambda c)^2} \tag{1}$$

in which $a \approx 0.1$, $\lambda = \sqrt[4]{\frac{k}{D}}$, k = coefficient of subgrade reaction and D = flexural rigidity of the PCC. The SIF for any other symmetrical load system may be obtained from

$$K_I = \int_0^c (\sigma(x))h(c,x)dx \tag{2}$$

in which $\sigma(x)$ = bending stress in the AC and h is given by [4]:

$$h = \frac{0.65c\pi \, (+c^2 - ac^4\lambda^2 - 0.5cx + 0.739ac^3\lambda^2x + 0.19\,x^2 + 0.952ac^2\lambda^2x^2)}{(1 + a\lambda^2c^2)c^3\sqrt{\pi(c - x)}} \tag{3}$$

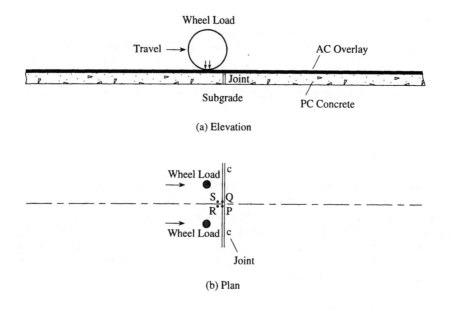

(a) Elevation

(b) Plan

Figure 1. AC Overlay on a Jointed Rigid Pavement.

3. Bending and shear stresses in the overlay under vehicular loading

The bending and shear stresses in a plain-jointed rigid pavement with an AC overlay may be found by superposing the stresses in a pavement caused by the transverse joints on the stresses in the pavement *without a transverse joint*. The bending and shear stresses in the concrete slab are first determined from CHEVRON multilayered computer program for a pavement without any transverse joints. Then the stresses on the crack faces are nullified by superposing equal and opposite stresses thereby creating the joint. For the load at any point P (Figure 1) the deflection is

$$w_P = \frac{2}{PE(1-v^2)} \int_0^{c'} (K_I^2)_P dc \tag{4}$$

in which $(K_I^2)_P$ is the SIF with the load at P. For a typical pavement, the cracking modes of interest are the opening Mode I, the in-plane sliding mode II, and the tearing Mode III. By analogy [5], the stress intensity factors for Mode II and III are of the form given for Mode II:

$$K_{II} = \int_0^c (\tau(x))h(c,x)dx \tag{5}$$

For combined modes I and III, the equivalent SIF is

$$K_{Ieq} = 0.5K_I + (1+2v_2) + 0.5\sqrt{K_I^2(1-2v_2)^2 + 4K_{III}^2} \tag{6}$$

so that the deflection at point Q can be obtained using K_{Ieq} instead of K_I. From the reciprocal theorem in elasticity, the deflection w_{PQ} at P due to the wheel load at Q, is equal to the deflection w_{QP} at Q due to the load at P. Furthermore, the superposition principle is applicable, so that for equal wheel loads are applied at both P and Q

$$w_P + w_Q + 2w_{PQ} = \frac{2}{PE(1-v^2)} \int_0^{c'} (K_{IeqP} + K_{IeqQ})^2 dc \tag{7}$$

The deflection at Q due the load at P is therefore

$$w_{QP} = \sqrt{w_P w_Q} \tag{8}$$

The shear and bending stress in the y-direction in the AC are respectively

$$\tau_{xy} = -\frac{E_{ac}h}{2(1+v)} \frac{\partial^2 w}{\partial x \partial y} \tag{9}$$

$$\sigma_{yy} = \frac{-E_{ac}h}{2(1-v_{ac}^2)} \left(\frac{\partial^2 w}{\partial y^2} + v \frac{\partial^2 w}{\partial x^2} \right) \tag{10}$$

in which E_{ac} and v_{ac} = the Young's modulus and Poisson's ratio of the AC in the y -direction.

4. New anti-reflective cracking design

The concept is simple. The reflective crack in the AC cannot propagate if the joint movement is negligible. The design is illustrated in Figure 2. It consists of continuous steel reinforcement of about 0.7% of the gross cross-sectional area of the concrete, with 6 cm diameter pipe dowels. A fiberglass layer 0.03 cm thick is bonded to the leveling course and AC to enhance the tensile strength of the AC and to prevent the entry of water through the joint and corrosion of the steel. Alternately the steel can be coated with epoxy in the vicinity of the joints to prevent erosion. A mechanistic analysis follows.

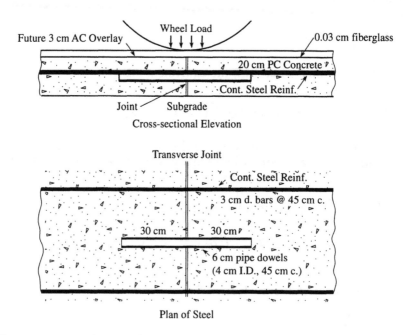

Figure 2. New Anti-Reflective Cracking Rigid Pavement / Overlay Design.

5. Thermal Stresses

5.1. Uniform Temperature Drop
5.1.1. *Conventional pavement*

Consider a uniform temperature drop ΔT in a rigid pavement with an AC overlay but *no transverse joints*. If the pavement is imagined to be clamped in the location where the real joints are, then the stresses in the AC and PCC induced by a temperature drop of ΔT are:

$$\sigma_{ac} = \frac{1}{1-v_{ac}} \, \alpha_{ac} E_{ac} \Delta T \tag{11}$$

$$\sigma_c = \frac{1}{1-v_c} \, \alpha_c E_c \Delta T \tag{12}$$

in which E = creep modulus, v = Poisson's ratio and α = coefficient of thermal contraction and the suffixes $_{ac}$ and $_c$ refer to the AC and PCC respectively. The joint in the concrete is now simulated by removing the clamps and nullifying the stresses on the faces of the joint in the concrete. However, if the interface between the AC and the PCC is fully bonded, the stress along the interface is an order of magnitude greater than the major principal stress in the AC (Zak and Williams, [7]). But the interface bond strength is less than 70 kPa. Consequently, the bond will be broken in the vicinity of the joint for a distance of about 30 cm and the local stresses so determined would be in considerable error. However, the stress on the interface away from the joint can be found, from which the stress in the AC can be determined. The bond between the concrete and the AC overlay is proportional to the relative displacement, with the maximum value of δ_u = 2 mm. Therefore the thermal stress in the AC is

$$\Delta\sigma_{ac} = \frac{1}{H_{ac}} \left(f_a (l - l_b) \frac{\delta_c}{\delta_u} \right) \tag{13}$$

in which H_{ac} = thickness of the overlay, f_a = interface bond strength, lb = unbonded length, lb = unbonded length and δ_c the movement of the concrete given by

$$\delta_c = \frac{2\,\sigma_c H_c}{2\,E_c \dfrac{H_c}{l} + 2\,f_a \dfrac{l}{du} + f_s \dfrac{l}{du}} \tag{14}$$

in which fs = frictional resistance of the subgrade, assumed proportional to the movement between the PCC and the subgrade with a maximum frictional coefficient = 1 mobilized at a displacement of 2 mm. The bond between the concrete and the AC overlay is proportional to the displacement, the maximum value being limited to 10 psi at a displacement of 2 mm. The length l_b over which the bond is broken is then about 30 cm on either side of the joint, with a temperature drop of $22°$ C.

5.1.2. *Elastic-joint pavement*

Continuous reinforcement across the transverse joints are provided to minimize the movement of the PCC slab. The additional frictional force on the AC when the stresses on the concrete joint faces are nullified are then considerably smaller than given by Eq.(13). The ultimate bond stress is approximately τ_b = 12 N/mm^2 at a slip of 1 mm. In the vicinity of the joint, the bond is broken and the unbonded length $_u$ is given by Bazant [6]) If F_s = the steel force per unit width, then the net force contributing to the frictional drag on the AC is

$$\Delta\sigma_{acs} = \frac{P_c - F_s}{P_c}\Delta\sigma_c \tag{15}$$

The total steel force is

$$F_s = \frac{1}{s}\left(\alpha_s A_s E_s \Delta T + f_s l \frac{\delta'_c (l - l_u)}{2\delta_b}\right) \tag{16}$$

in which δ'_c = concrete movement, α_s, A_s and E_s = the coefficient of expansion, the area and modulus of steel. The joint opening in the concrete is then

$$\delta w_c = \delta'_c + \frac{P_c - F_s}{P_c}Z l \tag{17}$$

in which Z = coefficient of shrinkage = 5 (10^{-4}).

5.2. Nonuniform temperature drop
5.2.1. *Conventional overlay*
During the daytime, the slab may be curled upwards and during the night it may be warped downwards. In the former case, the bending stress at the at the bottom of the AC over the transverse joint is tensile, and in the latter it is compressive, so that the daytime curl is more critical. Let the ends be clamped so that the nonuniform heating with a temperature gradient of θ, generates uniform bending moments in the PCC and AC, given respectively by:

$$M_c = D_c(1 + v_c)\alpha_c \theta \tag{18}$$

$$M_{ac} = D_{ac}(1 + v_{ac})\alpha_c \theta \tag{19}$$

The transverse joint is now introduced by nullifying the bending moment in the concrete. The slab then curls upward. The resultant deflection at the center due to the temperature curl and the weight is

$$w_R = 0.0257 \, w \, l^4 k - 0.5\theta\alpha(l^2 + b^2) \tag{20}$$

in which in which θ = temperature gradient. $k = 2b^4/(b^4 + l^4)$ is a correction factor for obtaining the deflection for a rectangular slab from that of a square slab (Ugural, [8]) and b and l are the half-width and half-length of the slab, respectively. The additional stress in the AC from the temperature gradient is

$$\Delta\sigma_{ac} = \frac{6}{H_{ac}^2}\left(-D_{ac}(1 + v_{ac})\alpha_c\theta\, l\frac{w_R}{w_T} + M_{ac}\right) \tag{21}$$

5.2.2. Elastic-joint overlay
An internal bending moment must be generated in the steel to make the slope in the steel zero. Therefore the steel couple M_{sT} is obtained from

$$\frac{M_{sT}\delta w_c}{E_s I_s} + \frac{M_{sT}l}{E_c I_c + E_{ac}I_{ac}} - \Phi = 0 \tag{22}$$

The effect of this couple on the K_1 in the AC is evaluated as before.

6. Vehicular Stresses
6.1. Elastic-joint overlay
The bending and shear stresses from the wheel loads in the conventional pavement has already been discussed. The effect of the steel reinforcement and the steel dowel bars are now evaluated. Without the steel, the slope of the pavement at the transverse joint θ_c is obtained from REFLEX (available on request). With the steel across the joint the slope of the steel must be zero from symmetry. Therefore

$$\frac{M_s\theta_c}{M_c} + \frac{M_s\delta w_c}{E_s I_s} = \theta_c \tag{23}$$

in which M_s = internal moment in the steel. The additional stress in the AC from the vehicular load on the pavement with elastic joints is

$$\Delta\sigma_{acs} = \sigma_v\left(1 - \frac{M_s}{M_c}\right) \tag{24}$$

The total stresses and deflection from vehicular loading, shrinkage and temperature changes are then obtained from a computer program REFLEX. From the bending and shear stresses the rate of crack propagation is given by [5]:

$$N_f = 0.033\frac{K_{Ic}^2}{f_y^2}\left[\ln\left(1 - \frac{K_o^2}{K_{Ic}^2}\right) - \ln\left(1 - \frac{K_I^2}{K_{Ic}^2}\right) - \frac{K_I^2}{K_{Ic}^2} - \frac{K_o^2}{K_{Ic}^2}\right] \tag{25}$$

in which f_y = flexural tensile strength, K_{Ic} = fracture toughness, K_I = SIF from thermal and vehicular loading and K_o ≈ $0.05K_{Ic}$ = threshold SIF dependent on the ratio of the minimum and maximum stresses.

Finally the crack spacing is checked by an empirical formula by McCoullough [9] to ensure that the minimum spacing between cracks is 4.57 m.

7. Overlay design

An 20 cm thick concrete pavement (E_c= 44800 MPa, v = 0.15) on a subgrade (E_f= 31 MPa, v = 0.45) is designed with an AC overlay with modulus ranging from 1724-5171 MPa, v = 0.35, and thickness 3.6-7.4 cm. The design is illustrated in Figure 2. The stresses from vehicular and temperature drop of 22°C and a temperature gradient of 0.65°C/cm are presented graphically in Figures 3 and 4. The fatigue life of the conventional design and the elastic-joint design are presented in Figure 5.

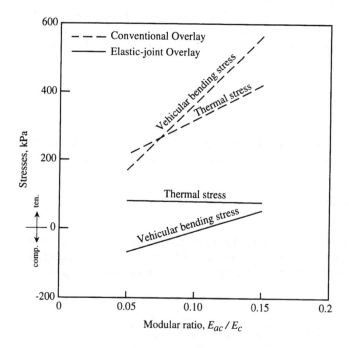

Figure 3. Vehicular and Thermal Stresses in AC Overlay, 7.6 cm thick.

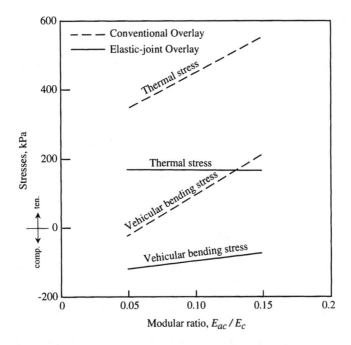

Figure 4. Vehicular and Thermal Stresses in AC Overlay, 3.6 cm thick.

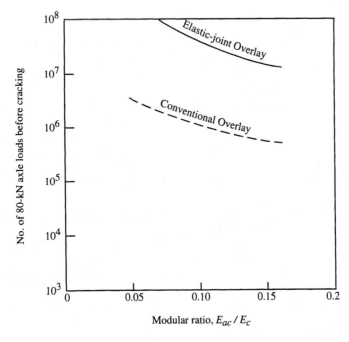

Figure 5. Number of 80-kN Axle Loads for Reflective Cracking of Overlay.

8. Conclusions

There is significant unbonding at the interface of the AC overlay and the concrete due to stresses from vehicular loading and temperature changes. Unbonding increases the tensile stress in the overlay, and may lead to traction failure. The dominant mode of crack propagation is the tensile bending stress at the base of the AC except for very thin overlays, when the shear stress controls.

The deployment of 0.7% of steel and 6 cm pipe dowels can reduce the tensile stress in the overlay to zero, the shear stress to about 5% and the thermal stress to about 30%. Further reduction of the thermal stress by increasing the amount of steel may generate intermediate cracks between the transverse joints, and is not recommended. The overall result is that the new design can delay reflective cracking 20-30 times longer than that of a conventional design. and it permits the use of stiffer AC mixes to minimize rutting.

The use of elastic joints reduces the deflection of a 20 cm thick rigid pavement under an 80 kN load from 0.055 cm to 0.042 cm, so that the pavement would be smoother riding. The reduction of the joint width from a typical value of 0.14 cm to 0.05 cm, means less chance of pumping at the joints.

The present model is simple and captures the fundamental behavior of AC overlays on rigid pavements. The new design is much superior to the conventional one.

9. References

1   Ramsamooj, D. V. (1973).   Prediction of reflection cracking in pavement overlays *Transportation Research Record*, 773, 1973.
2   Sherman, G. B. (1974).   Reflection cracking. *Pavement Rehabilitation*, Proc. of a Workshop, FHWA.
3.   Folias, E. S. (1970) On a plate supported on an elastic foundation and containing a finite crack, *Int. Journal of Fracture Mechanics*, 6:3, pp 257-263.
4.   Ramsamooj, D. V., Lin, G. S. and Ramadan, J. (1995). Stresses in pavements with joints and cracks. Submitted to *Engineering Fracture* 4.5.
        Sih, G. and Liebowitz, H.   (1968). Mathematical Theories of brittle fracture. *Fracture. An Advanced Treatise*. Vol. II, pp. 1-190. *Mechanics*, November.
6   Bazant, Z. P. (1985).   Fracture in concrete and reinforced concrete. *Mechanics of Geomaterials*, John Wiley. pp 259-303.
7.   Zak, A. R. and Williams, M. L. (1963).   Fracture Mechanics of Composites. *Fracture. An Advanced Treatise*. Vol. VII, pp. 675-769.
8.   Ugural A. C. (1981). *Stresses in Plates and Shells*. McGraw Hill Book Co.
9.   McCoullough, F. R. (1988).   Continuously Reinforced Concrete Pavements. *Concrete Pavements*. pp 279-318.
10.   Ramsamooj, D. V., (1990). Prediction of fatigue life of AC beams. ASTM Journal of Materials & Testing.

The dualism of bituminous road pavements cracking

D. SYBILSKI
Road and Bridge Research Institute, Warszaw, Poland

Abstract
The mechanism of the cracking of bituminous layers on the cement bound subbase consists in two processes: thermal cracking and mechanical cracking under a traffic load. Both modes of failure may perform separately or simultaneously. In the most cases actual cracking mechanism, its initiation and propagation are caused by tensile stresses exceeding local tensile strength of bituminous material. The analysis consisted in tensile stresses calculation at the bottom and top of bituminous layers regarding wheel and thermal loads at various temperatures and bituminous layers thicknesses. Tensile stresses were compared to tensile strength of bituminous materials. As a conclusion three cases were defined: **Case A** - crack initiation at bottom of the subbase due to tensile radial stresses under a wheel load, **Case B** - crack initiation at top of bituminous layers as a result of thermally induced tensile stresses exceeding tensile strength of bituminous layers, **Case C** - superposition of load conditions, i.e. the thermal stresses exceeding the tensile strength of a bituminous material on the pavement surface and the radial stresses exceeding the tensile strength of a cement bound material in the bottom of the subbase.
Keywords: bituminous pavement, mechanical and thermal cracking, stress analysis

1 General remarks

Cracking of bituminous layers on the cement bound subbase is one of the distress

Reflective Cracking in Pavements. Edited by L. Francken, E. Beuving and A.A.A. Molenaar. © 1996 RILEM. Published by E & FN Spon, 2–6 Boundary Row, London, SE1 8HN. ISBN 0 419 22260 X.

modes of the road pavements and one of the most discussed ones these days. The mechanism of the failure consists in two processes:

- thermal cracking
- mechanical cracking under a traffic load.

The two listed modes of failure may perform separately or simultaneously.

Mechanical cracking may be created by the three possible modes of the crack displacement (and stresses, respectively), i.e.: normal tension, normal shear, parallel shear [1]. The particular interest is however devoted to the normal tension mode of the failure as it is believed that in the most cases actual cracking mechanism, its initiation and propagation are caused by tensile stresses exceeding local tensile strength of bituminous material.

Thermal cracking as well as mechanical cracking involve a fatigue phenomenon, when process has a repetitive character. In such case a repetitive stress (or strain) level is lower than that, which leads to failure in static conditions.

Stress analysis discussed in the paper has been based on certain assumptions:

- the normal tension created by radial stresses in the pavement under a wheel load is a reason of "mechanical" cracking
- both thermal and mechanical cracking are analyzed in static conditions, i.e. one load to the failure.

2 Assumptions

2.1 Pavement construction and load conditions

Fig. 1 shows the considered pavement construction variants. The stress analysis has been done for the different construction and temperature conditions:

- bituminous layers total thickness varied from 2,5 to 30 cm,
- subbase made of a cement soil stabilization or a lean concrete, which means the different stiffness module and tensile strength
- two variants of temperature conditions: low temperature (-28,4°C) and high temperature (+20°C).

The total amount of different cases of pavement and temperature conditions regarded 20 variants.

Vehicle load was assumed single standard axle load 100 kN with tire pressure 0,65 MPa. Assumed values of material properties for subbase materials and bituminous layers are given on Fig. 1. It was assumed that stiffness modulus of cement bound materials was independent from temperature. For bituminous layers various values of stiffness modulus was considered for low and high temperatures.

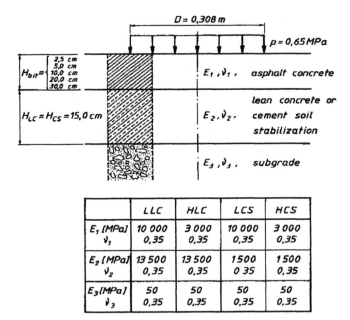

Fig. 1. Variants of pavement construction, wheel load and temperature conditions: LLC - Low temperature, Lean Concrete, HLC - High Temperature, Lean Concrete, LCS - Low temperature, Cement soil Stabilization, HCS - High Temperature, Cement soil Stabilization

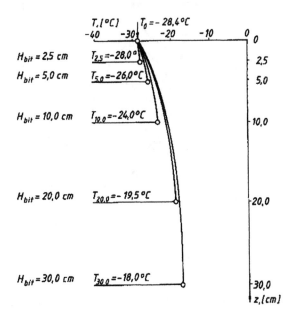

Fig. 2. Temperature distribution within bituminous layers

2.2 Temperature of pavement layers

Temperature within pavement layers were obtained from nomograph of Southgate et al. [2]. Calculations of pavement's temperature were done only for low temperature variants (Fig. 2), assuming that at high temperature +20°C there is not thermally induced stress in bituminous layers. Relaxation times are enough short.

2.3 Thermally induced stresses and tensile strength of materials

It is assumed that the stiffness and the tensile strength of cement bound materials are not dependent on temperature in the concerned range. On the contrary rheological properties, e.g. stiffness modules, of bitumen and bituminous layers are strongly dependent on a temperature. Thereupon, thermally induced stresses in bituminous layers and their tensile strength vary with temperature, which was comprehensively analyzed by Fabb [3] or Arand [4]. In the analysis presented, results of tests conducted by Eulitz [5] at the Technical University Braunschweig, Germany, were used. Fig. 3 shows results of tensile strength and thermally induced stresses chosen for the stress distribution analysis.

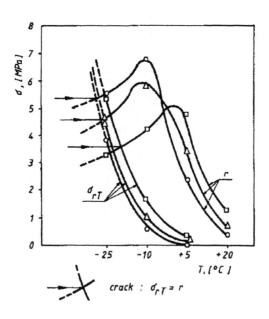

Fig. 3. Tensile strength and thermally induced stresses versus temperature (values from Eulitz [5])

3 Analysis of stress distribution

3.1 Mechanical, horizontal stresses

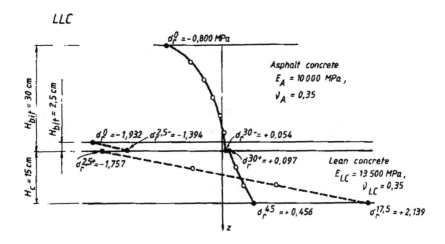

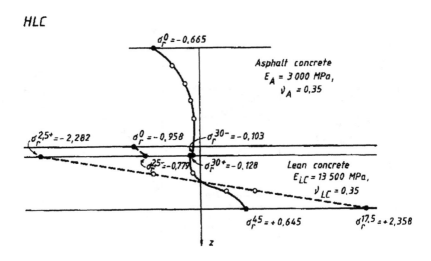

Fig. 4. Radial stresses under a wheel load within bituminous layers and lean concrete subbase at: low temperature (LLC) or high temperature (HHC)

Stresses in a pavement were calculated for the concerned conditions with the computer program ELSYM5. Stresses were calculated under a wheel load in bituminous layers and in a subbase at few depths. Fig. 4 shows horizontal, radial stress values in bituminous layers and in lean concrete subbase for respective cases: 30 or 2,5 cm total thickness of bituminous layers and for low (LLC) or high (HLC) temperature. Horizontal stresses for the above conditions but for variants with cement soil stabilization subbase (lower stiffness modules) have a similar distribution within layers, but with some significant value differences.

Fig. 5 shows the relationship of radial stresses at the top and at the bottom of subbase depending on bituminous layers' total thickness for respective cases. Tensile stresses exceed tensile strength of cement bound subbase at its bottom:

- lean concrete subbase would crack when total thickness of bituminous layers was lower than 15 cm at low temperature and lower than 24 cm at high temperature
- cement soil stabilization would crack when total thickness of bituminous layers was lower than 9 cm at low temperature and lower than 14 cm at high temperature.

More inconvenient stress conditions are at low temperature.

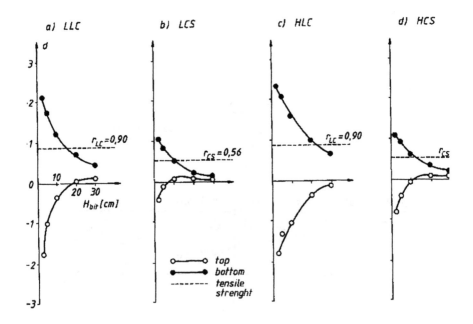

Fig. 5. Radial stresses under a wheel load at the top and bottom of cement bound subbase versus thickness of bituminous layers: a) Low temperature, Lean Concrete, b) Low temperature, Cement soil Stabilization, c) High temperature, Lean Concrete, d) High temperature, Cement soil Stabilization

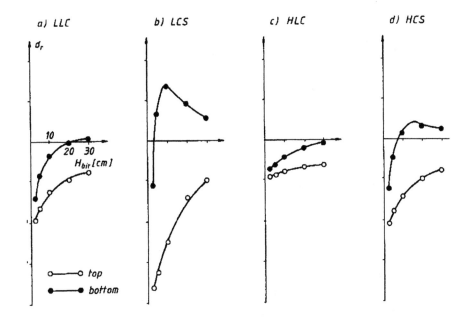

Fig. 6. Radial stresses under a wheel load at the top and bottom of bituminous layers versus their thickness

Fig. 6 presents the same relationship but at the top and bottom of bituminous layers. Thin layers are subjected to about twofold higher values of stresses at low temperature than at high temperature. In a case of cement soil stabilized subbase, the most critical conditions at low temperature are when total thickness of bituminous layers is 10-12 cm.

3.2 Thermally induced stresses
Temperature, which is not homogenous and constant within bituminous layers, causes that thermally induced horizontal tensile stresses and tensile strength of bituminous material varies along a depth of layers. Using Eulitz's results, both properties were determined at the top and at the bottom of bituminous layers for the assumed temperature conditions (Fig. 7). The variance of thermally induced stresses at the top and at the bottom of bituminous layers with their thickness is given on Fig. 8.

3.3 Tensile strength reserve
A difference between tensile strength (r) and horizontal stress (σ_r) is designated as a tensile strength reserve (Δr). Depending on the nature of horizontal stress, i.e. mechanistic (under a wheel load), thermally induced or total (superposition of the both), tensile strength reserve takes different values and different relationship with the

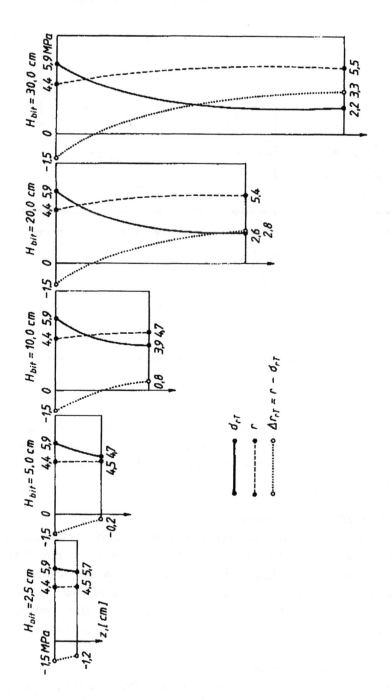

Fig. 7. Thermally induced stresses in bituminous layers, tensile strength and tensile strength reserve of bituminous layers at low temperature

increasing thickness of bituminous layers (Fig. 7). Fig. 9 shows a relationship of total horizontal (radial and thermal) stresses and tensile strength reserve at the top and bottom of bituminous layers versus their thickness determined for low temperature conditions for two cases of subbase material.

Instead of examining the relation of a stress versus tensile strength, a tensile strength reserve may be considered (Fig. 10). Negative values of a tensile strength reserve indicates that a material in that particular point would crack. It is interesting to notice that, in a case of cement soil stabilization, tensile strength reserve towards radial stresses under a wheel load (and thereafter total horizontal stresses) shows a significantly irregular relationship exhibiting a minimum when total thickness of bituminous layers is about 6 cm. Positive action of traffic may be concluded on the top of bituminous layers which creates compressive stresses superimposing tensile thermal stresses in thin bituminous layers, it is opposite at the layer's bottom. Value of tensile strength reserve calculated for static load conditions informs whether a material provides a reserve against the specific phenomenon as a stress concentration due to a notch creation and a material fatigue.

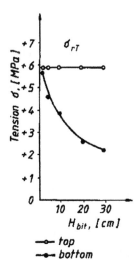

Fig. 8. Thermally induced stresses at the top and bottom of bituminous layers versus their thickness

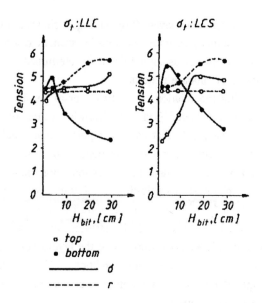

Fig. 9. Total horizontal stresses and tensile strength at the top and bottom of bituminous layers versus their thickness

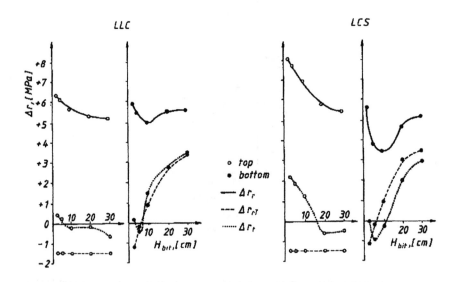

Fig. 10. Tensile strength reserve at the top and bottom of bituminous layers versus their thickness: Δr_r - tensile strength towards radial stresses under a wheel load, Δr_{rT} - tensile strength towards thermal stresses, Δr_t - tensile strength towards total horizontal stresses

4 Conclusions

The analysis of a dualism of crack initiation mechanism indicates that the most appropriate illustration may be as that given on Fig. 11. The mechanism contains three cases depending on the nature of horizontal stress responsible for a crack initiation. Fig. 11 considers a particular case of one winter frosty day and night.

- **Case A** consists in a crack initiation at the bottom of subbase due to action of tensile radial stresses under a wheel load, the stress exceeding the tensile strength of a cement bound material of subbase. The most inconvenient conditions for this case exhibit at relatively high pavement temperature (here it is presented for a winter midday).

- **Case B** consists in a crack initiation at the top of bituminous layers - on pavement's surface. Thermally induced tensile stresses exceed the tensile strength of bituminous layers causing a crack creation. It should be noticed that case B crack initiation does not depend on traffic conditions and pavement construction. It does depend on the initial tensile strength of surface material and, of course, on pavement's temperature. The lower the pavement temperature, the higher possibility of crack initiation. Case B concerns pavement service conditions in a winter midnight assuming no traffic and significant fall in a temperature.

- **Case C**, at last, shows a superposition of load conditions, i.e. thermal stresses exceeding a tensile strength of bituminous material on pavement surface and radial stresses exceeding a tensile strength of cement bound material at the bottom of subbase. Such conditions may perform in a winter morning, when pavement temperature is sufficiently low and a heavy traffic starts. This is the most critical case of crack creation conditions.

The existence of a tensile strength reserve in a bituminous material provides a transfer of specified stresses due to material fatigue or a stress concentration in a notch, when a crack tip was already created. Increase in total thickness of bituminous layers may be an effective, though expensive, method of crack pavement rehabilitation. It may be noticed that considerable tensile strength reserve increment is observed up to a total thickness of bituminous layers of about 20 cm. Thicker layers do not give any more significant growth in tensile strength reserve.

More convenient stress distribution is provided, when subbase is made of a cement soil stabilization than of a lean concrete. Lower stiffness modulus causes that radial tensile stress reaches lower value.

Value of thermally induced stress in bituminous layers, at a sufficiently low temperature, is significantly higher than of that created as a response to a traffic load.

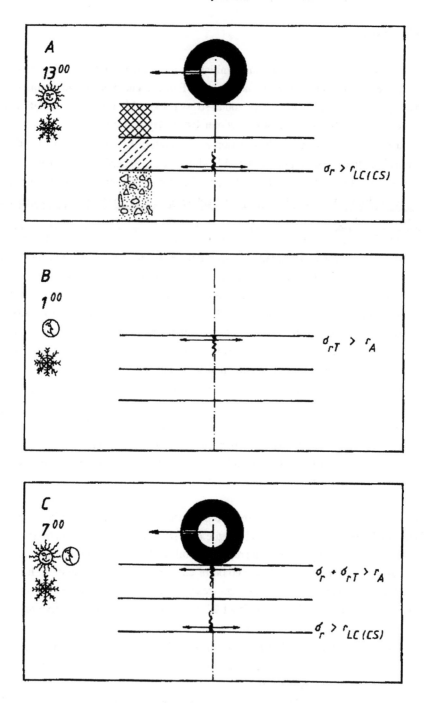

Fig. 11. Pavement crack initiation mechanism depending on the load and environmental conditions

Crack initiation on a bituminous pavement surface does not depend on pavement construction. A creation of crack tip depends on pavement surface temperature and properties of bituminous material. It is believed that the most important role for tensile strength of a bituminous mixture at low temperatures plays bituminous binder. The softer the bitumen is, the lower the failure temperature. Monismith et al [6] also mentioned that bitumen properties have a decisive importance for pavement resistance to low temperature cracking.

Considering the aforementioned crack initiation mechanism it should be noticed that the existence of any microcracks in a bituminous pavement surface, as for example those created during the bituminous layer compaction, may accelerate the crack initiation working as a notch, where the stress concentration takes place. Any other discontinuities in a pavement surface, as for example transverse joints, may play the same role of a crack initiation.

5 References

1. Haas R., Ponniah E.: *Design oriented evaluation of alternatives for reflection cracking through pavement overlay*. International RILEM Conference on Reflective Cracking in Pavements. Université de Liegé. march 1989.
2. Southgate H., Deen R.: *Temperature Distribution Within Asphalt Pavements and Its Relationship to Pavement Deflection*. Highway Research Record 291, 1969.
3. Fabb T.R.J.: *The influence of the mix composition, binder properties and cooling rate on the asphalt cracking at low temperatures*. Proc. AAPT 43, 1974.
4. Arand W.: *Einfluss der Zusammensetzung von Walzasphalt auf das Verhalten bei Kälte*. Strasse und Autobahn 8, 1987.
5. Eulitz H.-J.: *Kälteverhalten von Walzasphalten. Prüftechnische Ansprache und Einfluss kompositioneller Merkmale*. Helf 7, Schriftenreihe: Strassenwesen. Institut für Strassenwesen. Technische Universität Braunschweig.
6. Monismith, C.L., Hicks, R.G., Finn, F.N.: *Evaluation of the Tests for Asphalt-Aggregate Mixtures Which Relate to Field Performance*. Proc. of the 4th International RILEM Symposium. Budapest, 1990. Chapman and Hall.

Reflective cracking control via reinforcing systems: FE modelling of reinforced overlays

A. SCARPAS, A.H. DE BONDT and G. GAARKEUKEN
Delft University of Technology, Delft, Netherlands

Abstract
Overlaying is one of the most popular and cost effective techniques of rehabilitation of cracked pavements. The placing of reinforcement between the overlay and the top layer of the cracked pavement is currently being utilized as a technique for delaying the development of cracks into the overlay. In order to enable the road designer to quantify the contribution of reinforcement in carrying tensile forces across cracks in overlayed pavements, CAPA, a user-friendly, PC based, finite elements system has been developed. By means of CAPA analyses of an actual Dutch pavement profile it is shown that adequately bonded reinforcement can reduce the speed of crack propagation into the overlay and hence prolong the economic lifetime of the pavement.
Keywords: Finite elements, grids, nets, overlay deterioration, pavement rehabilitation, reinforced overlays, wovens.

1 Introduction

The placing of reinforcement between the cracked pavement and the overlay is one of the currently proposed techniques for delaying the development of reflective cracking. From basic principles, the presence of adequately anchored reinforcement enables the transfer of tensile forces across the crack faces, Fig. 1, and, as a result, it enhances the moment carrying capacity of the cracked section. In addition, by limiting crack opening, it reduces the stress concentration at the crack tip and hence the speed of crack propagation.

In addition to enhancing the flexural response of a cracked section, reinforcement can also contribute significantly to the shear response by enabling the development of the aggregate interlock mechanism, Fig. 2.

Reflective Cracking in Pavements. Edited by L. Francken, E. Beuving and A.A.A. Molenaar. © 1996 RILEM. Published by E & FN Spon, 2–6 Boundary Row, London, SE1 8HN. ISBN 0 419 22260 X.

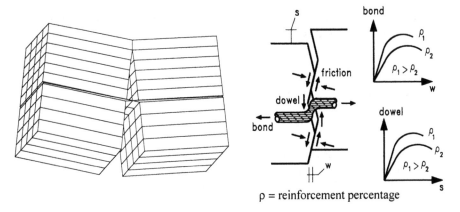

ρ = reinforcement percentage

Fig. 1 Flexural action of reinforcement

Fig. 2 Reinforcement enhances the shear response of a cracked section

Nevertheless, in the recent past, concerns have been raised as to the effectiveness of reinforcement in bituminous overlays. By enabling the detailed study of the local phenomena and mechanisms that may affect the interaction of reinforcement with the surrounding pavement materials, the finite elements method can assist the designer in his evaluation of different reinforcing systems.

2 Micro-mechanical modelling of bond

Because of the wide range of products which are currently used for overlay reinforcing purposes, micro-mechanical analyses were undertaken with CAPA [1] in order to determine the mechanisms contributing to the development of bond in each category of products. Due to the detailed nature of the investigation, the analyses concentrated on a small section of the pavement. Both the asphaltic material and the reinforcing strands were modelled by means of ordinary finite elements while interface elements were utilized to simulate the bond regions between the reinforcing strands and the surrounding material, Fig. 3.

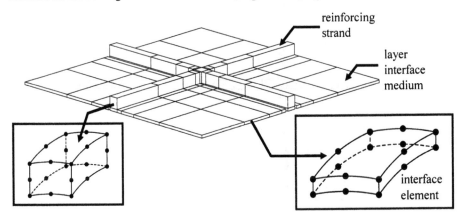

Fig. 3 Micro-modelling of bond between reinforcing strands and asphaltic material

For an interface element, the constitutive relation associating local stresses to relative local displacements can be defined as :

$$
\left\{ \begin{array}{c} \tau_{tt} \\ \tau_{ss} \\ \sigma \end{array} \right\} = \left[\begin{array}{ccc} D_{tt} & 0 & 0 \\ 0 & D_{ss} & 0 \\ 0 & 0 & D_{nn} \end{array} \right] \cdot \left\{ \begin{array}{c} s_{tt} \\ s_{ss} \\ w \end{array} \right\} \tag{1}
$$

By properly adjusting the shear stiffness terms D_{tt} and D_{ss} of the interface elements connecting the reinforcing strands to each other in orthogonal directions, various junction characteristics can be simulated. In a similar manner, the adhesion characteristics of the strands with the surrounding medium can be specified.

On the basis of the strength and stiffness characteristics of the junctions connecting the reinforcing strands in orthogonal directions, three main categories of reinforcing products were identified : **wovens**, in which no significant forces can develop at the junctions, **grids**, in which the strength and the stiffness of the junctions are at least equal, if not higher, to those of the individual strands and **nets** in which the junctions have adequate strength but not rotational stiffness.

Fig. 4 indicates the response of a woven when a tensile force is applied at one of the strands. In this case, forces between strands spanning in orthogonal directions can be only transmitted at the junctions by means of friction. As a result only a minimum amount of the applied pullout force is resisted by the engagement of strands spanning in the cross direction. Most if not all of the applied force is dissipated in the surrounding asphaltic medium by adhesion and mechanical friction along the loaded strand. Depending on the situation, this type of force transfer may place severe demands on the strength and stiffness characteristics of the bonding medium resulting to excessive slip and eventual loss of load carrying capacity.

In case of grids and given the appropriate construction conditions, because of the rigid joining of the strands in the orthogonal directions, pullout forces in one direction can be transmitted through the junctions to the strands spanning in the cross direction and engage them

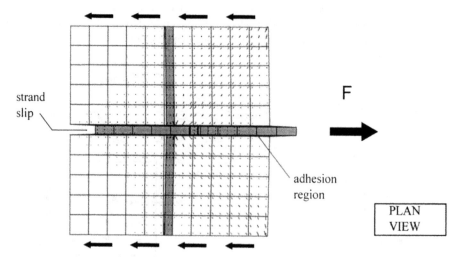

Fig. 4 CAPA simulation of bond between a woven and the surrounding medium

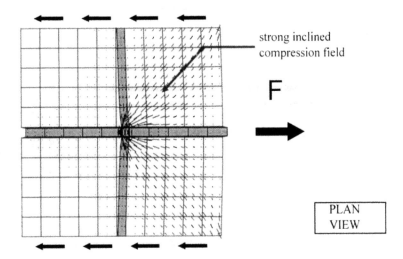

Fig. 5 CAPA simulation of bond between a grid and the surrounding medium

in resisting the imposed pull-out force by bearing against the surrounding medium. The resulting field of inclined compressive stresses, Fig. 5, constitutes a very efficient and definitely stiffer anchoring mechanism in comparison to the chemical adhesion and mechanical friction mechanism of the previous case. Nevertheless, before it can be relied upon, care must be taken during the paving process to ensure that the reinforcing layer is properly contained within the body of the surrounding asphaltic material.

3 Finite elements simulation of reinforced overlayed pavements

In actual finite elements analyses of pavements, micro-mechanical simulation of the interaction of the reinforcing strands with the surrounding medium would be impractical. Instead, average values of bond stiffness, representative of both the adhesion mechanism and the bearing mechanism (if any) can be determined by means of laboratory pull-out tests.

Once these are available then, the finite elements arrangement shown in the schematic of Fig. 6 can be utilized for the simulation of a cracked reinforced overlayed pavement.

Ordinary quadrilateral finite elements can be used for modelling the geometry of the pavement layers and bar type elements for modelling the reinforcement (see inset in Fig. 6). For the simulation of bond between the reinforcement and the surrounding pavement materials interface elements can be utilized. By varying the transverse stiffness various types of bond can be specified.

Because of construction difficulties during paving, the bond between the reinforcement and the top of the old cracked pavement may have completely different characteristics from that between the reinforcement and the overlay. By means of the arrangement of Fig. 6, different bond characteristics can be assigned to the interfaces on either side of the reinforcement.

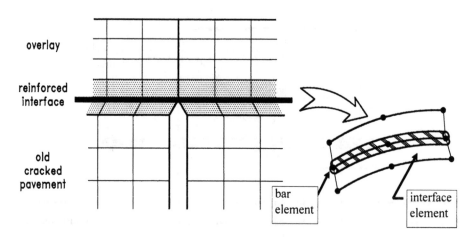

Fig. 6 FEM simulation of bond

4 Calculation of the lifetime of a reinforced asphalt concrete overlay

In a companion paper [1], a methodology was presented by means of which the lifetime of an asphaltic overlay can be computed. On the basis of the discrete cracking procedure incorporated in CAPA and by utilization of Paris' and Miner's rules, the initiation and propagation of cracks into the overlay can be traced. The methodology accounts also for the effects of debonding that may occur at the interface region between the overlay and the surface of the old pavement.

This methodology can be utilized also for the computation of the lifetime of a reinforced asphaltic overlay. The finite elements arrangement of Fig. 6 enables separate checks to be made for debonding between: (i) reinforcement and the surface of the old pavement and/or (ii) reinforcement and the bottom of the overlay. In the following and in order to demonstrate the application of CAPA in pavement rehabilitation studies, the case of an actual Dutch highway pavement, the A50 highway in Friesland, will be considered.

5 Case study

A schematic of the cracked pavement profile and the reinforced overlay is shown in Fig. 7. The influence on overall pavement response of the characteristics of the interface region between the overlay and the surface of the old pavement will be investigated by considering two different cases : (i) a low strength interface material, leading to local bond failure and hence debonding of reinforcement and, (ii) a high strength interface material in which debonding is prevented. For the latter case, the influence of bond stiffness on overlay performance will be demonstrated. Details of the analyses can be found in [2].

CAPA can handle any combination of traffic and/or environmental loads like temperature, subsidence, frost heaving etc. However, in the following, the case of traffic loading will be considered only. The passage of a wheel load is simulated by placing a distributed load of 0.707 MPa at different distances from the existing transverse crack axis.

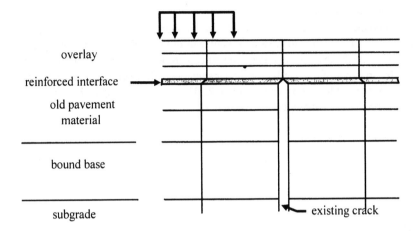

overlay

reinforced interface

old pavement
material

bound base

subgrade

existing crack

Fig. 7 A50 pavement profile

For a given crack length, the equivalent stress intensity factor K_{eq} is calculated for the various load positions. During the passage of a wheel, the interface region between the old pavement layer and the overlay is subjected to shear and normal stresses. The exact state of stress depends, among others, on the position of the load and the extent of cracking. Before the stress intensity factors corresponding to a given configuration can be computed, the normal stiffness D_{nn} and/or the shear stiffness D_{tt} of the interface elements must be adjusted so as to reflect the local physical conditions of the interface region over which they span (e.g. $D_{nn} = D_{tt} = 0$ for elements in which tensile fracture has occurred or $D_{tt} = 0$ for shear failure etc.). The characteristics of the pavement constituent layers are listed in Table 1.

Table 1 Pavement layers characteristics

	thickness [mm]	E [MPa]	ν [-]	D_{tt} $[(^N/_{mm})/mm^2]$	D_{nn} $[(^N/_{mm})/mm^2]$	EA [N/mm]
overlay	50	5500	0.35	-	-	-
top interface	1	-	-	5.61	5500	-
reinforcement	-	-	-	-	-	25000
bottom interface	1	-	-	1.0	5500	-
old pavement	200	3500	0.35	-	-	-
bound base	400	10000	0.2	-	-	-
subgrade	200	100	0.35	-	-	-

The shear stiffness characteristics D_{tt} of the interface region were chosen so as to reflect : (a) the good bond conditions typically resulting between the reinforcement and the overlay -due to embedding of the reinforcement at construction time within the surrounding hot asphalt- and, (b) the less favourable conditions between reinforcement and the surface of the existing pavement (i.e. bottom interface).

5.1 Weak interface material

In a companion paper [2] it was demonstrated that debonding at the interface between the overlay and the surface of the old pavement may lead to the initiation and propagation of secondary cracks into the overlay. In order to investigate the effectiveness of reinforcement in maintaining the integrity of the overlay even after some local debonding has occured, CAPA analyses were performed for the pavement of Fig. 7. The specified strength characteristics of the overlay and the interface materials are listed in Table 2.

Table 2 Strength characterisics

	σ_f [N/mm^2]	τ_f [N/mm^2]
overlay	1.0	5.0
interface	1.0	2.0

Because of the stress concentration introduced at the bottom of the overlay due to the presence of the existing crack in the old pavement layers, crack propagation along the axis of the existing crack was observed first. This crack is indicated as primary crack in Fig. 8. However, soon after, it was accompanied by failure and hence debonding of the adjacent interface region. The resulting double flexure mode of deformation (e.g. see Fig. 8) led to tensile fracture at the bottom of the overlay and the generation of a new crack, indicated as secondary crack in Fig. 8, at a short distance from the primary crack axis. In subsequent increments, this new crack became dominant and reached to the top of the overlay while the primary crack arrested for the rest of the loading history.

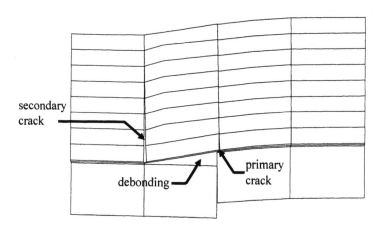

Fig. 8 Crack development in the overlay

The distribution of forces in the reinforcement (tension positive) along the interface region is shown in Fig. 9 and Fig. 10 for two different load positions and for different secondary crack propagation increments. In CAPA a **crack propagation increment** is defined as the number of finite element rows through which the simulated crack has propagated. In both simulated cases, the presence of the secondary crack at a distance of 75 mm from the primary crack axis

reinforced A50

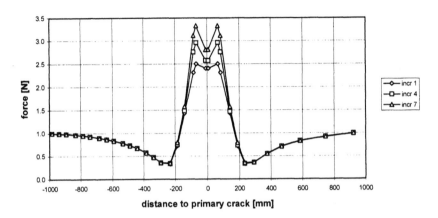

Fig. 9 Distribution of forces in the reinforcement for traffic load directly above
the primary crack axis

reinforced A50

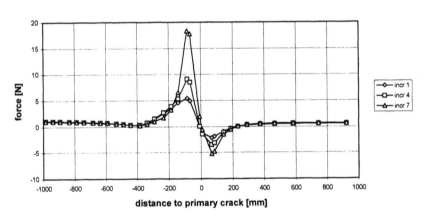

Fig. 10 Distribution of forces in the reinforcement for traffic load directly next
to the primary crack axis

can be identified by the high concentration of tensile actions in the reinforcement at this location.
It must be remembered that because of symmetry of actions with respect to the axis of the
primary crack, two symmetrically located cracks develop in the finite elements mesh. CAPA
allows any number of cracks to develop or propagate concurrently.

Propagation of the secondary crack into the body of the overlay results to a change of the
geometric characteristics of the structure and hence a gradual redistribution of actions. As a
consequence, the tensile forces in the reinforcement vary with crack length.

By enabling the determination of the magnitude and the distribution of the forces applied
on the reinforcement, the finite elements method enables the designer to quantify the contribution of a specific reinforcing system and hence to optimize his pavement design by choosing
between the available alternatives.

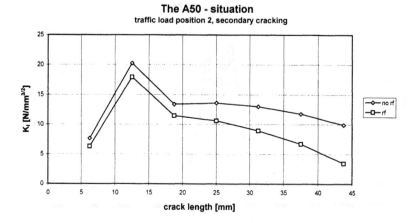

Fig. 11 The influence of secondary crack length on K_I

The contribution of reinforcement in prolonging overlay life can be justified on the basis of Fig. 11 in which the distributions of mode I stress intensity factors K_I at the tip of the secondary crack are compared for the reinforced overlay case studied in this contribution (indicated as "rf" in Fig. 11) and the unreinforced overlay case examined in [1] (indicated as "no rf" in Fig. 11). From this figure it can be concluded that, by preventing unrestrained crack opening, adequately anchored reinforcement can reduce, if not annihilate, the energy (as represented by K_I) available for crack extension and hence, because of Paris' law :

$$\frac{dc}{dN} = A \cdot \left(K_{eq}\right)^n \tag{2}$$

reduce the speed of crack propagation dc/dN into the overlay. In Eq. 2 A and n are overlay mix material parameters and K_{eq} represents the equivalent stress intensity factor for mixed mode crack propagation conditions. Values of $A=10^{-7}$ and $n=4$ have been utilized.

As shown in Fig. 12, in case of the A50 pavement, reduction of the speed of crack propagation due to the presence of reinforcement resulted to a significant increase in the life of the overlay. Details can be found in Gaarkeuken et al. [3].

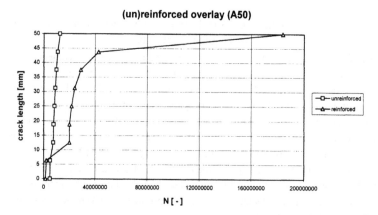

Fig. 12 Influence of the presence of reinforcement on overlay life

5.2 Strong interface material

The A50 pavement profile was also utilized in order to investigate the influence of reinforcement bond stiffness on overlay life. The strength characteristics of both the interface material and the overlay were chosen so as to exclude the development of secondary cracking. As shown in Fig. 13, a wide range of reinforcing products was simulated. The simulated bond stiffnesses D_{tt} span the whole range of bond possibilities.

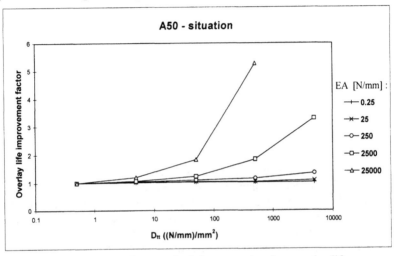

Fig. 13 Influence of reinforcement bond on overlay life improvement factor

As expected, an improvement of the bond characteristics enhances the capability of the reinforcement to limit crack opening and this, from the discussion in Section 5.1, results to an extension of the life of the overlay.

In general it must be realized that the beneficial effect of reinforcement is case dependent and that the pavement structural and material parameters have direct influence on the overlay life improvement factor.

7 Acknowledgements

Development of the CAPA system has been made possible by the financial support of the Netherlands Technology Foundation (STW).

8 References

1. de Bondt, A.H. (1995) *Theoretical Analysis of Reinforcement Pullout*, Report 7-95-203-16, Division of Road and Railway Construction, Faculty of Civil Engineering, TU-Delft, The Netherlands.
2. Scarpas, A., de Bondt, A.H., Molenaar, A.A.A. and Gaarkeuken, G. (1996) Finite elements modelling of cracking in pavements, in *Proceedings of the 3rd RILEM Conference on Reflective Cracking in Pavements*, (ed. L. Francken et al.), E & FN Spon, London.
3. Gaarkeuken, G., Scarpas A. and de Bondt, A.H. (1996) *The Causes and Consequences of Secondary Cracking*, Report 7-96-203-23, Division of Road and Railway Construction, Faculty of Civil Engineering, TU-Delft, The Netherlands.

Reflective cracking control via stress-relieving systems

A.H. DE BONDT and A. SCARPAS
Road and Railroad Research Laboratory, Faculty of Civil Engineering, Delft University of Technology, Netherlands

Abstract
The purpose of stress-relieving systems is a reduction of the shear stiffness of the overlay / existing pavement interface. These systems should enable the occurrence of considerable movements of the old layers without causing high stresses in the overlay. Stress-relieving systems mainly consist of a thick layer of (modified) bitumen sprayed onto the old surface before the overlay mix is layed down. A constitutive model is presented, capable of showing the effect of thickness and material properties of the applied bitumen as well as pavement surface geometry, on the stiffness value which can be "offered" by a stress-relieving system. A thermal overlay design example showed that, under specific circumstances and within the scope of the one-dimensional bar model utilized, a reduction of the interface shear stiffness by a factor 10 implies a factor 2 increase in lifetime of a 140 mm thick bituminous surfacing which is placed on top of a concrete slab.
Keywords: Bitumen, Crack Opening, Interface Shear, Stress-Relief, Temperature, Textile, Thermal Overlay Design, Transverse Cracking.

1 Introduction

Reflective cracking is a worldwide pavement problem which has generated a lot of questions and scientific interest. One of the oldest ways of treating the problem is the application of a so-called "stress-relieving system" (a layer of bitumen with or without a textile to ease construction), which enables sliding between overlay and old surface to occur without creating huge shear stresses.

It is clear that a stress-relieving system has to be able to provide a low shear stiffness of the interface between old (cracked) pavement and overlay; even for a considerable lateral displacement and after a great number of load repetitions. In the next section a constitutive law for a pavement layer interface is presented.

Reflective Cracking in Pavements. Edited by L. Francken, E. Beuving and A.A.A. Molenaar. © 1996 RILEM. Published by E & FN Spon, 2–6 Boundary Row, London, SE1 8HN. ISBN 0 419 22260 X.

2. Constitutive Law of a Pavement Layer Interface

The surface of a pavement structure is not flat; it shows undulations. Texture depth measurements showed typical values for the Mean Profile Depth (= MPD) ranging between 0.7 and 1.0 mm for different Dutch stone mastic asphaltic mixes (maximum size of the stones 6, respectively 11 mm) and 0.7 mm for a Dutch dense asphalt concrete with a maximum aggregate size of 16 mm [1]. The MPD is defined as the voids volume at the pavement surface divided by the area. It can be derived that the texture depth is twice the recorded MPD-value (see figure 1).

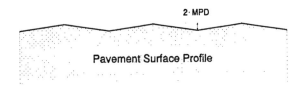

Figure 1   Sketch of Pavement Surface Profile

Figure 2 shows a sketch of a pavement layer interface loaded by a shear force S as well as a normal force T (tension positive). The undulation of the existing pavement is characterized by an angle δ. The bitumen thickness is denoted by t_0.

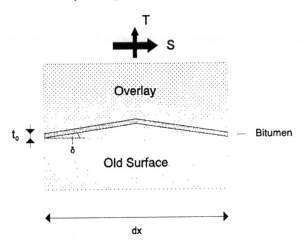

Figure 2   External Forces acting on a Pavement Layer Interface

It is obvious that extension of the bitumen occurs due to the presence of the normal force T. The increase of the bitumen thickness (Δt) can be determined from:

$$\Delta t = \frac{\sigma_N \cdot t_0}{E_{bit}}$$

in which σ_N is the normal stress ($= T/dx$) and E_{bit} represents the modulus of the bitumen. The latter parameter is rate and temperature dependent. All in all, it implies that due to the action of the normal force T the bitumen thickness has increased to the value t_0^* ($= t_0 + \Delta t$).

Of course, the interface also shows lateral movement due to the presence of the shear force S. Figure 3 presents a sketch of the shear deformation of the bitumen which is located between the overlay and the old surface.

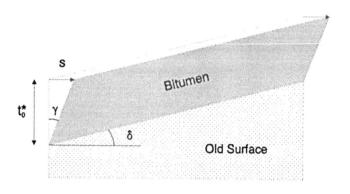

Figure 3   Shear Deformation of Bitumen in Pavement Layer Interface

It can be observed that shear deformation is allowed to occur as long as the angle γ is smaller than $(90°-\delta)$. Via the formula $\tau = G_{bit} \cdot \gamma$ the following relationship between shear stress τ and lateral displacement s can be derived:

$$\tau = \frac{1}{3} E_{bit} \arctan \left(\frac{s}{t_0 \left(1 + \dfrac{\sigma_N}{E_{bit}} \right)} \right)$$

In this equation the shear modulus of the bitumen G_{bit} has been replaced by $E_{bit}/3$; this is possible, because the Poisson's ratio of pure bitumen is equal to 0.5. The relationship is valid for lateral displacements s smaller than s_0^*, where:

$$s_0^* = \frac{t_0 \left(1 + \dfrac{\sigma_N}{E_{bit}} \right)}{\tan (\delta)}$$

If s is equal to s_0^* the protruding teeth of overlay and old surface touch.

Figure 4 presents τ versus s for several transverse stresses σ_N.

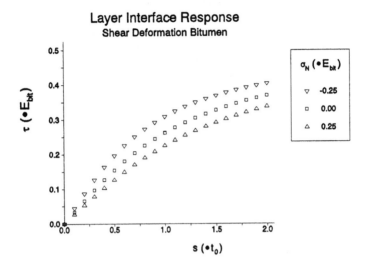

Figure 4 Shear Stress versus Lateral Displacement for Several Transverse Stresses

As expected, in compression a stiffer response is found than in tension. It can also be seen that by increasing the bitumen application rate ($\rightarrow t_0$ larger), it can be achieved, that a specific lateral displacement generates smaller stresses. Figure 5 shows that the lateral displacement at the moment of touching between overlay and old surface increases with decreasing undulation angle δ.

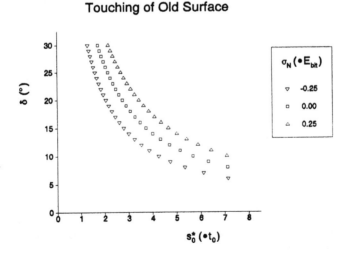

Figure 5 Lateral Displacement at Touching for Several Transverse Stresses

It is clear that the "free" sliding phase of the interface ends when the lateral displacement has generated an angle γ equal to $(90°-\delta)$. At this moment the protruding teeth of overlay and old pavement surface touch; the bitumen at the conract area is squeezed out. Further shear deformation implies that sliding occurs along a fixed angle $(= \delta)$ and "dry" friction is developed along the surface plane. Figure 6 gives the forces which are active in this "frictional" phase.

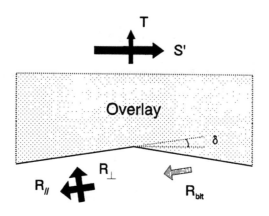

Figure 6   Forces active during "Frictional" Phase

If it is assumed that friction along the surface plane occurs according to the well-known dry friction rule $R_{//} = \mu \cdot R_{\perp}$, where μ is the friction coefficient, then the shear stress - lateral displacement relationship can be described as follows:

$$\tau = \frac{E_{bit}(s - s_0^*)}{2 t_0 \cos \delta (\cos \delta - \mu \sin \delta)} -$$

$$\left(\frac{\sin \delta + \mu \cos \delta}{\cos \delta - \mu \sin \delta} \right) \sigma_N + \frac{1}{2} G_{bit} \left(\frac{\pi}{2} - \delta \right)$$

Figure 7 presents for a specific undulation angle δ, friction coefficient μ and confining pressure σ_N, the shear stress τ versus the lateral displacement s for several bitumen application rates t_0.

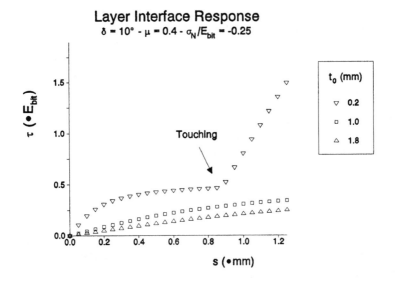

Figure 7   Effect of Bitumen Application Rate on Layer Interface Response

Interesting is that only for the low application rate the effect of touching between overlay and old surface is visible within the range of lateral displacements shown.

From this section it can be concluded that by using the stress-strain curves of the bitumen utilized, in combination with the geometric characteristics of the pavement surface, it is possible to determine the shear stiffness which can be "offered" by a stress-relieving system (of course, a check by means of proper interface testing is always recommended). It has become clear that the interface shear stiffness provided by a stress-relieving system decreases with increasing thickness and decreasing stiffness of the applied bitumen. The bitumen has to have a large elastic range to prevent irrecoverable lateral displacement to take place, because the latter would imply friction development (overlay teeth then touch old surface teeth). In case of the use of a textile, it should have a high ultimate strain to prevent potential disturbance of the "free" sliding process of the stress-relieving system [2].

In the next section the mechanism of temperature induced pavement movements is discussed. The effect of these type of movements is often treated via the use of stress-relieving systems.

3. Mechanism of Temperature Induced Pavement Movements

Virgin pavements, consisting of materials with cementitious components, often show soon after construction transverse cracks. This is due to the fact that temperature drops and frictional restraint at the support interface causes high tensile stresses [3]. The temperature drop can be quite high, because chemical induced heating can be followed by a cold period. All in all, it is clear that even long sections of pavement layers without joints gradually transform into slabs.

The presence of cracks/joints in pavements can be a problem if an overlay is required to improve e.g. serviceability. This because if the overlayed cracked/ jointed pavement structure is subjected to temperature drops [4] the slabs perform a plying action to the overlay. This action can have the effect that the cracks/joints in the existing pavement grow fast upwards in and through the overlay. Finally, the cracks reach the pavement surface and then the ingress of water is allowed. This phenomenon has become one of the classical pavement engineering problems.

Overlayed slabs which are subjected to temperature drops are restrained in their movements by friction at the existing slab - support interface and at the overlay - existing slab interface [3]. The relative slab size $\lambda*L$ determines the magnitude of the axial force which will be developed in the centre of the existing slab. If the shear stiffness of the overlay - existing slab interface is much higher than the

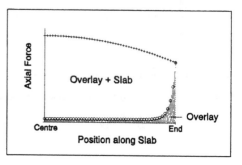

shear stiffness of the existing slab - support interface, then reasons of equilibrium cause that the axial force in the centre of the existing slab is transferred to the overlay right above the crack/joint (see adjacent figure). This implies that huge stresses occur at this location, because of the high axial stiffness of the existing slab and the small thickness of the overlay.

Figure 8 shows a sketch of the extremes which can occur when an overlayed slab suffers from temperature drops. The effect of friction between slab 1 and the support is neglected (→ a worst case scenario for the full 'friction' situation).

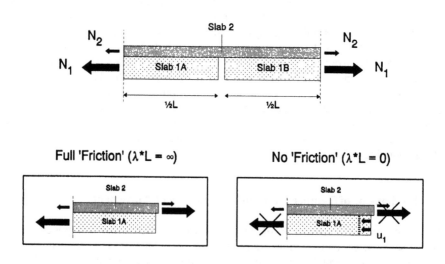

Figure 8   Extreme Situations for Overlayed Cracked/Jointed
Pavements Subjected to a Temperature Drop [3]

De Bondt [3] derived a model, based on elementary bar theory (see figure 9), which enables to get insight in the effects of frictional shear stiffness resistance at the support (c), slab size (L), axial stiffness of slab ($E_1 \cdot h_1$) as well as overlay ($E_2 \cdot h_2$), coefficient of thermal contraction of slab (α_1) as well as overlay (α_2), size of the temperature drop in slab (ΔT_1) as well as overlay (ΔT_2) and last but not least the shear stiffness of the interface between overlay and slab (k_{if}). This model can be utilized for a first evaluation of overlays with or without stress-relieving systems.

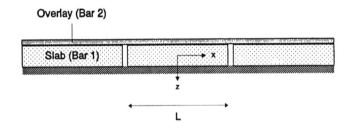

<u>Figure 9</u>   Model of Overlay on Slabs (= Cracked/Jointed Pavement)

The overlay stress (σ_2) above the crack/joint ($x = \frac{1}{2}L$) can be obtained from:

$$\sigma_{2,x=\frac{1}{2}L} = -E_2\,\alpha_2\,\Delta T_2\,\left(1 - \frac{ab\,\alpha_1\,\Delta T_1\,(\lambda_2 C_1^- C_2^+ - \lambda_1 C_2^- C_1^+)}{\alpha_2\,\Delta T_2\,(a\,\lambda_2 C_1^- C_2^+ - b\,\lambda_1 C_1^+ C_2^-)}\right)$$

p : $(c + k_{if})/(E_1 h_1)$	λ_1 : $\sqrt{((p + r)/2 + s)}$
q : $k_{if}/(E_1 h_1)$	λ_2 : $\sqrt{((p + r)/2 - s)}$
r : $k_{if}/(E_2 h_2)$	C_1^- : $\exp(\lambda_1 \cdot \frac{1}{2}L) - \exp(-\lambda_1 \cdot \frac{1}{2}L)$
s : $\sqrt{(\frac{1}{4}(p - r)^2 + qr)}$	C_1^+ : $\exp(\lambda_1 \cdot \frac{1}{2}L) + \exp(-\lambda_1 \cdot \frac{1}{2}L)$
a : $-r/((p - r)/2 + s)$	C_2^- : $\exp(\lambda_2 \cdot \frac{1}{2}L) - \exp(-\lambda_2 \cdot \frac{1}{2}L)$
b : $-r/((p - r)/2 - s)$	C_2^+ : $\exp(\lambda_2 \cdot \frac{1}{2}L) + \exp(-\lambda_2 \cdot \frac{1}{2}L)$

In the next section a thermal overlay design example is given.

4. Thermal Overlay Design Example

The classical problem of a bituminous surfacing on top of a slab, which is subjected to a surface temperature drop (ΔT_0), has been studied. The question is, which surfacing thickness h_2 is needed to ensure that a large number of temperature variations can take place before failure occurs. This classical case has become even more important the past years, because of the widespread use of recycled materials in cement treated bases in secondary roads. For these type of roads the temperature mechanism is often the predominant factor for surfacing (overlay) thickness design.

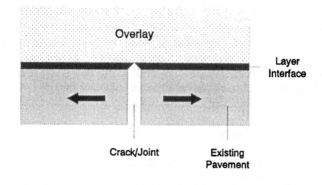

Figure 10   Classical Reflective Cracking Problem

First of all, temperature drop data is needed. Figure 11 shows values for a specific thermal diffusivity a, obtained from the equation given by Jayawickrama, et al. [5].

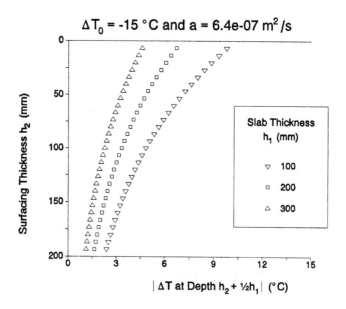

Figure 11   Computed Temperature Drops at Middle Height Slab

Using the temperature drop in surfacing as well as slab the overlay stress above the crack/joint can be determined (see figure 12). The temperature drops at middle height of the surfacing ($\frac{1}{2} \cdot h_2$), respectively the slab ($h_1 + \frac{1}{2} \cdot h_2$) were used. In the example c was set equal to 0.005 (N/mm)/mm, L to 10 m, α_1 to 1.1e-05 -/°C, E_1 to 10000 MPa, α_2 to 2.8e-05 -/°C and E_2 to 1000 MPa.

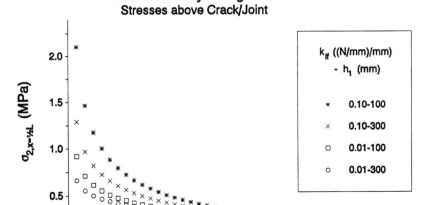

<u>Figure 12</u>   Computed Overlay Stresses

It can be seen that, as expected, the overlay stress increases with decreasing surfacing thickness h_2 and increasing layer interface shear stiffness k_{if}. It is interesting to see that the largest effect of a change of these parameters occurs in the lower thickness region. Values for k_{if} can be obtained from layer interface shear testing [2]. Of course, both monotonic as well as cyclic tests should be carried out for a range of temperatures and displacement rates. The displacement rate to which the interface will be subjected in the field, can be obtained by dividing the magnitude of the opening of the crack/joint (Δu) by the time which is required for the opening process [6]. The magnitude of the opening can be measured before overlaying by using nails located at both edges [7] or by means of the following equation if the generated friction at the slab - support interface can be neglected:

$$\Delta u = \alpha_{slab} \left| \Delta T_{slab} \right| \frac{1}{2} (L_A + L_B)$$

where A is the slab in front of the crack/joint and B the slab beyond the crack/joint. The procedure of using thermal movements determined before overlaying is allowed, because an asphaltic overlay hardly restrains thermal slab movements; it only reduces the temperature drop in the slab (insulation) [3].

Figures 13 and 14 (→ detail near origin) show the relative life of the overlay for the analyzed cases. The lifetime was obtained by using a fourth power damage law. It is clear that a softer layer interface (→ k_{if} low) is beneficial.

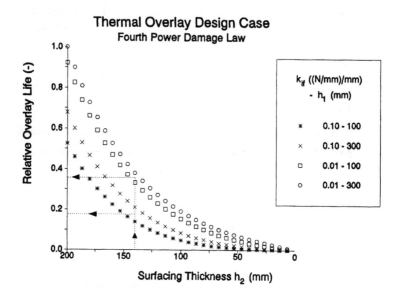

Figure 13   Computed Relative Overlay Life

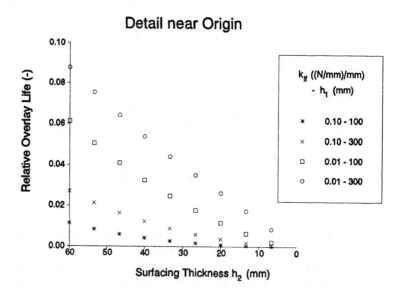

Figure 14   Detail near Origin

5. Conclusion

From information on applied bitumen and pavement surface geometry it is possible to determine the interface shear stiffness generated by a stress-relieving system; design equations then allow quantification of the effect of these systems in the field.

Acknowledgement

Both the Delft Research Project and this publication have been made possible by the financial support from the Netherlands Technology Foundation (STW).

References

1. Bennis, T. (1994) *Skid Resistance Measurements on Dutch Test Sections.* Internal Report, Ministry of Public Works and Transport, the Netherlands.
2. de Bondt, A.H. (1995) *Properties of Anti-Reflective Cracking Systems.* Report 7-95-203-22, Faculty of Civil Engineering, Delft University of Technology.
3. de Bondt, A.H. (1995) *Effect of Temperatures on Pavement Response.* Report 7-95-203-13, Faculty of Civil Engineering, Delft University of Technology.
4. de Bondt, A.H. and Steenvoorden, M.P. (1995) *Theoretical Analysis of Temperatures in Pavements.* Report 7-95-203-21, Road and Railroad Research Laboratory, Delft University of Technology, Delft, the Netherlands.
5. Jayawickrama, P.W., Smith, R.E., Lytton, R.L. and Tirado, M.R. (1987) *Development of Asphalt Concrete Overlay Design Equations.* Texas Transportation Institute, Texas, USA.
6. de Bondt, A.H. and Saathof, L.E.B. (1993) *Movements of a Semi-Rigid Pavement Structure.* RILEM Conference on Reflective Cracking, Liège.
7. de Bondt, A.H. and Steenvoorden, M.P. (1995) *Reinforcement Test Sections A50 (A6) Friesland.* Report 7-95-209-21, Road and Railroad Research Laboratory, Delft University of Technology, Delft, the Netherlands.

Ukrainian experience of system retarding reflective cracking

B. RADOVSKY, V.V. MOZGOVOY, I.P. GAMELYAK,
O.E. TSEKHANSKY, V.V. KOSTRITSKY and
O.G. OSTROVERHY
Transportation University of Ukraine, Ukraine
F.P. GONCHARENKO
Ministry of Road Construction of Ukraine, Kiev, Ukraine

Abstract

Problems related to the formation of reflective cracking of asphalt concrete pavements in Ukraine are tasked.

The experience goes back over 20 years of improving the stability of pavements against the formation of cracks during design, construction and maintenance of roads.

The technical and economical aspects of the effect of different methods of improving crack-resistance of asphalt-pavements are analyzed.

Impressive results are obtained from rational combination of constructive, technological and the regulation of the parameters of materials.

The use of dispersed reinforcement, polymers and crack obstructing layers effectively help in reducing the formation of cracks.
Keywords: Crack obstructing layers, dispersed reinforcement, formation of cracks, polymers.

1 Problems connected with the formation of reflective cracking in asphalt pavements in Ukraine

Ukraine is a great country in the center of Europe occupying an area of 603.7 thousand square km with population of 52,2441 million. Density of road network is 285 km for a thousand sq. km. Ukrainian road network (excluding urban roads) consist of 172 thousand km, of this 94,4% are roads with hard surface. Presently the road network is not being expanded and main resources are used in the maintenance and repairs.

Reflective Cracking in Pavements. Edited by L. Francken, E. Beuving and A.A.A. Molenaar. © 1996 RILEM. Published by E & FN Spon, 2–6 Boundary Row, London, SE1 8HN. ISBN 0 419 22260 X.

Taking into consideration the integration of Ukraine to the economy of Europe there appears a trend of wider use of bases treated with binder materials inplace of untreated crushed and sand stone. It is considered that treated bases with binder materials increase the bearing capacity of pavements in general. The analysis of Ukrainian experience confirms the existence of two different ways of the design of pavements with treated bases.

In the first method, pavements with thick bases of asphalt concrete and other materials of organic nature are widely used [1]. These type of bases have a lot of advantages: high resistance to thermal cracks due to the closeness of the thermomechanical and thermophysical properties of the materials of the base and the surface; little stone content due to high bending stiffness(rigidity) of the base; structure technology due to the few number of structural layers. However the disadvantage of such method is the deficiency of petroleum bitumen in Ukraine.

The second method appeared due to the availability of developed level of cement industry in the country [2] and is expressed by trying to reduce to the minimum the thickness of black surface instead of using bases from stones and soils treated with mineral binders. In this case, the increase in bending stiffness is achieved due to high modulus of stiffness of the treated base that allows the reduction in thickness of pavement and economy of bitumen. However, asphalt concrete pavements with bases consisting of inorganic binders possess little resistance to thermal cracks.

Experience has shown that the main disadvantage of strengthening existing pavements of cement concrete, reinforced concrete slabs and flexible pavements having developed network of cracks with asphalt concrete is the formation of refective cracks after 1-4 years in the new pavement. The sealing of these cracks is a very complex and wasteful procedure.

Reflective cracks are one of the most widespread types of destruction of asphalt concrete pavements. Their formation lead to the weakening of the structure of the whole pavement. Due to the breach of the entity of the pavement its distribution capability is reduced which causes over strain of sublayers and base soil in the area of the cracks and lead to their rapid destruction under the action of transports. Further destruction is intensified by the penetration of precipitates through the cracks. Reflective cracks are one of the major reasons of the reduction of the durability of asphalt concrete pavements and airdrome surfaces in general. With the formation of thermal cracks, the evenness of the surface is gradually worsened, there appears the danger of "break off" of the destroyed particles of the pavement surface, which considerably reduces comfortability, leads to the increase in transport expenses and reduces traffic safety. Thermal cracks in asphalt concrete pavements require a great deal of resources for their repairs.

2 Analysis of the experience of increasing the stability of pavement against the formation of reflective cracks

In the seventies a trial had been made in the Ukraine to analyze national and foreign experience of asphalt concrete pavements construction on cracked-blockbase pavements. Researches also had been organized aimed at developing practical procedures for increasing crack resistance of pavements. Had been developed for this purpose method of evaluating crack resistance of a pavement, the features of which are as follows.

The pavement is considered as a visco-elastic layer, rheological properties and strength of which depend on temperature. The relationships arrived at are used for determination of stress in pavement and limiting condition depending on relaxational and thermomechanical properties of materials, thicknesses of pavement and base, lengths of block-base and distance between the blocks, coefficient of friction between pavement and base, temperature changes in time etc [3-7].

The influence of different factors on the formation of reflective cracks was analyzed. Practical recommendations were suggested based on the analyzes carried out.

One of the ways of improving thermal cracking resistance of pavement is to regulate the composition and properties of asphalt mixture. Apart from familiar methods (polymer additives, dispersed reinforcement with synthetic fibres, reducing the difference between the coefficient of linear deformation of bitumen and crushed stone) it is expedient to call for optimal bitumen content as to thermal cracking resistance, use bitumens with PI varying from -0,5 to 0,5, increase adhesion between the mineral material and bitumen at low temperature of asphalt concrete (particularly using surface-active agents) [4, 5, 8].

Another way of reducing the danger of thermal cracks occurrence on asphalt concrete pavement is by means of judicious pavement design. In this sense it is expedient to pave crack-interrupting interlayers (granular, elastic) between the asphalt pavement and crack-blocked base, provide friction between the base and the underlying layer, increase the length of section with decreased friction between asphalt pavement and the cracked-blocked base near the joint, decrease the Young's modulus of the base and the length of its blocks [5-7].

The third way of reducing the danger of thermal cracking of asphalt pavements is by implementing a suitable engineering measure: paving the cement concrete base at low air temperature, providing the lowest temperature of the asphalt concrete by the end of compacting, hindering the cooling of asphalt concrete by means of temporary heat-proofing, laid on the pavement, particularly when paving asphalt concrete at low air temperature.

Many of the solutions developed had been used in practice. The survey of different asphalt concrete pavements shows substantial effect of the asphalt concrete composition on its crack rezistance. Particularly the modification of bitumen and reinforcement of asphalt concrete with micro-dispersed fibers and also micro-reinforcement greatly increase crack resistance [6,7]. Forecasting effective properties of bitumen-polymer compositions as well as asphalt concrete is possible on the basis of the correlation of structural theory of self reinforcing medium [15]. The survey of asphalt concrete pavements shows the need of providing some relations between the thicknesses of pavement and block base with account of material properties and natural conditions [8-11].

Constant surveys were carried out on some asphalt pavements using different above measures. From 1972 surveys conducted on a particular road in central Ukraine confirm the known data and the results of theoretical analysis of earlier and more intensive crack formation in asphalt concrete over cement concrete base in comparison with bases of asphalt concrete or granular base (table 1).

The first section of the pavement has the following structural composition: fine granular asphalt concrete h=3,5 cm; coarse granular asphalt concrete h=5 cm; black crushedstone h=6 cm; black sand h=3 cm; stone residue h=10 cm; soil- sandy loam. Structural composition in the second section is as follows: Fine granular asphalt concrete h=3,5 cm; coarse granular asphalt concrete h=5 cm; black crushedstone h=6 cm; fractional crushedstone h=30 cm; stone residue h=10 cm; soil- sandy loam.

Table 1. Indices of crack-formation of pavements

Number of sections observed	Conc. brand of base.	Month of placing conc.	Life span with-out crack	No. of cracks for 1 km for a lifeterm of, year 5	7	12	Medium interval between cracks,m (12yrs)	Crack index (12yrs)
1	190	V-VI	1	106	108	110	9.1	0.55
	190	II-III	1	–	46	98	10.2	0.49
	185	III	1	–	45	104	9.6	0.52
	190	IV	1	–	70	96	10.4	0.48
	190	V-VI	1	–	–	134	7.5	0.67
	260	IV	1	–	–	126	7.9	0.63
	250	I-II	1	–	–	110	9.6	0.55
	260	IV	1	90	–	125	8.0	0.63
	260	I-II	1	–	–	73	13.7	0.37

Results show that the placement of cement concrete in cool period enables a decrease of twofold in the seventh year of service of the quantity of transverse cracks in asphalt pavement that is laid over the joints of the base. After 12-14 years cracking stabilizes.

Such technique of using black crushedstone with thickness h=6-10 cm between a treated base and a single or double layer of asphalt concrete is widely used during the construction of new roads as well as during reconstruction projects. However, experience has often shown that these types of construction are not stable against the formation of reflective cracks in pavements.

As results of analysis carried out have shown, one of the ways of increasing the stability of pavements against the formation of reflective crackings is the use of the so called granular crack-interrupting layer in a treated base. This layer serves as a hinge layer.

Eventhough, little studies were carried out on the performance of such pavements both positive and negative results were observed. As a result of that special comparisonal experiments and theoretical analysis of pavements with dense asphalt concrete and base made up of lean concrete containing an intermediate layer of white crushedstone or black crushedstone were carried out. These experiments were carried out with the help of circular test bench in Kiev [12,13].

The pavements compared have the same type of base consisting of an underlying layer of granite residue (h= 15 cm), laid over clayey soil and a bearing layer of cement concrete (brand 150). A layer of crushedstone is laid over the base: in the first section- black crushedstone (h=8 cm), in the second, crushedstone (h=13 cm), while in third section consist of untreated crushedstone (particle size 0-40 mm and thickness 8 cm). The pavement in the first and third sections consist of two layers of : a) coarse granular asphalt concrete (h=5 cm) and b) fine granular asphalt concrete (h=4 cm). While the pavement in the second section has a single layer of fine granular asphalt concrete (h=4 cm).

Tests were conducted using three types of transports. Two leading trolleys serving as loads and two supporting trolleys with different wheel system serving as loads from heavy transports were used. Trolleys of this type have the front axle as supporting while the rear axle as leading. The size of their base is 3880 mm in width. Every axle in cross-sectional direction has a couple of wheels with distance between them being 1910 mm. The front axle has single while the rear axle has twin tyres of the type 320-508D. The distance between the centers of the imprints of the twin tyres is 370 mm. Load from the front axle on the road is 44 kN while load from the rear axle changes from 60-120 kN.

The heavy trolleys are double-axled. Each axle has one wheel. One of the trolleys has a single axle wheel, with tyre size of 18.00-24. The second trolley has twin wheels (size 15.00-20).

The centers of imprints of the twin tyres are spaced at 900 mm from each other, the distance between the axles of the heavy trolleys in longitudinal direction changes from 1700 to 3000 mm.

The load acting on the single wheel varies from 40 to 90 kN and on the twin wheel from 40 to 120 kN. The pressure in the tyres of the single wheel is 0,5 MPa, while in the twin wheel is 0,55 MPa.

The speed used during the test varies from 1 to 25 km/h. Static loads are applied during long stops of the trolleys within the interval of 12-24 hrs.

The pavement is tested for a period of two seasons. Each period lasted from March to October. For the whole period of the test, 10000 circles of the test bench with trolleys of different loads and regime of movement were carried out. 6000 circles of which are with maximum loads and 1500 circles are with minimum loads. And in every test section at least 150 static loadings were carried out. During testing, vertical deflection and vertical normal stresses were measured. Results received proved the equivalence of all the three variants of test in terms of their rigidity and distribution capacity. Towards the end of the test in the last season (Sept.-Oct.) final test of the pavements were carried out (table 2).

The deflections of the surface of crushedstone layers measured during punching-press tests with the help of dynamic loading in every test section have almost the same value. This shows that the conditions of service in such pavements are ensured by the equivalent rigidity (stiffness) of the layers.

At the end of the tests, visual check of the surface of the test sections was carried out. No signs of destruction, damage or deformation were found out. Due to the accumulation of residual (permanent) strain, insignificant rutting is observed in highly loaded lanes in all the three sections. The depth of the rutting does not pass 20 mm. The least rutting depth of 11 mm, is found in the second section with a layer of black crushedstone (h=13 cm) and single layer surface coating. The values of rutting depth in sections one and three are similar. This proves that the use of untreated crushedstone inplace of black crushedstone does not affect shear resistance under the action of heavy transports.

For a detailed comparison of the stress-strain conditions of the pavements in question, stresses and displacements in them were calculated using the method of Privarnikov for stratified semi space [14].

In calculating the stresses and displacements, a speed of 20 km/h of vehicles and loads of 100 kN on the rear axle at 20°C pavement surface temperature are used.

Results show that according to the values of: a) vertical deflection u_z of the pavement surface, b) horizontal normal stresses in the cement concrete and c) vertical normal stresses in the layer of residual granite and in the soil being studied are equivalent. Notably in all the three variants, black and white intermediate

Table 2.   Results of deflection and rutting measurement

Index	Index values in sections		
	I	II	III
Surface deflection of pavement under the center of twin wheel of the back axle of the test trolley with load 37 kN and tyre pressure 0,55 MPa and pavement temperature 17°C, MM	0.26	0.23	0.24
Rutting depth, mm	17	11	15

crushedstone layers are subjected to the action of compressive (not tensile) horizontal normal stresses. This circumstance ensure a favourable condition for their work in the pavement. In the compressive zone, the top layer of asphalt concrete pavement of all the three pavements is subjected to compressive forces as well as the top fiber of the second layer of asphalt concrete pavement of the first and third variants.

While at the same time the lower fiber of asphalt concrete pavement of all the three pavements are subjected to the action of tensile horizontal normal stresses. From the results, it is clear that replacing the lower layer of asphalt concrete pavement by black crushedstone as done in the second test section in comparison with the first leads to the elimination of the action of tensile stresses in asphalt concrete pavement.

Tensile stresses in the lower fiber of the second layer of asphalt concrete pavement increas from 0.07 to 0.4 MPa by replacing black crushedstone (in section one) with ordinary untreated crushedstone (section three). However, this value (0.4 MPa) is considerably lower than the Standard value of asphalt concretes, which in principle allows the replacement of black crushedstone with white crushedstone in the interlayer between the treated (strengthened) base and asphalt concrete pavement. Considering the fact that by replacing black crushedstone with white one tensile stresses increase in value, some additional calculations were carried out in order to assess the influence of thickness and modulus of elasticity of intermediate crushedstone layer on tensile stresses in lower fiber of the pavement situated above them.

Numerical analysis of horizontal normal stresses in the lower fiber of asphalt concrete pavement confirm that tensile stresses from bending considerably depend simultaneously on the modulus of elasticity of intermediate granular layer and its thickness. As a result of this, permissible combination of these parameters for use in intermediate layers of crushedstone is established.

The use of untreated crushedstone as crack-interrupting interlayer between a stabilized (strengthened) base and asphalt concrete pavement is recommended. This is widely use in Ukraine and pavements with this type of interlayers have lifeterm from 1 to 15 years.

3 Prospective problem

At present it is proposed the solution of practical problems aimed at reducing the use of materials and energy during the constructing and maintaining of road network with account of reflective cracking. These problems include:
— creating a Standard method for the design of asphalt concrete pavements that resistance to the formation of reflective cracks,
— developing of catalogue of rational design using cement based materials with optimal thermo-rheological properties of pavements and bases,
— developing of design properties of composite materials based on probability-geometrical model of compaction of particles mixture and packing of particles,
— implementation of effective crack-interrupting layers using new materials.

4 References

1. Mozgovoy, V. V. (1978) Construction of thick-layered asphalt concrete base for road construction in Kiev region. *Inform. paper of the Min. of Road construction of Ukraine,* Kiev.
2. Goncharenko, F.P. (1995) Prospective of using inorganic binders in road construction in Ukraine. *Highway Journal of Ukraine,* Kiev, No. 4. pp. 38–39.
3. Mozgovoy, V.V. (1981) Determination of stresses in asphalt concrete as in viscoelastic medium during pavement temperature decrease, *Journal of roads and road construction,* Kiev, No. 29. pp. 60–63.
4. Mozgovoy, V.V. (1983) Design of flexible road pavements that are resistant to cracks, *Journal of roads and road construction,* Kiev, No. 31. pp. 19–22.
5. Radovsky, B.S. (1986) Thermal stresses in asphalt concrete pavements, *Improving the durability of road pavements. Collection of scientific works.* Soyuzdornii, Moscow, pp. 18–24.
6. Mozgovoy, V.V., Tsekhansky, O.E. (1988) Influence of transverse cracks and joints of stabilized base on thermal crack-resistance of road pavements, *Journal of roads and road construction,* Kiev, No. 42. pp. 66-72.
7. Radovsky, B.S., Mozgovoy, V.V. (1989) Ways to reduce low temperature cracking of asphalt pavements, *4th Eurobitume symposium,* Madrid, pp. 571–575.
8. Gokhman, L.M., Basurmanova, I.V., Radovsky, B.S., Mozgovoy, V.V. (1989) The use of polymer-bitumen binder on the basis of DST,

9. Borovoy, M.V., Gnativ, N.Y., Gochman, L.M., Davidova, A.R. (1989) Complex organic binders, *Highway Journal of Ukraine*, Kiev, No.3. pp. 40-43.

10. Borovoy, M.V., Gnativ, N.Y., Merzlikin, A.E., Gamelyak, I.P., Yurchuk, A.P. (1990) Dispersed reinforced asphalt concrete in road construction, *Highway Journal of Ukraine*, Kiev, No. 2. pp. 30-31.

11. Kovalchik, Y.P., Gnativ, N.Y. (1989) The use of metallic mesh in reinforcing road pavements, *Highway Journal of Ukraine*, Kiev, No. 4. p.30.

12. Malevansky, V.V., Radchenko, L.A., Malevansky, G.V. (1968) Test bench for testing road construction materials, *Highway Journal*, Moscow, No. 6. pp. 8-10.

13. Radovsky, B.S. (1980) Experimental research of stress-strain condition of pavements considered as multilayered viscoelastic base under the action of moving load, *Applied Mechanics*, Moscow, Vol. 16, No. 4. pp. 131-135.

14. Privarnikov, A.K. (1973) Three dimensional deformation of a multilayered base, *Resistance and strength of elements of structures*, Dnepropetrovsk. pp. 27-45.

15. Kostritsky, V.V. (1993) Structural theory of self-reinforcing composites, *Transaktion YIII International Conference On Mechanics Of Composite Materials*, Riga, Latvia. pp. 98-110.

Reflective cracking in asphaltic overlays on rigid pavements can be delayed

A.H. VAN DE STREEK
Infrastructural Facility Assessment & Management Department, Unihorn Infrastructure Consultants, Scharwoude, Netherlands
B.J.A.M. LIESHOUT
Quality Department Civil Engineering Division, Ooms Avenhorn, Scharwoude, Netherlands
I.H. WOOLSTENCROFT
International Sales and Marketing Department, Bayex, St. Catharines, Ontario, Canada

Abstract

The most important problem of asphaltic overlays on portland cement concrete or rigid pavements is reflective cracking. This paper describes the approach for establishing the maintenance measure to be taken on Highway 99, the way the measure was implemented in 1993 and the current pavement condition. Highway 99 is a heavily trafficked two lane road. Due to its age and the loss of bearing capacity the rigid pavement was severely damaged.

On the basis of construction and traffic data, visual condition surveys and falling weight deflection measurements the structural condition of the pavement was assessed. A number of options were considered and for various design criteria the life time cycle was evaluated. In this evaluation costs and the practicability of the measure were taken into consideration. The project was executed under a pre-established quality assurance plan.

After completion of the work, falling weight deflection measurements and visual condition surveys were executed regularly to monitor the structural condition of the pavement. As to date the pavement performance is satisfactorily. No reflective cracking has been identified and the life expectancy matches the predicted level.

This paper also discusses other projects where reinforcement grids, SBS modified asphalt and/or SAMI's have been applied succesfully on both rigid and flexible pavements.
Keywords: crack controlling products, maintenance, pavement performance and analysis, reflective cracking.

1   Introduction

Reflective cracking in rigid (as well as in flexible) pavements is a well known phenomenon. Although well known, processes inducing cracking and effective ways for controlling it were only partly understood in the early 90's. To gain more insight, not only theoretically but also practically, we decided to start a number of projects. These projects were meant to apply our theoretical concepts into practice, to see whether the predicted results and pavement performance could be achieved in reality. The projects were set up in such a way that new developments and insights in the field of reflective cracking could be incorporated to enhance our knowledge and understanding of the failure mechanisms in pavements for further modifying of our performance model. The following N99 Highway project is our main study project.

Reflective Cracking in Pavements. Edited by L. Francken, E. Beuving and A.A.A. Molenaar. © 1996 RILEM. Published by E & FN Spon, 2–6 Boundary Row, London, SE1 8HN. ISBN 0 419 22260 X.

2   Case 1: Highway 99

2.1   History

Highway 99 is nowadays a heavily trafficked two-lane road. Part of this road is situated on a sea embankment (constructed mid 20's) between the former island of Wieringen and the city of Den Helder in the far north of the Province of Noord-Holland in the Netherlands.

In the 1930's the existing paving stone was replaced by a cast in place concrete pavement. The pavement thickness varied from 18 cm in the centre part to 23 cm at the longitudinal slab edges. At these edges three 16 mm reinforcement bars were placed. Slab dimensions are 3 by 15 metre. The longitudinal construction joint was probably doweled, the transversal expansion joints were doweled. All joints were (originally) filled with asphalt emulsion. Later on special sealants were used.

The concrete pavement lies on a layer of sand with a layer thickness varying from 0.3 to 2.0 metre. Under the sand lies a clay layer with a thickness varying from 0.1 to 1.0 metre. The natural subbase consist of layers with clay sands, peat and boulder clay. On both sides of the road are shoulders, consisting of 18 cm asphalt on sand. The ground water level is found at about 3 metre below pavement level.

In the 1980's the pavement manager noticed an ever detoriating pavement condition of the section between km. 10.2 and 11.8. Around 1990 the condition was starting to drop below a critical level. The pavement condition then could be characterized by:

*   transversal cracking about every 4 metre (plate length is 15 metre);
*   cracking at slab corners, some times partially or even complete loss of concrete material. Repairs with (cold) asphalt in bad condition;
*   loss of finer material of the pavement surface, the surface looked weathered;
*   opening of the longitudinal joints (up to 1 cm) and the transversal joints (up to 5 cm);
*   most of the joints (and the wider cracks) have been sealed, but the quality of the seal is poor to bad; water can easily ingress;
*   at several locations vertical displacements at joints or cracks during the passage of heavy commercial traffic can be noticed;
*   although pumping was expected, no visible signs (like deposits of very fine graded material adjacent to the joints) could be noticed at the time of the survey;
*   unevenness due to irregular settlements.

2.2   Basic assumptions and limiting conditions

The visual condition survey showed that the road had to be upgraded at a (very) short notice. The road maintenance management decided to consider all maintenance options, provided that the measure to be taken would have a design life period of 10 years (in case of maintenance) or 20 years (in case of complete reconstruction).
Other assumptions were:

*   number of vehicles per lane per day   : 3.800
*   percentage commercial traffic   : 5
*   number of 100 kN axle loads per lorry   : 1.3
*   increase in traffic intensity   : 3 %

Apart from that the following limiting conditions were posed:

*   due to local circumstances a detour (15 km over very small local roads) was impossible, at least one lane had to be open to the traffic permanently;
*   traffic disruption had to be minimized, construction time had to be as short as possible;

- due to the presence of adjacent works (drainage system, crash barrier) increase in pavement level should be minimized;
- salvaged materials (if any) had to be re-used as much as possible.

2.3 Investigation: set-up and results

First of all construction and maintenance data were gathered. Construction data were available (see 2.1), no major maintenance had ever been executed. No information was available on the concrete quality at the time of construction or at the moment of investigation. Therefore it might be needed to take cores for further testing. A visual condition survey had already been performed, the results are described in 2.1. The distresses were equally distributed over the total section. The first 200 - 300 metres showed more (asphalt) repairs.

Falling weight deflection measurements were used to provide information on the bearing capacity of the subbase and subgrade, the structural condition of the concrete pavement and the measure of load transfer at joints and cracks. In July 1992 per lane every 50 metres measurements at a 50 kN load level where taken in both wheeltracks and in between these tracks. Both lanes were measured. Additionally load transfer measurements at transversal joints and (some) cracks were executed at random chosen locations. Traffic conditions did not allow load transfer measurements at the longitudinal joints.

The results of the "50 metres" measurements were analyzed with the help of the Sustentational Energy Response Pavement Evaluation Model (SERPEM) [1]. On basis of subgrade modulus Es, equivalent layer thickness heu and sustentational energy response W, the uniformity of the sections in longitudinal and transverse direction were judged. Although variation in pavement characteristics in longitudinal direction was found, there was no need for breaking down in subsections. Of interest was the variation of the pavement characteristics in transverse directions, as presented in table 1. In this table the mean values per track over the total section length are presented. From the figures it can be concluded that the west bound lane has a higher subgrade modulus (favourable), a lower equivalent layer thickness (unfavourable) and a somewhat higher sustentational energy response (unfavourable).

Table 1.   Mean values of some pavement characteristics of Highway 99

Location	Subgrade modulus (MPa)	Equivalent layer thickness (m)	Sustentational energy response (J/m3)
West bound lane			
Right wheel track	164	2.49	39
Lane axis	195	2.17	39
Left wheel track	181	2.46	36
East bound lane			
Right wheel track	127	3.21	35
Lane axis	175	3.19	28
Left wheel track	128	3.34	32

The results of load transfer measurements can be summerized as follows. At the transversal joints and clearly open cracks the load transfer was minimal to non-existent. At the joints the dowels did not seem to function anymore. At the "closed" cracks the load transfer was adequate.

Based on the investigation results, the residual pavement life was considered nil. Maintenance or reconstruction had to be executed as quickly as possible. As the pavement condition was very poor, further investigation into the concrete quality was found of no use.

2.4   Measure determination

Because of the lack of support of the plates locally (becoming visible in the vertical movements of the plates), the loss of load transfer at joints and (open) cracks, the wide opening of the transversal joints, the bad quality of the more than 50 years old concrete (evident from cracking and loss of material at the pavement surface) and the results from the falling weight deflection measurements, complete reconstruction of the pavement was thought to be the technically most viable solution. However, due to the limiting conditions as described in 2.2, this solution was not practicable if not impossible. Therefore it had to be investigated whether an overlay was possible and if so, what materials with suitable material engineering properties could be used.

In the 1989 RILEM-conference "Reflective cracking of pavements" a number of authors, a.o. Haas and Joseph [2], pointed out the importance of establishing the cause(s) of the problem(s), meanwhile not forgetting the practicability of the measures chosen. From the results of the investigations carried out it was clear that, although the plates were cracked, vertical plate movements at joints and cracks were occuring (mode 2 cracking). As the (cracked) plate lengths amounted about 4 metre on average, horizontal movements due to temperature variations could also be expected (mode 1 cracking). After all years of service, shrinkage problems were deemed quite unlikely. As the prevailing crack type was transverse cracking, mode 3 cracking was thought unlikely.

To minimize crack propagation problems, it was decided to pre-crack the concrete plates. By subsequently rolling the cracked pavement, the plates could be embedded into the subbase thus restoring contact between layers and improving bearing capacity of the (cracked) pavement. Thus mode 1 and 2 cracking were thought to be minimized.

At the time the overlay had to be determined, only the results of the falling weight deflection measurements at the existing (not yet pre-cracked) pavement were available. To be able to determine the overlay thickness on the pre-cracked pavement, the approach of Van Gurp and Molenaar [3] viz. reducing the effective moduli of the (pre-)cracked pavement, was incorporated in the SERPEM-model. Using only standard asphalt products, effective (excl. levelling/profiling) overlay thicknesses varying from 15.5 for the better parts to 17 cm for the worst parts were found.

Subsequently a number of alternative products were considered. In [4] Molenaar stated that to prevent crack propagation there are two options, depending on the crack mode. The first option is the bonded overlay, reducing stresses and strains in the existing pavement and subgrade, depending on thickness and relevant material properties of the overlay. The second option is the unbonded overlay, according to Molenaar an excellent solution to retain horizontal movements, but not applicable in the presence of high shear stresses.

Though Molenaar was very decisive about the incompatibility of both options, we felt that a combination of these options could be advantageous in the present case, provided that the materials were carefully selected. Our idea was to apply first a bonded overlay on the pre-cracked concrete pavement. As crack propagation was thought inevitable, subsequently as a kind of security measure to retard further crack propagation, a debonding measure was sought. Such a measure was thought to be advantageous by a.o. Rigo, et.al. [5].

The desired characteristics for the material of a bonded overlay, viz. low A- and n-values as used in Paris law, were pointed out by Molenaar [4]. Sealoflex® SBS polymer modified as-

halt fulfilled these demands quite satisfactorily (see figures 1 and 2 and table 2) [6]. As for
le debonding measure, we agreed with Molenaar that conventional debonding measures
uld nullify the bonding measure. Therefore we looked for a (modified) bituminous pro-
uct for use in a stress absorbing layer (SAMI), which had excellent adhesive properties,
w temperature susceptability and high shear strength. To further create what could be
illed a controled debonding, we looked for a material with which we could reinforce the
AMI. The idea was to permit a limited degree of debonding to prevent crack propagation.
his debonding effect however had to be controled by some kind of reinforcement in order
) prevent strains in the layer above the SAMI, to delay crack propagation. What we were
)oking for was a reinforcement material displaying high strength and low strain characte-
stics. To our opinion only glass fibre or metal grids could fulfill these demands. As it was
nown that good placement of the materials was of paramount importance for the success of
le measure, we also looked for easy to place, low-failure prone materials.

igure 1.

tress intensity factor
ersus
rack growth rate of
andard and SBS
ense asphalt concrete.

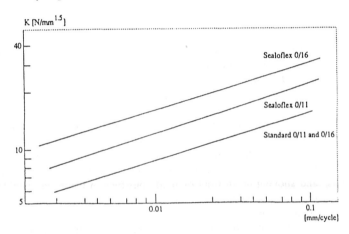

igure 2.

rack growth
haracteristics of
andard and SBS
ense asphalt concrete.

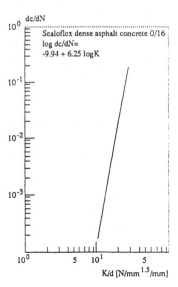

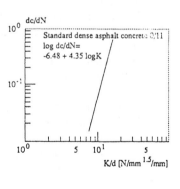

Table 2.   Characteristic properties of Sealoflex® SFB5-50 and SAM.C-30® binders.

Binder	Penetration	$T_{R\&B}$	T_{FRAASS}	Elastic Recovery	Ductility	
					at 5°C	at 15°C
	(0.1 mm)	(°C)	(°C)	(%)	(cm)	(cm)
Sealoflex®	40 - 60	>90	< -17	90	15	65
SAM.C-30®	90 - 130	>70	< -20	90	90	90

Table 3.   Characteristic properties of GlasGrid® 8501.

Material	Tensile force across		Modulus of elasticity	Grid size
	length (kN/m)	width (kN/m)	(MPa)	(mm x mm)
Glass fibre strands, covered with polymer modified coating, self-adhesive at one side	100	100	69000	12.5 x 12.5

Selected were SAM.C-30®, a SBS polymer modified sprayable (at 180°C) bituminous binder, for use in the SAMI and GlasGrid® 8501 as reinforcement grid. Their relevant characteristics are given in table 2 resp. table 3. It should be noted that the self-adhesive property of the reinforcement grid was considered of major importance, as such aspects as crinkling and shoving up in front of road construction equipment could be prevented. An important aspect was the watertightness of the pavement construction with SAMI, even in case cracks would propagate through.

The combination of measures, needed to fulfill the pavement design life under the traffic conditions as mentioned in 2.2, were determined with the SERPEM-model. Finally the following combination of measures was accepted by the road authority as being at least equal in life time expectancy to the overlay using standard products. Comparison of costs of the standard versus the alternative measures showed that the alternative measures were less costly. As since then the costs of alternative products/measures, due to scale enlargement and placement experience (compared to standard products/measures) have diminished, this surely will be more so today.

Applied measures:
1. Cracking and seating of the existing pavement. Remaining plate size: 1 m2;
2. Local repairs:
   - existing asphalt repairs (especially those with flux-basis asphalt) were replaced by hot mix asphalt repairs;
   - joints and cracks were filled in with a non flux-basis sealant;
2. Double prime coat (first layer: 0.3 kg/m2, second layer: 0.2 kg/m2) to ensure optimal adhesion between concrete and asphalt levelling/base layer;
3. 5 cm. SBS polymer modified Sealoflex° STAB 0/16 (base course);

4. Controlled debonding measure (over the total pavement surface); consisting of:
   - placement of self-adhesive glass fibre reinforcement GlasGrid° 8501, on top of which was placed;
   - Stress Absorbing Membrane Interlayer (SAMI) constructed by spraying 2.5 kg/m2 SBS polymer modified SAM.C-30® and spreading about 7 kg/m2 aggregate 8/11 on top of the SAMI;
5. 4 cm. open graded asphalt concrete 0/16.
6. 5 cm. drain asphalt 0/16.

Figures 3 to 7 give an impression of the N99-project.

Figure 3.

Original pavement condition.

Figure 4.

Pavement condition after cracking and seating.
On the other lane the new construction, except for the drain asphalt layer, has already been applied.

382 *A.H. Van De Streek* et al.

Figure 5.

Mechanized application of GlasGrid reinforcement at the base course.

Figure 6

SAMI applied on reinforcement grid, notice the open aggregate spreading.

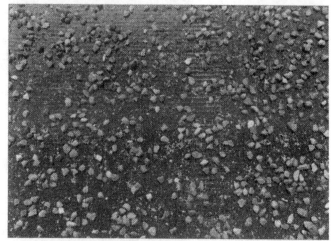

Figure 7

Application of open graded AC over the reinforcement/ SAMI-system.

2.5 Research program, initial results

The determination of the maintenance measures has partly been based on assumptions, especially in case of the pavement condition after cracking and seating the pavement and after applying the different stages of the maintenance measure. Of course it is of great interest to check whether the predicted pavement life can be realized in practice too. In order to confirm our assumptions and check the pavement life we decided to execute and analyse falling weight deflection measurements at the following stages during construction and service life:

1. Immediately after cracking and seating the concrete pavement in September 1993;
2. Immediately after applying the SBS STAB base course in September 1993;
3. Immediately after applying the temporary top layer of open graded AC on top of the reinforcement grid and SAMI in September 1993;
4. After applying the drain asphalt top layer in April 1994;
5. In September 1995 on the complete construction.

Due to limited accessability of the project site the falling weight deflection measurements during the stages 1 to 4 could only be executed on the (poorer) west bound lane.

The results of the measurement in terms of unit equivalent layer thickness heu and energy response W are presented in figure 8 (for presentational reasons only the results of the right wheel track of the west bound lane are presented here). Measurement results of the 'old' concrete pavement are incorporated in this figure too. From the results one can clearly see the expected change in heu en W during the different stages.

Special attention is requested for the results obtained in stage 3. When these results are compared to those of stage 2, a slight decrease in heu and a strong increase in W is noticeable. Here the effect of the (controlled) debonding layer is made visible.

From the collected data the residual life time of the pavement for the different construction stages and after completion during service life have been calculated on basis of SERPEM. The results are presented in table 4.

Table 4 Residual life time of the pavement during construction and after completion.

Stage	Residual life (years)
Existing concrete pavement (July 1992)	none
Cracked and seated concrete pavement (September 1993)	none
After applying SBS base course (September 1993)	5
After applying temporary top layer on reinforcement/SAMI (September 1993)	9
After applying drain asphalt top layer (complete construction, April 1994)	12
On complete construction (September 1995)	12

During all stages the pavement has been visually surveyed. After applying the base course, at two transversal joints a transversal crack in the asphalt layer had been propagated. After applying the temporary top layer and the drain asphalt layer no damages were visible. The same goes for the 1995 survey.

Cores taken from the pavement showed excellent adhesion between all layers.

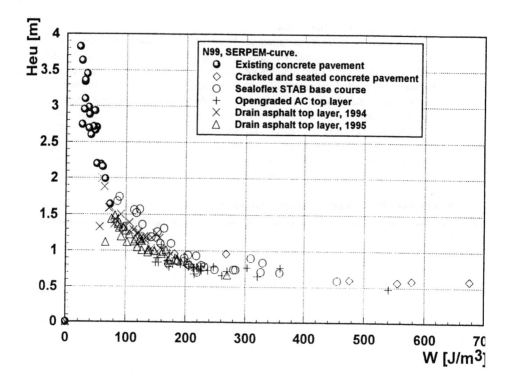

Figure 8   Results of the Falling weight deflection measurements according to SERPEM
              for the various stages during and after construction.

2.6   Future research program

It is intended to follow the course of the pavement condition during its service life by yearly executing a detailed visual condition survey. As a result of this survey, the location, size and severity of each visible distress will be recorded on maps. This enables one to compare distress development over the years. Additionally in 1996, 1999 and 2002 falling weight deflection measurements will be executed and analyzed to determine the structural pavement condition. Taking and examining cores could become part of the research program if found useful (e.g. examining crack growth).

We do realise that the research approach chosen for this project is performance based. It is also true that nowadays analytical methods based on laboratory experiments related to crack propagation problems are available. We intend to bring in a number of these new concepts to complement our research efforts (e.g. CAPA [7]). These should help us in providing a better insight in establishing the exact nature of failure mechanisms as well as providing us better tools for taking decisions regarding the choice for products to be used and measures to be taken in crack prevention.

3 Case 2: Melkweg, Purmerend

The Melkweg is a 5 to 5.5 m wide country road. The construction of the road: 8-10 cm asphalt (several layers) on 10-15 cm rubble subbase on a clay and peat subgrade. The width of the shoulders is about 0.5 m. Immediately next to the shoulder, on both sides of the road, is a ditch. The water level in the ditch lies about 0.5 metre below pavement level.

Traffic intensity on the road is low, but as the road was and is being used as a short cut by concrete hauling trucks to a major housing development project, commercial traffic loads are high and relatively frequent.

In 1990 the pavement showed over the total length wide longitudinal cracks, especially near the pavement edges, moderate to severe ravelling and locally complete collapse of the road. The transversal and longitudinal profile of the road was locally unacceptable.

As due to the minor importance of the road on one side and lack of sufficient funds on the other side, complete reconstruction of the road, thought to be necessary from the technical point of view, was not possible. In selecting the maintenance measure, the main problems identified were lack of bearing capacity, lack of support of the roadside and crack propagation. On the basis of theoretical considerations, in which the existing pavement was taken as a cracked semi-rigid subbase, the maintenance measure was determined as follows:

1) 4 cm (minimum) STAB 0/22 as a levelling/base course;
2) self-adhesive glass fibre reinforcement GlasGrid® 8502 with on top a SAMI, consisting of 2,5 kg/m2 SBS polymer modified SAM.C-30® and 7 kg/m2 aggregate 8/11 (both over the total pavement surface);
3) 4 cm open graded AC 0/16;
4) 4 cm dense AC 0/16.

The basic idea of this approach was to improve the bearing capacity of the road by the application of 12 cm of asphalt and to prevent crack propagation by means of the SAMI/reinforcement-layer. As the longitudinal cracks were quite wide, but no clear height differences between both sides of the cracks could be established, reinforcement of the pavement to prevent the road sides from "toppling into the ditch" was thought possible. After all, the toppling mechanism which induced horizontal strains and stresses in a transversal direction, thus creating the longitudinal cracks, would be arrested by applying a reinforcement measure that could be effective in the horizontal plane. As GlasGrid® 8502 has a double quantity of reinforcing material in transversal direction, it was thought very useful for this project.

As to the date (February 1996) no distresses can be observed, the pavement is performing as expected.

4 Case 3: Europaweg, Helmond

The Europaweg is a dual carriageway (2 lanes each), consisting of a 23 cm concrete pavement on a 15 cm tar/lime stabilisation on sand. The pavement was constructed in 1967. The average traffic intensity was assumed to be 3500 80 kN SAL per lane per day.

Over the years a process of faulting at the transverse joints developed. Besides an ongoing decrease in riding comfort (locally very serious), also an increase in noise annoyance level became evident. The road authority looked for a low cost solution by which the existing pavement could be used to its full remaining capability.

As the overall bearing capacity of the road was deemed sufficient, it was concluded that the maintenance measure should be aimed at providing an even surface with high riding comfort. For this purpose a thin overlay would suffice. However, crack propagation had to be prevented during the design life (15 years) of the maintenance measure. To limit the layer thickness (and the cost) the following measures were chosen.

1. Cracking and seating of the existing pavement, remaining plate size: 1 m2;
2. Cleaning and filling of joints;
3. Double prime coat (0,3 and 0,2 kg/m2) on the concrete;
4. SAMI, constructed by spraying 2,5 kg/m2 SBS polymer modified SAM.C-30® and spreading about 7 kg/m2 aggregate 11/16 on top of the SAMI, over the total surface;
5. 4 cm STAB 0/16;
6. Self-adhesive glass fibre reinforcement GlasGrid® 8501, over the total surface;
7. 3,5 cm stone mastic asphalt 0/11.

These measures have been applied in the summer of 1992. To this date (February 1996) no distress has been observed so far.

5. Conclusions

In this paper the approach for the determination of the maintenance measures as well as the application of specially selected products for a number of projects, where crack propagation was thought to be the main problem, has been described.

Subsequently the pavement performance has been followed (for 3 to 5 years) until now. So far we have found that the pavement performance is in compliance with our predictions and expectations.

Nowadays new (analytical) methods, like CAPA, for the analysis of crack propagation problems are available. We will bring in these methods in the described projects to obtain even better insight in failure problems and to provide better tools for taking decisions regarding the choice for products to be used and measures to be taken.

6. References

1. Srivastava, A. (1989) *Pavement Evaluation bij SERPEM approach*, IVth Eurobitume Symposium - Madrid, pp
2. Haas, R., Ponniah, E.J. (1989) *Design oriented evaluation of alternatives for reflective cracking through pavement overlays*, RILEM-conference - Luik, pp 23-46;
3. Gurp, C.A.P.M. van, Molenaar, A.A.A. (1989) *Simplified method to predict reflective cracking in asphalt overlays*, RILEM-conference - Luik, pp 190-198;
4. Molenaar, A.A.A. (1989) *Effects of mix modifications, membrane interlayers and reinforcements on the prevention of reflective cracking in asphalt overlays*, RILEM-conference - Luik, pp 225-232;
5. Rigo, J.M., et al (1989) *Laboratory testing and design method for reflective cracking interlayers*, RILEM-conference - Luik, pp 79-87;
6. Srivastava, A., Hopman, P.C., Molenaar, A.A.A., (1990), *Sealoflex polymer modified asphalt binder and its implications on overlay design*, Paper submitted at 1990 ASTM-conference, San Antonio, Texas;
7. Scarpas, A., Blaauwendraad, J., Bondt, A.H. de, Molenaar, A.A.A., (1993) *CAPA: a modern tool for the analysis and design of pavements* RILEM-conference - Luik, pp 121-128.

Fatigue improvement of asphalt reinforced by glass fibre grid

D. DOLIGEZ
Chomarat, Mariac, France
M.H.M. COPPENS
Netherlands Pavement Consultants, Hoevelaken, Netherlands

Abstract

When we look at the different fibres which are used to combat crack propagation, and their manufacturing (non woven, grids ...), it is worth checking which kind of movement happens on the road in order to recommend the best material for the best use at the best price.

In the last ten years, glass fibre grids with different strength levels and different open structures have been tested.

In this paper, we shall emphasize the use of glass fibre grids to improve the behaviour of flexible structures and avoid crack propagation due to traffic load repetition.

Keywords : Glass fibre grid, Fatigue, Asphalt Reinforcement, Dynamic bending test.

1 Introduction

To summarize in a non exhaustive way the main types of cracks we can find and some of the possibilities to maintain the road, we may state :

- The problems with thermal cracks of rigid structure can be solved using a bituminous interlayer between the cracked road and the new overlay. This is where we can use S.A.M.I., non woven, or better, non woven reinforced by glass fibre grid, impregnated in place by bitumen. The stress of the crack is diffused into the bituminous interlayer and cracks can take 2 or 3 times longer to propagate through the overlay. In addition,

Reflective Cracking in Pavements. Edited by L. Francken, E. Beuving and A.A.A. Molenaar. © 1996 RILEM. Published by E & FN Spon, 2–6 Boundary Row, London, SE1 8HN. ISBN 0 419 22260 X.

this bituminous membrane is considered good for watertightness.

The possibility to use a composite non woven' reinforced by a glass fibre grid allows the use of a thick layer of bitumen, about 1 mm thick, without any problem of slippage of this interlayer and thus to improve even more the life time of the new overlay.

When there is vertical shearing movement between slabs, amplitude must be considered in order to check which material that can be used. It is very difficult to find a solution to this phenomenon.

- Cracks due to widening the old road
  If the differential movement is too big, it is not possible to prevent the crack propagation. In the other cases, a lot of road widening contracts have been completed, using grids with reinforcing capacity to make the horizontal connection between the two structures. Impregnated non wovens put just below the overlay can only be used as a bituminous interlayer to waterproof the cracks which could appear.
- Fatigue cracks of flexible structure
  These are the cracks due to traffic load repetitions. The heavier the traffic or the higher the deflection of the road, the better the performance of the glass fibre grid.
  This is the subject which will be developed hereafter, when glass fibre grids are put below the hinge point to improve the fatigue resistance of the structure.

Will be developped :

- the main caracteristics of the glass fibre compared to other materials like asphalt mix
- then, experimental works that have been done at the NPC laboratory in the Netherlands
- and finally, a few case histories from France and Sweden.

2 Main characteristics

What performes raw material ?

It is good to compare the elasticity modulus of the main raw material which are used in the road market.

These values are the general values that we find in literature.

E. Modulus
- Glass   70000 MPa
- Polyester   8000 to 20000 MPa
- Polypropylene 4000 MPa
- Asphalt mix   5000 to 10000 MPa

When stress is applied to the composite, asphalt mix + reinforcement, the higher elasticity modulus material is the best. This is why glass has been chosen.

3 Cyclic bending tests on asphalt beams

To study the effect of crack propagation, several series of tests were run on asphalt beams with different reinforcing grids. The method for preparing asphalt slabs and beams is described in [1],[2].

The beams –600 x 180 x 80 mm³– were loaded in four-point bending mode in a servohydraulic test rig at 30 Hz. Both reference beams and reinforced beams comprised a 30 mm asphalt binder layer and 60 mm dense asphaltic concrete as an overlay. The grids were applied on the lower layer using normal bitumen emulsion. To simulate a crack in the old pavement, a sharp notch was sawn in the 30 mm asphalt layer. For the test at 5°C, the beams needed a 4500 N load to arrive at practical testing time, ranging from one hour to one day. At –20°C, maximum load was 8500 N. During the complete test runs it was possible to follow crack propagation visually through windows in the climatized room because the edges of the beams had been polished and specially painted.

The unreinforced beams were tested in threefold to get a reliable reference level, the grid reinforced beams were tested in twofold. Although fatigue crack propagation were different for the reference and the reinforced beams, the time of failure could be established without problems. The very important influence of the reinforcing grid can be seen in fig. 1, in which the total number of load repetitions until failure is given for the unreinforced beams and several types of glass fibre grids.

Crack development rate is much lower for reinforced asphalts.

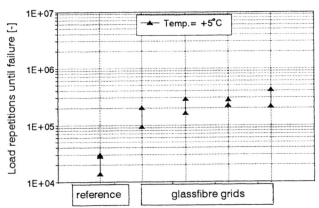

Fig.1. Results of dynamic bending test

4 Case histories

4.1 RN 139, Port-la-Nouvelle, France
Road authority : Department of Aude
The works are located at the entrance of the city, on each
side of a railway. They include the road junction between
the railway and the city.
.   Traffic class is T0 (750 lorries per day).
.   The road had major problems of structural fatigue with
    a big rutting about 100 mm deep, alligator cracks and
    pot-holes.
Core samples have been taken. They show that there
were only between 60 to 120 mm asphalt mix. Characteristic
deflection (Benkelman beam) was 1,73 mm (C5) coming-back
to 0.2 to 0.4 mm. From the *"French design guide"*, one
should have milled up 0.50 m of the road and then built up
a new road with :
.   0.34 m   lean mix (cement bound aggregates)
.   0.12 m   bituminous base course
.   0.04 m   overlay.
But a new city by-pass was to be built shortly after
which would remove a lot of lorries from this road. It
was therefore decided to build the road as following :
.   milling-up 60 mm (there was no structure at all in some
    areas)
.   levelling from 20 to 50 mm with 0/10 aspahlt mix
.   laying of the open mesh glass fibre grid
.   70 mm ovelay 0/10.
This work site has been followed up every year and has
shown the following results :
.   October 1993: deflection (Benkelman beam)
    2.2 mm coming back to zero
.   December 1994: deflection (Benkelman beam)
    1.3 mm coming back to zero
.   December 1995: deflection (Lacroix deflectometer)
    0.8 mm ($\bar{x} + 2\sigma$).

    ($\bar{x} + 2\sigma$ = the mean value plus twice the standard
    deviation).
Interpretation :
It is interesting to see that the road is evolving very
positively. Deflections are getting better and better
every year. Visually, nothing has appeared. « It seems
that the elasticity effect of the glass fibre distributes
the load stress and helps the structure of the soil to
become stable » the authority engineer said.

4.2 Avenue Gustave Eiffel, Narbonne, France
The road is located in an industrial estate.
.   Traffic : 150 lorries of more than 5 tons per day in
    each direction.

. The construction was 400 mm aggregate and 80 mm asphalt (base course quality).
. Before the works in May 1994, deflection was 1.5 mm ($\bar{x} + 2\sigma$).
. Mean value : 0.9 mm
. σ : 30.

It was decided to lay the glass fibre grid ROTAFLEX and to cover it with 50 mm asphalt mix 0/10. This solution had an expected life of 15 years with 4 % traffic increase per year.

After the works in September 1994, deflection was measured and the results were 0.5 mm (mean value). In January 1996, deflection (Lacroix deflectometer) was 0.46 mm.

4.3 Rue Alfred Catel, Amiens, France
Contractor : COLAS
. Traffic : 200 lorries per day in each direction.
. The construction was 250mm aggregates 50mm asphalt mix.
. Deflexion was 1.05 mm (x + 2σ)

The works were done in July 1994 : the road has been reinforced by the glass fibre grid (open mesh) covered by 50 mm asphalt mix.

This solution had an expected life of 12 years without crack propagation, with 4 % traffic increase per year.

4.4 Route Departementale 1015, Bouttencourt, France
Contractor : COLAS
. Traffic : 100 lorries in each direction.
. The road was built with 150 mm aggregates, 100 mm stones cobbles(40/70) and 0.5 mm asphalt mix.
. Deflection was 1.5 mm ($\bar{x} + 2\sigma$)

The works were done in September 1993 : levelling, applying the glass fibre grid covered by 80 mm high modulus asphalt mix.

This solution had an expected life of 15 years without crack propagation, with 4 % traffic increase per year.

Later deflections ($\bar{x} + 2\sigma$) :
. Dec. 1993 : 0.80 mm
. June 1994 : 0.95 mm.

4.5 Bus street, Växjö, Sweden
. Location : the University area in Växjö
. Road authority : City of Växjö
. Contractor : Skanska.

Bus street on the University area, only intended for city buses (< 100 buses/day). Due to much too inadequate, almost non-existing, superstructure, very severe deformation and large crocodile cracks had appeared. The deformation in ruts and cracks was measured to more than 100 mm.

The works were done in September 1992.
. Machine screeding to restore the crossfall
. MAB 12 mm 0-100 mm
. Tack coat 0.3 kg/m² BE65R emulsion
. Reinforcement with the glass fibre grid
. New top 80 kg MAB12 approx. 30 mm on all the surface
. Compaction.
Three years later (1995), no crack and maximum depth in ruts approx. 3 mm. Without grid the street would have declined to its old appearance after a couple of years. Another year later (February 1996) there is still no crack and the ruts are 4-5 mm deep (mainly due to wear). No deformation can be noted.

Thus, up to now, all these flexible structures are a lot improved by using Rotaflex far below the hinge point. After several years of traffic loading, no cracks appeared even in very hard conditions.

5   Conclusion

We have developed a full range of products to combate crack propagations, and we have found that the best material to give real reinforcement of asphalt mix is open mesh grid made in glass fibers.
We recommand much attention to the application procedures which must be followed to get the expected results.

6   References

1. Godard, E., Coppens, M.H.M. and Doligez, D. Asphalt reinforced by a glassfibre grid : fatigue properties. *Revue Générale des Routes et Aérodromes* (Dec. 1993).

2. Coppens, M.H.M., van de Ven, M.F.C. Dynamic bending tests on asphalt beams with reinforcement. *Wegbouwkundige werdagen CROW* (March 1994).

Reflective cracking – design and construction experiences

F.A. SLIKKER
Ballast Nedam Grond en Wegen B.V., Amstelveen, Netherlands
A.R. NATARAJ
Netherlands Airport Consultants B.V., The Hague, Netherlands

Abstract
Two almost identical flexible pavement constructions with a pre-cracked cement treated base were constructed at geographically separated sites under similar environmental conditions. During construction, reflection cracks appeared in the asphalt layers at one of the sites only.

An investigation into material properties and crack movement concluded that the asphalt properties and large thermal coefficient of the aggregate in the rigid base in combination with the extreme diurnal temperature variations were the main causes for the early crack development on this site.

Remedial works comprised optimalisation of the asphalt material, an additional asphalt layer and the inclusion of a bitumen impregnated geotextile interlayer. To date, three years after the event, cracks have not reappeared.
Keywords: Aggregate, cement treated base, geotextile, interlayer, pre-cracking, reflective cracking, temperature variations

1 Introduction

A part of the works carried out by Ballast Nedam in one of its turn-key design & construction projects in the Middle East included flexible pavements for two airfields on the Arabian Peninsula. NACO carried out the design on behalf of Ballast Nedam and provided the technical support during construction. This document summarizes the experiences with regard to reflective cracking in the asphalt pavement that occurred during construction on one of the sites.

Reflective Cracking in Pavements. Edited by L. Francken, E. Beuving and A.A.A. Molenaar. © 1996 RILEM. Published by E & FN Spon, 2–6 Boundary Row, London, SE1 8HN. ISBN 0 419 22260 X.

2 Design

The design of the flexible pavements for the 15 m wide taxitrack system on both sites was based on a single wheel load of 16,750 kg at a tyre pressure of 2.35 MPa, representing the design aircraft. The resulting pavement construction is shown in Fig. 1. The design included a rigid base, whereby the following precautions were included in anticipation of the occurrence of reflection cracks:

- Pre-cracking of the Cement Treated Base (CTB) into a panel size of 4 x 4 m using the cutting wheel of a pneumatic wheeled roller while the CTB was still green, refer to Fig. 2. The CTB was constructed in two layers, each layer separately cut to a depth of 1/3 of its thickness. This method has been very effective, as evidenced by the core samples taken at the location of precrack.
- Extra thickness of asphalt layer compared to the recommended thickness by FAA [1] and PSA [2] methods

Asphalt Surface Course	ASC	50 mm	
Asphalt Concrete Base 2	ACB-2	60 mm	
Asphalt Concrete Base 1	ACB-1	60 mm	
Cement Treated Base	CTB	260 mm	
Subbase		150 mm	(Site B)
		300 mm	(Site A)
Compacted Subgrade			

Fig. 1. Flexible pavement construction

3 Problem

In May 1993, the beginning of the hot season, the first small cracks appeared in the asphalt base layers of Site A, sometimes as soon as a few days after paving. Initially only longitudinal cracks appeared but later on transverse cracks as well, matching the pattern of the pre-cracks in the CTB. Cracks appeared initially at almost 100% of the locations with one layer of base course and at 30% of the locations with two layers of base course. Only a small area of pavement was completed with surface course asphalt and at that date no reflection cracks were observed. At Site B there was no occurrence of reflection cracks, although diurnal temperature variations were of the same magnitude.

Although these reflection cracks would not influence the structural integrity of the pavement, it would increase the maintenance effort from the user. Having a particular low maintenance expectancy, the client was unlikely to accept extensive crack repair, hence, a remedial had to be found that would be effective in the long term.

Fig. 2. Cutting device

With construction activities going on and a fixed hand-over date of the project, a solution had to be found at short notice that would allow incorporation of works already executed and preferably at low cost.

As a consequence, the production was limited to the asphalt base layers, thus allowing work to continue at a reduced pace, thus providing more time for an investigation into the cause of the problem and possible remedial measures.

4 Investigation

4.1 General
The investigation comprised the works on both sites and included on-site temperature and crack measurements, as well as extensive research into the properties of the paved material and production procedures.

4.2 Material properties
Material from already laid pavement layers were core sampled and partly shipped back to Holland for laboratory testing. The asphalt production and construction procedures were reviewed. The results as summarized in Tables 1 and 2 gave indications that the following factors could have influenced the rapid appearance of cracks:

- Aggregate absorbtion   - 0.43% against 1.23%
- Mix proportions   - affecting the binder film thickness
- Proportion of dune sand   - 11% against 7% could affect the tensile strength
- Mixing time at the plant   - 45 sec. against 30 sec. (ageing)
- Compaction techniques   - better compaction improves performance
- Mix properties   - air voids and VFB influencing ductility

Table 1. Summary of material properties

Material	Item	Site A	Site B	Unit	Comment
Bitumen	Penetration	70	62		fresh
		36	43		recovered
	Softening Point	49.5	51.5	°C	fresh
		57.5	54.5	°C	recovered
	Penetration Index	-0.5	-0.3		fresh
		-0.3	-0.5		recovered
Asphalt cores	Resilient modulus	13.7	19.3	GPa	@ 0 °C
		4.1	6.7	GPa	@ 15 °C
	Tensile strength	2.5	3.4	Mpa	@ 0 °C
		1.7	1.8	Mpa	@ 15 °C
	Horizontal strain	3.3	3.3	mm/m	@ 0 °C
		10.6	8.3	mm/m	@ 15 °C
	E-modulus	584	1414	MPa	@ 0 °C
		240	366	Mpa	@ 15 °C
	Toughness	7.0	5.1	N/mm	@ 0 °C
		7.5	9.5	N/mm	@ 15 °C

Table 2. Asphalt mix properties

Property		Site A		Site B		Unit
Aggregate	type - proportion	Basalt (Wadi)	52	Limestone	61	%
Sand	type - proportion	Crusher sand	40	Crusher sand	34	%
		Dune sand	11	Dune sand	7	%
Filler	type - proportion	Crusher dust	2.9	Crusher dust	3.1	%
	type - proportion			Cement	2.0	%
Asphalt absorbtion		1.23		0.43		%
Bitumen	proportion	5.5		5.5		%
	film thickness	6.7		9.5		µm
Mixing	cycle	45		30		sec.
	temperature	160-165		150-160		°C
Compaction technique		no vibration		vibration used		
Marshall	job mix density	2449		2350		g/cm^3
	stability	12200		11680		N
	flow	2.9		3.5		mm
	air voids	4.0		2.1		%
	VFB	70		85		%
	VMA	15.2		14.0		%

Table 3. Crack movement measurements

Location	Layer	Average movement (mm)
Site A	CTB	0.70
Site A	ACB	0.35
Site B	CTB	0.22
Site B	ACB	not cracked

4.3 Site measurements

Lacking advanced measuring equipment on these remote sites, crack measurements were taken at regular intervals during the day at several locations, either with a vernier calliper, measuring the distance between two fixed points (nails) on either side of the crack, or by means of crack monitor. The average values as presented in Table 3 show a significant difference between the two sites.

The movement measurements were accompanied by temperature measurements in various layers of different stages of construction. The results are presented in Table 4 and 5.

Table 4. Temperature registration Site A

Construction layer and position		Maximum (°C)	Minimum (°C)	Difference (°C)	
Ambient		45	27	18	
CTB	at 60 mm depth	53	33	20	(100%)
Ambient		43	25	18	
ACB-1	middle	65	32	33	
CTB	at 60 mm depth	52	38	14	(70 %)
Ambient		42	23	19	
ASC	middle	62	30	32	
ACB-2	middle	59	35	24	
ACB-1	middle	52	39	13	
CTB	at 60 mm depth	49	43	6	(30%)

Table 5. Temperature registration Site B

Construction layer and position		Maximum (°C)	Minimum (°C)	Difference (°C)
Ambient		41	26	15
CTB	top	47	27	20
CTB	at 65 mm depth	43	28	15
CTB	at 180 mm depth	38	31	7

Table 6. CTB material

Location	Coefficient of thermal movement (E^{-6}/°C)	Type of aggregate
Site B	6.3	Limestone
Site A	14.6	Basalt

Based on the collected data, the coefficient of thermal movement of the CTB materials were estimated and found to be significantly different in both sites due to the different type of aggregate used, refer to Table 6.

4.4 Conclusion

The poorer mix properties of Site A and the high temperature coefficient of the aggregate in the CTB in combination with the local large diurnal and seasonal temperature variations were identified as the main contributors to the early crack development.

5 Remedial Measures

Whatever improvements were possible pertaining to the mix properties were made immediately, which were mainly the adjustments of grading, increase of binder content and better compaction, all at Site A. These improvements were later confirmed by means of a similar test regime. It was felt that the improvements in isolation were not sufficient. Based on a literature study, which included earlier publications of this conference, various additional remedial measures were considered:

- additional asphalt layer
- interlayer, i.e., SAMI / impregnated geotextile / reinforcing grids
- crack widening and structural fill
- polymer modified asphalt
- sand asphalt
- replacement of the CTB by a crushed aggregate base
- any combination of the above

Because both crack width and movement were relatively small and service records of application of polymer modified bitumens in the country or as a SAMI in airfield pavements were lacking, the choice was made to increase the asphalt thickness by 50 mm and to introduce an unmodified bitumen impregnated geotextile. This solution allowed incorporation of works already completed. Further the implementation of the remedial measures could follow the staged completion of the pavement works. The final remedial measure adopted is shown in Fig. 3.

Additional asphalt concrete layer	50 mm
Asphalt Concrete Surface Course	50 mm
Impregnated Geotextile interlayer	
Asphalt Concrete Base Course 2	60 mm
Asphalt Concrete Base Course 1	60 mm
Cement Treated Base Course	260 mm

Base construction unchanged

Fig. 3. Revised pavement construction Site A

6 Construction

Material and construction specifications for the geotextile interlayer were prepared and the non-woven textile AM2 from Fibertex was selected. Fibertex was able to provide the relevant reference projects and the technical support. The geotextile was mechanically unrolled by means of the manufacturers provided un-rolling equipment mounted on a adjusted agricultural tractor onto a layer of hot-sprayed bitumen of approximately 0.90 kg/m^2.

The start of the execution was supervised by an experienced foreman, arranged through Fibertex, which clearly made the difference. Especially in hot and windy conditions, tight control on the amount of sprayed bitumen is crucial. With too little bitumen, the geotextile does not stick to the surface and blows away, whilst too much bitumen will cause the geotextile to be picked up by the wheels of the asphalt trucks.

Construction activities resumed at full speed late 1993. Today, three winters later, it has been confirmed that cracks have not reappeared.

7 Conclusions

Design of flexible pavement using a semi-rigid base has a number of performance benefits especially at airports with heavy wheel loads and high tyre pressures. Further, good quality bases can be produced using relatively lower quality granular materials which offer attractive cost benefits. However, proper attention to the design and specification stage is necessary to prevent the occurrence of reflection cracks. Areas requiring particular attention are:

- Assessment of crack movement based on the expected thermal properties of the available materials at site in relation to the local temperature variations.
- Pre-cracking of a semi-rigid base is a must and the pattern of pre-cracking should be related to the assessed crack movement.
- Specify a simple, effective and practical, cheap pre-cracking method such as shown in Fig. 4. Since then this system has been used extensively at Schiphol Airport Amsterdam.
- Specify asphalt mixes which are ductile and creep resistant.
- Incorporation of crack growth mitigation or arresting layer(s) into the structural design of the pavement.

In case impregnated geotextile is included, a well experienced foremen/crew is essential to a smooth and good execution without headache.

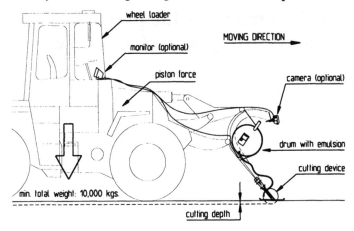

CUTTING DEVICE FOR C.T.B. DEVELOPED BY NACO

Fig. 4. Improved cutting device

8 Acknowledgement

W. De Wijs, Ballast Nedam Grond en Wegen B.V., for initiating and editing this contribution.

9 References

1. FAA Advisory Circular 150/5320-6C (1978) *Airport Pavement Design and Evaluation.*
2. UKMOD Property Service Agency (PSA) (1990) *A guide to Airfield Pavement Design and Evaluation*

Steel reinforcement for the prevention of cracking and rutting in asphalt overlays

J.R.A. VEYS
N.V. BEKAERT S.A., Zwevegem, Belgium

Abstract

In 1981 trials were first carried out using Mesh Track as reinforcement in asphalt overlays. Following these succesful trials Mesh Track steel reinforcement has been used regulary to prevent cracking and rutting in many problem areas. After 15 years of widespread practical experience the opportunity exists to make a full evaluation of the effectiveness and the durability of steel reinforced asphalt. This paper evaluates the product itself within a number of application areas, e.g. overlays on peat, on concrete slabs, underdesigned roads etc., and also focusses on recommendations for laying the product taking into account the latest developments and trends within road construction (e.g. slurry sealing, porous asphalt mixes, recycling potential of asphalt). Keywords: cracking, rutting, application areas, laboratory testing, project evaluation, practical advice, recycling, steel reinforcement, slurry sealing.

1 Introduction

Over the past fifteen years Mesh Track steel reinforcement nettings have been used as a solution for a number of problem areas :

  -asphalt overlay on cement concrete roads
  -prevention of both primary and secondary rutting on asphalt roads
  -asphalt overlays on peat roads (weak subbases)
  -asphalt overlays in hilly areas (slippage problems)
  -prevention of cracking problems due to freeze/thaw cycles
  -prevention of cracking problems due to shear (e.g. widened roads)
  -reinforced overlay for underdesigned asphalt roads
  (alternative to reconstruction)

Reflective Cracking in Pavements. Edited by L. Francken, E. Beuving and A.A.A. Molenaar. © 1996 RILEM. Published by E & FN Spon, 2–6 Boundary Row, London, SE1 8HN. ISBN 0 419 22260 X.

The purpose of the present paper is to show the efficiency and durability of steel reinforced asphalt for a range of application areas. These statements are based on large field experience of installing several million of square meters of Mesh Track on different application areas, on field evaluations of several of these projects and on extensive laboratory research.

The present paper highlights the following interesting topics :

- application areas of steel reinforcement nettings
- results from laboratory tests, showing the efficiency of steel reinforcement nettings and the importance of the fixing technique (nailing versus slurry seal)
- results from long term pavement performance with recommendations concerning laying procedures and prior repairs
- recycling experience

2 Application areas

A comparative analysis was carried out in 1992 by a working group of the CROW in the Netherlands (ref. 2). Tables 1 and 2 give an overview of the technical possibilities to solve different aspects of dammage on the road.

Notation : ++ good solution / + possible solution / - no solution

Table 1 : Technical possibilities to solve different aspects of dammages

	Grid		Netting	
	Steel	Plastics	Steel	Plastics
Reflective Cracking				
- K1-mode	++	++	++	++
- K2-mode	++	+	++	+
Widening	++	++	++	++
Long. cracks				
- edge interlock	++	++	++	++
- fatigue	++	++	++	++
Transv. cracks	++	++	++	++
Slippage	++	+	++	+
Deformation				
Rutting				
primary	++	+	++	+
secondary	+	-	+	-
Corrugation	+	-	+	-
Increase bearing capacity	+	+	+	+

Table 2 : Technical possibilities to solve different aspects of dammages

| | Asphalt overlay | Woven | Sami | | Fibres |
			Nonwoven	Bitiminous inlay	
Reflective Cracking					
- K1-mode	+	+	++	++	++
- K2-mode	-	-	-	-	+
Widening	-	+	-	-	-
Long. cracks					
- edge interlock	-	+	-	-	-
- fatigue	+	+	+	+	++
Transv. cracks	+	+	+	+	+
Slippage	-	-	-	-	-
Deformation					
Rutting					
primary	++	-	-	-	-
secondary	+	-	-	-	-
Corrugation	+	-	-	-	-
Increase bearing capacity	++	-	-	-	+

The above tables evaluate the performance of different inlays in 3 major problem areas. From these it is possible to highlight the main differences/advantages of a steel netting within these different domains.

Anti-cracking properties
Because of its rigidity (resistance to shear and deflection) and tensile strength Mesh Track (steel netting) performs better in solving damage due to K2-mode tension intensity factors (e.g. widened roads).
The efficiency of Mesh Track in preventing cracking due to slippage is because of the products design. Lateral slippage (creep of the asphalt) is prevented through the interlock of the asphalt aggregates into the large hexagonal meshes.
The above characteristics also form the basis of Mesh Track's excellent performance in solving the problems of cracking caused by K1-mode tension intensity factors, fatigue and edge interlock.

Deformation
The compartmentalisation of the aggregates into the Mesh Track meshes is the ideal way to prevent both primary rutting and corrugation (due to creep of the asphalt).
Because of its high Young's modulus (200 MPa) Mesh Track offers high flexural stifness to the reinforced overlay. It ensures an optimal distribution of the load and limits the tension and deformation in the subbase (secondary rutting).

Bearing capacity
3-point bending tests have proven that Mesh Track reinforced asphalt provides extra toughness to the overall structure. These tests were carried out by NPC in 1988 and the results were presented for the first Rilem-Congres in 1989 (ref. 7).

3 Results from laboratory testing

The BRRC developed a test device to assess the effectiveness of anticracking interlayers, more particularly when used on concrete slabs subject to thermal expansion and contraction (ref. 1). Figure 1 gives a schematic view of the test .
In this test, cyclic variations of the crack opening in the cement concrete base are performed at very small rates in order to simulate thermal movements of the underlayer. The onset and propagation of the crack in the asphaltic overlay are continuously followed during the test in order to investigate the effectiveness of the interlayer system.

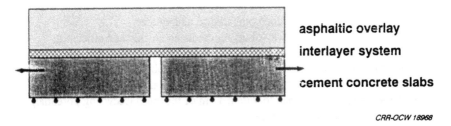

Figure 1 : Schematic view of the BRRC thermal cracking test (ref. 1)

3.1. Effectiveness of interlayer systems

Figure 2 below represents the development of crack length observed in a bituminous overlay during tests carried out with various interface systems.

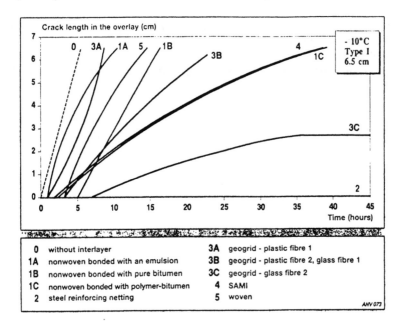

Figure 2 : Results of thermal cracking tests with various types of interlayer

From fig. 2 it is clear that :
- No cracks were observed in the case of steel reinforcement nettings, whereas a reference sample without interlayer fails completely after three hours of testing.
- All interlayers except steel reinforcing netting control/delay the speed of crack growth. Only steel reinforcing netting (Mesh Track) prevents crack growth.

3.2.   Influence of the fixing technique : nailing versus embedding in a slurry seal

In 1988 slurry sealing was introduced as an alternative to nailing for fixing Mesh Track to an existing road.
The main advantages of slurry sealing compared to nailing are :
- it is a user-friendly, economical way of laying the mesh quickly and efficiently
- slurry sealing gives a strong uniform bond to the existing pavement
- slurry sealing prevents water from penetrating into the sub-base
- it enables the thickness of the top layer to be reduced by at least 1 cm
- asphalt layers can easily be planed to within 1 cm of the mesh (recycling potential)
- it makes laying the top layer very easy
These benefits were found on a variety of different job sites.

Because of the increasing popularity of slurry sealing and the move towards modified bitumen the BRRC examined the combined efficiency (particularly regarding thermal cracking) of Mesh Track with a (polymer) modified slurry seal by thermal cracking tests and compared the results with those of nailed Mesh Track. From this research three key areas were highlighted (ref. 3) :

- the effect of the extra elasticity in the interlayer offered by the modified bitumen slurry seal (strain absorption interlayer)
- the bond offered by the slurry seal
- the degree of reinforcement offered by the Mesh Track (stress absorption interlayer).

Extra elasticity : Figures 3 and 4 show the result of the force exerted on the concrete block during the test in order to provide the cyclic variations of the crack opening.

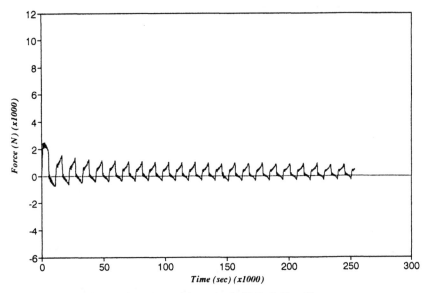

Figure 3 : Forces on the concrete base (slurry sealed Mesh Track)

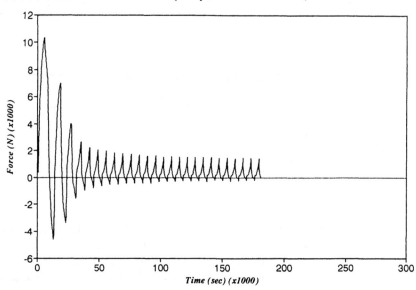

Figure 4 : Forces on the concrete base (nailed steel mesh)

It is clear that with slurry sealing the forces are much lower than with nailing. This is due to the strain absorption offered by the modified bitumen slurry seal. This absorption function is at a maximum at the beginning of the test.

Additionly a strong bond is given by slurry sealing the Mesh Track. This is shown in Figure 5 which indicates minimal slippage of the overlay to the underlayer.

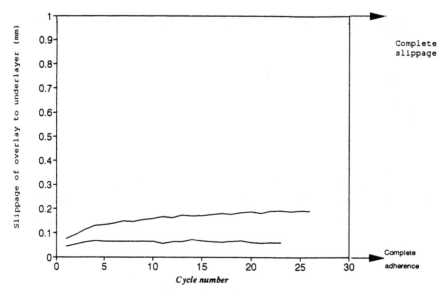

Figure 5 : Slippage of overlay to underlayer for slurry sealed Mesh Track
<u>Finally, the stress absorption potential or reinforcing aspect of the steel mesh</u> is
shown in table 3.

Table 3

	Nailed Mesh Track	Slurry sealed Mesh Track	No interface
Max. strain asphalt (mm) - 1st cycle	0.146	0.017	0.46
Max. strain asphalt (mm) - 23rd cycle	0.047	0.011	1.0

The deformation/strain of the asphalt (measured above the cracked concrete base) is
lowest for a slurry sealed Mesh Track. This is due to its dual function : the modified
slurry seal acts as a strain absorption interlayer whilst the Mesh Track acts as a
reinforcement to take up the tensile stresses at the bottom of the overlay.
It can be concluded from the tests of the Belgian Road Research Centre which prove
the dual function (strain + stress absorption) of a slurry sealed Mesh Track as an anti-
cracking interlayer that this is the recommended method for fixing a steel mesh to an
existing pavement.

4 Long term pavement performance of steel reinforcement nettings

In 1994 an evaluation was made by the Belgian Road Research Centre of different
projects with steel reinforcement nettings (ref. 4). Ten projects, of which the time
after renovation varies between two and five years, were selected in order to assess
the effectiveness and durability of Mesh Track in several problem areas :

- Five projects concern overlays on cement concrete slabs; two projects concern overlays on lean-mix concrete and three projects are overlays on cracked/rutted asphalt roads.
- On six projects nails and hooks were used to fix the Mesh Track ; on four projects the Mesh Track was slurry sealed.

Table 4 gives an overview of the different projects, with their year of repair, the reason of repair, the structural design and the results of the evaluation.
From table 4 it is clear that in terms of reflective cracking Mesh Track was very effective :
- Three projects showed no reflective cracking
- Two projects showed one reflection crack
- For project No. 6 (N499) reflective cracking was observed : one on fifteen cracks/joints reappeared after three years of repair. The cracks are fine and do not lead to disconfort for the user yet. The higher percentage of reflective cracking is due to the limited 4 cm overlay thickness, combined with large vertical movements ("rocking") at the edges of the concrete slabs. Measurements have been performed of the vertical movements at the edges of the slabs before overlaying : they were of the order of 1 mm. Better results undoubtedly could be obtained by cracking and seating the "rocking" slabs before overlaying and/or by increasing the overlay thickness (min. 50 mm).
With rutting the results were different for primary and secondary rutting. Mesh Track prevented secondary rutting, because it offered extra bearing capacity to the overall structure. Mesh Track was seen to limit the degree of primary rutting due to the interlock of the aggregates into the large meshes of the netting. However if the rutting is strictly non-structural (creep of the asphalt mix), the solution lays initialy in the asphalt mix itself. If a revised asphalt mix (e.g. drainasphalt or SMA) does not solve the problem the addition of Mesh Track will assist.
From these evaluations and from practical experience it is recommended that following practical points should be taken into account:
- crack and seat unstable ("rocking") concrete slabs before overlaying
- apply a regulating layer if required (This avoids the appearance of cavities between the steel reinforcement nettings and the underlayer. These holes can lead to badly compacted spots or may even remain at the stage of overlaying. They give rise to potholes already very shortly after repair.)
- the use of a wheeled tyre roller on the Mesh Track ensures the flatness of the netting (no undulations)
- a minimum overlay thickness above the Mesh Track is needed to guarantee an even surface (5 cm on slurry sealed Mesh Track, 6 cm on nailed Mesh Track)
- ensure good drainage of water alongside the road
- slurry sealing will guarantee a good uniform bond of the Mesh Track to the existing surface.

Table 4 : Project evaluation

Nr	Project id. Year of repair	Reason of repair	Sub-base + asphalt overlay	Evaluation
1	N 31 Bruges 1989	Rutting	lean concrete 21 cm asphalt (2 diff. types for wearing course)	good, positive effect from use of MT on rutting clearly to be seen
2	E17 Beervelde 1991	Rutting	gravel 20 cm asphalt	good, positive effect from use of MT on rutting clearly to be seen
3	Zwevegem Bellegemstreet 1990	Rutting + Cracking	section 1 : gravel + 10 cm asphalt section 2 : concrete slabs + 10 cm asphalt	very good, one reflective crack in section 2 has been observed
4	Bruges Lange Rei 1989	Cracking	lean concrete 5 cm asphalt	excellent, no damages
5	N 415 Zwalm 1991	Cracking	concrete slabs 14 cm asphalt	excellent, no damages
6	N 499 Maldegem 1991	Cracking	concrete slabs 4 cm asphalt	good, one on fifteen joints did reflect
7	Tournai Mont St Aubert 1989	Cracking	concrete slabs 8 cm asphalt	very good, initiation of one reflective crack, one pothole observed
8	Aalter - Nijverheidslane 1992	Cracking	concrete slabs (cracked + seated) 9 cm asphalt	excellent, no damages
9	N368 Beernem 1991	Cracking	gravel 12 cm asphalt	very good, initiation of one pothole
10	Grimbergen Welvaartsdijk 1992	Rutting + Cracking	gravel 12 cm asphalt	fair because of drainage problems , one pothole, 10- 15 mm of rutting

5 Recycling possibilities of steel reinforcement nettings

An evaluation was made in 1995 by the CROW in the Netherlands on the recycling possibilities for Mesh Track reinforced asphalt (Ref. 5).

A number of trial sections of different inlays were installed in order to evaluate the recycling potential of several products. The Mesh Track was slurry sealed onto an existing asphalt road and covered with a thin 60 mm asphalt overlay.

The working group gave the following general recommendations on recycling steel netting : "When working with steel netting as a reinforced inlay the overlay should be planed down before pulling out the mesh. The two separate products can be fully recycled as the asphalt doesn't contain any foreign substances from the mesh. There are no obstacles to warm recycling. Because of the high steel quality it is not recommended to plane through the mesh." The practical recommendations from their recycling testing are : "The steel mesh lays flat on the surface so it is safe to plane down to 1 cm above the netting without touching the mesh. It is no problem to pull the mesh from the surface using a hydraulic crane. As such it is easy to separate the reinforcement mesh from the asphalt." These tests proved that recycling Mesh Track is possible. However when renewing the overlayed surface as a company we recommend to plane down the overlay to 1 cm above the mesh. Due to the durability and corrosion resistant characteristics of the heavily galvanised mesh (ref. 6) it is recommended that the mesh track be left in place. The old mesh will still serve as an anti-cracking and anti-rutting interlayer for the new overlay.

6 Conclusions

The fifteen years of practical experience combined with extensive research have proven the effectiveness and the durability of Mesh Track within a number of application areas.

The main development has been proving the dual functionality of a slurry sealed (based on modified bitumen) Mesh Track. The extra elasticity offered by the modified slurry seal reduces the tensile strain at the bottom of the overlay. This has made it possible to adapt the existing Mesh Track and introduce a new steel mesh called Mesh Track 2 designed specially for renovating cracked asphalt roads.

7 References

1.   Belgian Road Research Centre (3/1995) *Anticracking Interlayers*, OCW Bulletin by Mr. L. Francken and Mrs. A. Vanelstraete

2.   CROW (1993) *Asfaltwapening : zin of onzin ?*, Publikatie 69, pp. 44-45

3.   Belgian Road Research Centre (1996) *Test report EP3775*, Thermal cracking test on slurry sealed Mesh Track

4.   Belgian Road Research Centre (1995) *Report EP 3563*, Evaluation of Mesh Track Projects

5.   Stichting CROW (1995) *Geen vrees voor de frees*, Publikatie 95

6.   Moens J. (1989) *Corrosion of galvanized wire reinforcement in asphalt coatings*, Ref. nr 046040/002/89 (RES 013455)

7.   Molenaar, A.A.A. (1988) *Reinforcing asphalt pavements with Mesh Track*

Geotextile anti-cracking interlayers used for pavement renovation on Southern Poland

W. GRZYBOWSKA and J. WOJTOWICZ
Institute of Roads, Railways and Bridges, Cracow University of Technology, Cracow, Poland

Abstract
In the paper, three selected projects of pavement renovation using nonwoven geotextiles as the interlayer retarding the thermal reflective crackings, are presented. The first part refers to the reconstruction of 1 km section Bielsko-Cieszyn of the national road No 1 Warsaw-Cieszyn, loaded with very heavy traffic. On the some stretch located on upgrade 6-7 %, the distresses manifested in the form of thermal crackings, as well as ruts, also with crackings. The water bleeding was seasonally observed. The rehabilitation works were realized in 1993.

The second project concerns the renovation of main street in the province capital -Bielsko, heavily loaded with traffic, at poor subgrade condition. The works were performed in 1994.

The third project had the pilot character, as its objective was to gain the first experiences at double surface treatment with application of geotextiles. The sections length of 600 m were situated on the national road No 778 Kraków - Wolbrom. Part of it were constructed in the late summer 1993, the rest in July 1994. The project included also the laboratory tests determining the optimal contents of the components of the surface treatment system. Constructed sections are under observations, until now no reflective crackings are noticed.
Keywords: nonvowen geotextiles, thermal cracks, overlay, surface treatment.

1 Introduction

The first attempts of experimental applications of nonvowen geotextiles to a renovation of bituminous pavements in the South of Poland were undertaken in the late eighties, when these materials were used to cover the longitudinal joint between

Reflective Cracking in Pavements. Edited by L. Francken, E. Beuving and A.A.A. Molenaar. © 1996 RILEM. Published by E & FN Spon, 2–6 Boundary Row, London, SE1 8HN. ISBN 0 419 22260 X.

the carriageway and the paved shoulder. The asphalt concrete was then mechanically laid on the whole width of the road (carriageway + shoulder) and compacted. The experiences gained during those works are related to the way of pretreatment of the road, manual or mechanical geotextile unrolling systems, longitudinal and transversal overlaps, repairing of the folds, and controlling the run of paver on the geotextile layer. As a result of those works, it has been found, that when the difference between the bearing capacity of basic traffic lane and the paved shoulder was substantial, the nonvowen geotextiles did not retard the appearing of the cracks. The values of deflection measured with the Benkelman Beam on sections with and without geotextiles proved, that the presence of the nonvowen geotextile layer did not bring about the effect of the strengthening. On the base of the practical knowledge, as well as the results of theoretical analyses (Salamon, 1994) [1], several projects on the application of nonwoven geotextiles against the reflective thermal cracking were realized. The obtained results are very good. Typical three projects realized in the South of Poland, with application of the Polish geotextiles are discribed in this paper.

The first presented project refers to the reconstruction in 1993 the national road No1 Warsaw-Cieszyn, loaded with very heavy traffic.

The second project concerns the renovation of main street in the province capital - Bielsko, i.e.street Bystrzańska, submitted to the very heavy traffic, at poor subgrade condition. The works were completed in 1994.

The aim of third project was to gain the first experiences at double surface treatment with application of the geotextiles. The section was situated on the national road No 778 Kraków - Wolbrom. Part of it was constructed in the late summer 1993, the rest in July 1994.

2 Renovation of the section Bielsko-Cieszyn of the national road Nr 1,

2.1 The condition of the existing pavement

The renovated section was originally constructed in years 1970-75 as the bypass of the town Skoczów, nearby the South frontier with Czech Republic. The pavement structure was designed for the heavy traffic, and comprised the bituminous layers, 16 cm of lean concrete with contraction joints and 12 cm of cement stabilized layers. In 1987 the next bituminous layers were laid in order to strengthen the structure and improve the longitudinal and transversal evenness, but any means against the reflective crackings were not applied. In 1993 the damages on the cut under the consideration, i.e. from km 617+405 to km 618+347, were already severe, and the transveral cracks reflected from the stiff base were accompanied by deep ruts and other defects. Some fatigue cracks were also registered on the 40 m long cut, because of unsatisfactory dranaige at that place and, in the consequence, low bearing capacity of the pavement structure.

Surveys of crackings were conducted in June 1993, at the high temperature, and, what was obvious, many microcracks in bituminous surface were then closed, but even though, 30 transveral cracks with openings >5 mm were identified. The cracks were randomly spaced and the distances were as follows:

- less than 25 m   - 75 % of the total number of cracks
-               25-40 m   - 7 % "   "   "   "
- more than 40 m   - 20 %.

The openings of cracks were different, in majority wider than 5 mm and at the pavement surface even wider than 40 mm (crumbled edges).

2.2 The concept of the repair works

The prevailing thermal character of cracks, encouraged the designers to apply the nonwoven geotextile interlayer to the overlay structure. In their opinion, the long lifetime service of the pavement gave the substantial probability, that thermal cracking processes already have been completed, and the majority of internal cracks is visible also on the surface. Above assumption allowed to plan the laying the geotextile layer only on the discontinuities of the pavement, not on the whole area. The concept of repairing of the pavement included the following works;

1. Dranaige of wet places in the subgrade of pavement, i e., reconstructing the whole thickness of the pavement structure with the subgrade in the weak spots, to the depth lower than the frost limit, that means about 1.2 m.
2. Milling the upper 8.5 cm layers, in order to cut the ruts and profiling to obtain a regular surface pattern under the geotextile layer.
3. Filling the cracks with openings wider than 3 mm in the following way:

- opening 3-10 mm : filling with the cationic emulsion
- opening >10 mm : after spraying the cationic emulsion inside the crack, filling with the fine graded bituminous mixture hot or cold

4. Thorough cleaning of the pavement surface, as no dirt or loose pavement material should be left after.
5. Spraying down the bitumen (asphalt $70°$ P) on the areas of discontinuities, with a quantity 1.1 kg/m^2, at the width of spraying 60 cm (+15 cm of bitumen surplus) on both sides of a crack, at the temperature of asphalt equal $140-150°$ C. The quantity of asphalt should be thoroughly controlled, because its deficit can entail an insufficient bond to a lower layer and an inadequate impregnation of the geotextile with bitumen. Excess of it can cause the slipping of the geotextile inside the overlay structure.
6. Manual unrolling geotextile along the crack, covering a both sides of a crack symmetrically, with the total width   of 120 cm. Unrolling should be done as quickly as possible after spraying, when bitumen is still hot. The places of laying geotextiles were marked very thoroughly at the road side.
7. Driving over geotextile layer with rubber roller or hand steel roller in order to improve the adhesion of the geotextile to lower layer.
8. Putting the asphalt mixture with the finisher onto the whole, repaired areas.

Detailed specifications for the above listed   works   were enclosed to the documentation [2].

The scheme of the applied overlay structure is presented in Fig.1.

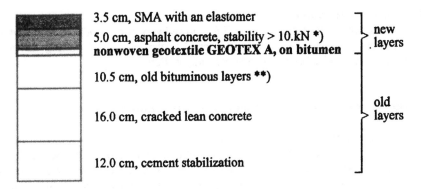

3.5 cm, SMA with an elastomer

5.0 cm, asphalt concrete, stability > 10.kN *)
nonwoven geotextile GEOTEX A, on bitumen
} new layers

10.5 cm, old bituminous layers **)

16.0 cm, cracked lean concrete
} old layers

12.0 cm, cement stabilization

*) stability according to the Marshall method

**) after milling of 8.5 cm

Fig. 1. The scheme of the overlay structure with nonwoven geotextile
as anti-termal cracks layer (road section Bielsko-Cieszyn)

2.3 Properties of the nonwoven geotextile applied as anti-termal cracks layer
The applied nonwoven geotextile GEOTEX A is the product of Polish Textile Company WATINA in Łęczyca town. Its properties were as follows:

- type of the material - polipropylen, quilted with a polyester thread, the resistance to the temperature - 155° C
- tensile strength: 6.1 kN/m
- elongation at break: 37 %
- unit weight: 220 g/m^2
- thickness : 1.2 mm

2.4   Properties and quantity of the bitumen applied   to impregnate the nonwoven geotextile
For impregnation the nonwoven geotextile as well as to bond it to the low layer, the asphalt cement 70° P was applied, with the following properties:
- softening point (ring and ball method): 55° C
- break point (Fraass method): -20° C

An amount of asphalt was calculated according to the formula developed by the Task Force 25, Method 8 [3]:
$$Q_{ef} = Q_o + Q_s \pm Q_c \qquad (1)$$
where:

Q_{ef} - effective amount of binder

Q_o - amount of binder to regenerate of the old surface ($0.2 \div 0.3$ kg/m^2)

Q_c - amount of binder to be absorbed by the geotextile, determined
experimentally, or according to formula Task Force 25 [3]

Puting corresponding data to the formula, one obtains the calculated amount of asphalt equal to 1.1 kg/m^2.

2.5 Results of observations
Observations carried out till the April 1996 showed very good condition of the whole repaired section, without any reflective cracks (the places of putting down of geotextiles were marked), only with very small, shallow ruts.

3   Renovation of the street Partyzancka in voivoda town Bielsko

3.1 Condition of the existing pavement
The new overlay had been designed for the section of 3 km length, for the very heavy traffic. The pavement was in the very bad condition, with a great number of thermal and fatigue cracks, "sickle" cracks (caused by the inadequate compaction of bituminous layer with the incorrect composition), potholes, ruts and other deformations, giving the evidence of the low mixtures stability. Those observations were confirmed by the laboratory tests, which showed, that the Marshall stability of examined samples was equal 55-65% of the required stability. Besides, bearing capacity investigations with Benkelman beam proved, that the elastic deflections do not meet the requirement for the heavy traffic, so the pavement structure needs strengthening. In the subgrade, mostly the cohesive soils occured, sporadically mixed with the burnt colliery shales. The ground water table was upraised only at the end of the chosen section.

3.2 The concept of the repair work
Numerous cracks of the different origin required applying of stress absorbing interlayer, which in that case, in the opinion of designers, should have been the nonwoven geotextile, reinforced with the grid. A composite material ought to have high strength, low elongation and high temperature resistance. Because of the lack of such high quality material (and also because of economic reasons), designers decided to perform the full strengthening with bituminous mixture, using the nonwoven geotextile on almost whole area of repaired section, in order to retard at least the propagation and reflection of thermal cracks. The concept of the pavement repairing included the following works:

1. Dranaige of wet places in the subgrade of pavement, see point 2.2.
2. Milling the upper 5.0-6.0 cm layers, in order to cut the ruts and profiling the old pavement, to obtain a regular surface pattern under the geotextile layer.
3. Filling the cracks, see point 2.2.
4. Cleaning the pavement surface, see point 2.2.
5. Spraying down the bitumen (asphalt 70° P) on whole appointed area with a quantity 1.1 kg/m2, see point 2.2.
6. Unrolling geotextile mechanically; it should be done as quickly as possible after spraying, when bitumen is yet hot. The longitudinal folds were less than 15 cm.
7. Driving over geotextile layer with rubber roller or hand steel roller to take sure about good adhesion the geotextile to lower layer.
8. Putting the asphalt mix with the paver onto whole, the repaired area.

Detailed specifications for listed above works were enclosed to the documentation [4]. The scheme of the applied overlay structure is given in Fig.2.

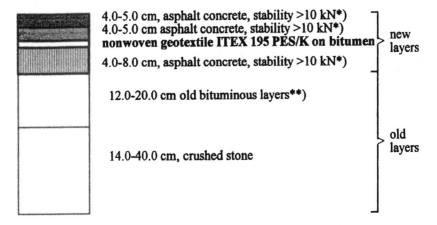

4.0-5.0 cm, asphalt concrete, stability >10 kN*)
4.0-5.0 cm asphalt concrete, stability >10 kN*)
nonwoven geotextile ITEX 195 PES/K on bitumen
4.0-8.0 cm, asphalt concrete, stability >10 kN*)
} new layers

12.0-20.0 cm old bituminous layers**)

14.0-40.0 cm, crushed stone
} old layers

*) stability according to the Marshall method
**) after miling of 5.0-6.0 cm

Fig.2. The scheme of the overlay structure with the nonwoven geotextile
(Partyzancka street in Bielsko)

3.3 Properties of the nonwoven geotextile
The applied nonwoven geotextile ITEX 200 PES/K was the product of Polish Textile
Institute in Łódź. Its properties are as follows:

- type of the material - 100 % polyester
- tensile strength:- longitudinal 7.3 kN/m
                           - transversal   10.3 kN/m
- elongation at break: - longitudinal 84.6 %
                                     transversal 83.8 %
- unit weight: 214 g/m^2
- thickness: 1.58 mm

3.4 Properties and quantity of the bitumen applied   to impregnate the nonwoven geotextile
For impregnation the nonwoven geotextile as well as to bond it to the low layer, the
asphalt cement 70° P was applied, with following properties:

- softening point (ring and ball method): 55° C
- break point (Fraass method): -20° C

The recommended amount of asphalt equal 1.1 kg/m^2 was calculated
accordingly to the formula developed by the Task Force 25 [3] and modified by the
authors in respect of its fluffy character (the thickness > 1.20 mm [4],[5].

To determine the adequate bitumen amount, the conception of the *"Fibers
concentration index"* was introduced [5]:

$$FCI = \frac{FG}{d \cdot \gamma \cdot G \cdot 1000} \quad [cm^3] \qquad (2)$$

where: FCI - fibers concentration index $[cm^3]$

FG - unit mass of geotextile $[g/m^2]$

thickness of geotextile   [mm]

γG - geotextile specific density $[g/cm^3]$

The corrected thickness for the fluffy geotextilies with thickness more than *1.2 mm* is suggested to be calculated accordingly to the formula:

$$d_{cor} = \frac{FCI \cdot d}{FCI_{ref}} \qquad (3)$$

where:

d_{cor} - corrected thickness of the given geotextile [mm]

FCI - fibers concentration index for the given geotextile $[cm^3]$

FCI_{ref} - fibers concentration index for the reference geotextile.

3.5. Results of observations

Observations have been carried out to the spring 1996 and showed good condition of the repaired section, without any reflective cracks.

4 Pilot section on double surface treatment with usage of geotextiles

4.1. The aim of the pilot section

In summer 1993, the Institute of Road and Railway Engineering, Cracow University of Technology, together with the Regional Directorate of Public Roads in Cracow, undertook the realization of the pilot section on old national road pavement. The technology of double surface treatment was applied, with an interlayer made of the Polish nowoven geotextile, saturated with binder.

The goal of this experiments was to verify the performance of the structure improved with geotextiles.

Predicted traffic was a heavy one, while, according to the deflections values measured with the Benkelman Beam, the bearing capacity of the section have met the requirements only for the medium-light traffic loading. Nevertheless, the separate, strengthening bituminous layer was not applied, (besides the profiling layer of average thickness 4 cm), in order to make faster the possibly appearance of reflective crackings and to gain the quicker estimation of the geotextile influence on retarding the crack propagation.

The more detailed description of this technology is given in the paper [5].

The conception of this technology is presented in Fig.3.

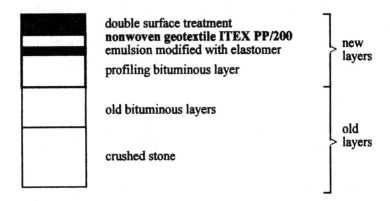

double surface treatment
nonwoven geotextile ITEX PP/200
emulsion modified with elastomer

new layers

profiling bituminous layer

old bituminous layers

old layers

crushed stone

Fig.3. The conception of application of geotextile as interlayer in double surface treatment, realized on the pilot section at the road Kraków-Wolbrom, km 21+400 ÷ 22+000

4.2 Properties of the nonwoven geotextile

The applied nonwoven geotextile ITEX 200 PP is the product of the Polish Textile Institute in Łódź, developed in collaboration with Cracow University of Technology. Its properties were as follows:

Type of material: 100 % polypropylene, the mixture of fibers of various dimensions:

- 18 dtex, length of 120 mm - 66 %

  6 dtex, length of 60 mm - 34 %
- unit mass: 180 g/m^2, ±10 % (in this case 195 g/m^2)
- density: 0.92 g/cm^3
- thickness : 2.7 mm
- tensile strength (width of the strip 5 cm): longitudinal: 4.7 kN/m
  transversal :7.0 kN/m
- elongation at break: longitudinal : 86 %
  transversal :83 %
- temperature resistance: 120 °C

4.3 Properties and quantity of the asphalt emulsion applied to impregnate the nonwoven geotextile

For saturating the geotextiles and bonding it to lower layer, an asphalt cationic emulsion was applied. Its properties were as follows:

- density: 1.04 g/cm^3
- asphalt amount = 69 %
- Engler's viscosity in 20 °C ÷ 11 °E
- breaking time: 5 minutes
- modification with the elastomer SBS

The quantity of the emulsion was calculated according to the formula developed by the Task Force 25 Group [3], modified by the authors for fluffy, not calandered geotextiles (the thickness under 2.0 kPa > 1.2 mm), the way of modification is discribed in the point 3.4 [5].

The obtained quantity of asphalt emulsion is equal up to 1.5 kg/m^2. Details of the calculations are given in the paper [5], the obtained values are presented in the table 1.

4.4 Quantities of materials

Table 1. Quantities of materials determined for the pilot and reference sections in comparison to the Polish Specifications (without the materials of profilig layer)

Components for the double surface treatment with nonwoven geotextile	Specification requirements	Values determined for pilot sections
The pilot section		
• Asphalt emulsion for saturating nonwoven geotextile and bonding to the lower layer	-	1.5 kg/m^2
• Nonwoven goetextile	-	ITEX 195 PP
• Double surface treatment:		
1° layer of binder: asphalt emulsion	1.2 kg/m^2	1.6 kg/m^2
1° layer of chippings 12/16 mm	14÷18 kg/m^2	16 kg/m^2
2° layer of binder: asphalt emulsion	1.6 kg/m^2	1.8 kg/m^2
2° layer of chippings 6/10 mm	11÷12 kg	12 kg/m^2
The reference section		
• Double surface treatment:		
1° layer of binder: asphalt emulsion	1.2 kg/m^2	1.6 kg/m^2
1° layer of chippings 12/16 mm	14÷18 kg/m^2	16 kg/m^2
2° layer of binder: asphalt emulsion	1.6 kg/m^2	1.8 kg/m^2
2° layer of chippings 6/10 mm	11÷12 kg	12 kg/m^2

Laboratory obtained amounts of the asphalt emulsion for double surface treatment differ somewhat from ones given in the Polish Specifications, but they have been finally verified in the field as adequate for the conditions of the pilot and reference sections.

4.5 Results of observations and conclusions
Constructed sections are under observations all the time. Until spring 1996 the condition of them is excellent.

5. Final conclusions

The conducted experiments and gained experiences allowed us to formulate the following conclusions:

1. The efficiency of nonwoven geotextiles in retarding the reflective thermal crackings is confirmed for the period 2-3 years. The behaviour of this system in a longer time is not yet known, but the pilot sections are still under observations. It should be underlined that the last winter in Poland was exeptionally severe, and the weather conditions were permitting the occurence of thermal reflective crackings.
2. The allowable maximum elongation of nonwoven geotextiles should be limited to avoid the excesive rolling waves and folds under the wheels of trucks.
3. The Polish nonwoven geotextiles have good properties as an anti-thermal cracking interlayer. Some problems are how to diminish the unit mass of those materials to decrease in the same time the amount of binder to saturate them, as well as to reduce the elongation value.
4. In the case of surface treatment, the fact that the thickness of geotextile is greater than usually applied is profitable, because the chippings are better embedded in the material.
5. The amount of binder to saturate the geotextile depends not only on porosity of the geotextile but also on compressibility of this material. It is related especially to the nonwoven geotextile of thickness greater than *1.2 mm* (under the load *2 kPa*). The optimum amount of bitumen should ensure the water tightness for geotextile in order to avoid water penetration and provide good adhesion to the lower layer.
6. The formula developed by the *Task Force 25* for calculating the amount of binder to saturate the geotextile gives doubtful results for fluffy geotextiles of the thickness *d > 1.2 mm* and in these cases could be applied only with the correction factor.

6 References

1. Salamon, J.W. (1994) *State of stress and strain in strengthened, multilayers elastic structures*, Ph.D. Thesis (in Polish), Cracow University of Technology, Institute of Road and Railway Engineering.
2. Wojtowicz, J. and Grzybowska W. (1993) *Specifications for the modernization of the national road Nr 1 (E 75), km 617+405÷618+347,* (in Polish), not yet published
3. Jaecklin, F. (1991)*Geotextiles dans las revetements hydrocarbones, part 11 of Le Manuel des Geotextiles,* Association Suisse de Geotextiles, pp.11.25-11.26 (in French and German),
4. Wojtowicz, J. and Grzybowska W.(1994) *Specification for the modernization of the Bystrzańska-Partyzantów street in Bielsko-Biała town (in Polish),* not yet published.
5. Grzybowska, W. and Wojtowicz, J.(1994) *Surface treatment with applications of geotextiles - first experiences.* Proceedings of the Conference Road Safety in Europe and Strategic Highway Research Program, Lille, France, September 26-28, 1994

Temperature behavior of nonwoven paving fabrics

D.M. CAMPBELL
Hoechst Celanese Corporation, Spartanburg, SC, USA
M. KLEIMEIER
Hoechst Trevira GmbH & Co KG, Frankfurt/M., Germany

Abstract

Rehabilitation of damaged roads by incorporating a paving fabric as a crack retarder and water barrier into the new overlay or chipseal has been well established over the past 15 years. Successful use of such systems depends on the quality of the fabric laydown and the compatibility of the individual components. In this context the quality of the paving fabric with respect to the asphalt is essential. During installation the fabric has to meet extreme demands of which the temperature of the asphalt is most critical. Once the paver has passed over the fabric, it can experience temperatures of more than $130°C$ within a short period before cooling occurs. This is especially true in situations requiring extra hot asphalt [winter, special transportation conditions]. There are clear changes in the fabric's dimensions when exposed to high temperature. These changes are a direct result of shrinkage and the tension created by shrinkage, referred to as shrinkage force. Shrinkage values and the development of shrinkage forces are primarily dependent upon the paving fabric's raw material. Laboratory test have been developed to quantify these changes. Testing has been performed on a representative collection of commercially available nonwoven paving fabrics. Irrespective of the raw material, thermally induced changes occurred to all of the tested nonwovens. In contrast to polyester [PES] fabrics, the shrinkage force of polypropylene [PP] fabrics increases dramatically beyond $100°C$. Dramatic deformation occurs to PP fabrics at temperatures in the range of $150°C$ and $160°C$, creating distinct shrinkage forces within the fabrics. The unusual shrinkage forces built up in some paving fabrics under certain conditions are considered to be one reason of irregular cracking of fabric rehabilitated road surfaces.

Keywords: Hot asphalt, paving fabrics, reflective cracking, shrinkage, shrinkage force, temperature.

Reflective Cracking in Pavements. Edited by L. Francken, E. Beuving and A.A.A. Molenaar. © 1996 RILEM. Published by E & FN Spon, 2–6 Boundary Row, London, SE1 8HN. ISBN 0 419 22260 X.

1 Introduction

In the last 15 years, the use of paving fabrics to reduce or delay reflective cracking in asphalt overlays has increased dramatically. It is estimated that nearly 100,000,000 square meters are sold annually in the U.S.A. alone for this purpose. Specifications for fabrics used to retard reflective cracking are often inadequate. One fabric property in particular that is often neglected is linear shrinkage, which has been associated with cracks during pavement construction [1].

When wrinkles (or cuts) are present in a fabric during an overlay operation, tensile forces caused by fabric shrinkage can produce a significant displacement of the fabric normal to the wrinkle or cut. Shrinkage occurs while the asphalt concrete overlay is hot and without appreciable tensile strength; thus, the fabric displacement carries the hot overlay with it, resulting in a crack in the new overlay within the wrinkle or cut area.

Overlay cracking due to the fabric shrinkage is a particular concern where relatively high mat temperatures are used during placement. For example, in Germany [2] and Great Britain [3] mix temperatures out of the mixer usuly ranging from 120°C to 200°C (248°F to 392°F) are commonly encountered with temperatures sometimes reaching 220°C (428°F). Typical paving temperatures in Europe are significantly higher than those employed in the U.S.A.

The objective of this paper is to address the effects of shrinkage on paving fabrics used in overlays and the possibility of subsequent premature cracking in new overlay surfaces. A simple, easily performed test to quantify fabric shrinkage and shrinkage force will be examined.

The results indicate that paving fabrics of polypropylene construction have a higher probability of producing cracks in a new overlay than paving fabrics of polyester construction.

2 Materials and test procedures

2.1 Fabrics

Three commercially available nonwoven paving fabrics were used in the study. These included polypropylene continuous filament (PP-CF), polypropylene staple fiber (PP-SF) and polyester continuous filament (PES-CF). Table 1 contains a product description for each fabric used.

Table 1 - Physical property descriptions

Product code	Material	Bonding	Weight [g/m²]	Thickness [mm]
PP-CF	PP	mechanical	140	1.3
PP-SF	PP	mechanical / thermal	145	1.3
PES-CF	PES	mechanical	140	1.4

2.2 Hot air shrinkage test

Each of the fabric types were evaluated for the effects of hot air shrinkage in accordance to the established test method, ASTM D-2646-79 (part 27). Specimens were cut 330mm or 300mm in the machine direction by 330mm in the cross machine direction. Each fabric type was placed in a circulating air oven and exposed to the following temperatures; 135°C (275°F), 150°C (300°F), 157°C (315°F) and 177°C (350°F) for three minutes each. The specimens were remeasured and the amount of shrinkage was calculated by expressing the change as a percent of the initial specimen dimension.

2.3 Hot air shrinkage force test

A special test was developed to measure shrinkage force. The equipment consisted of a tensile testing machine with clamps measuring 10cm wide and having a gage length of 20cm and an environmental chamber ['pipe oven'] 120mm wide and 1000mm long with a 1000 watt heating element capable of heating at a rate of 4°C/minute from 20°C to 100°C. Sample specimens were cut 100mm wide and 260mm long [with a tested dimension of 100mm by 200mm]. The specimens were clamped in the tensile testing machine at ambient temperature and pretensioned to 20 N/100mm. [Note: Pretensioning is necessary to take out slackness in the fabric specimens during clamping and to establish an uniform starting baseline.] The environmental chamber is enclosed around the clamps and specimen and the heat is turned on. A pen plotter records the amount of relaxation followed by contraction and the shrinkage force that is exerted against the tensile testing machines load cell as the temperature increases.

3 Test results

3.1 Hot air shrinkage test results

The results of the hot air shrinkage test clearly indicate a propensity for polypropylene products to shrink excessively at paving temperatures exceeding 135°C [275°F], refer to Photo 1. The polyester product had only minimal shrinkage at temperatures up to 177°C [350°F]. Table 2 contains the actual test results for hot air shrinkage.

Table 2 - Hot air shrinkage test results

Product code	Temperature °C [°F]	Machine direction*, mm	Machine direction, Δ%	Cross direction, mm	Cross direction, Δ%
PES-CF	135 [275]	293	2.3	321	2.7
	150 [300]	292	2.7	321	2.7
	157 [315]	289	3.7	320	3.0
	177 [350]	284	5.3	315	4.5
PP-CF	135 [275]	298	9.7	319	4.2
	150 [300]	289	12.4	292	12.3
	157 [315]	213	35.5	188	43.5
	177 [350]	125	62.1	135	59.5
PP-SF	135 [275]	279	15.5	312	6.3
	150 [300]	246	25.4	277	16.8
	157 [315]	229	30.6	216	35.1
	177 [350]	100	69.7	87	73.9

Photo 1 - Effects of hot air shrinkage at various temperatures [*Note: original sample size PES: 300mm x 330mm, PP: 330mm x 330mm]

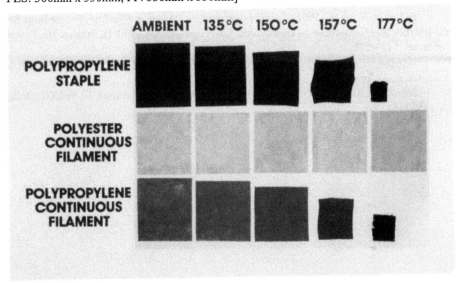

3.2 Shrinkage force test results

Initially during each test, all products exhibited a certain degree of relaxation prior to the effects of shrinkage. This relaxation appears as a downward slope on each curve [refer to the diagrams in Figures 1-6] and is believed to be a result of linear extension during the initial introduction of heat to the sample, a heat sink effect.

Depending upon polymer type, each test sample reached a different temperature in which a tension [*shrinkage force*] occurred. For the polyester samples, this temperature was approximately 75°C [170°F] and the maximum tension force observed was approximately 1 N/100mm at 90°C [195°F]. For the polypropylene samples, tension started at 120°C [250°F] to 130°C [265°F] and reached maximum tension forces of 10 to 20 N/100mm at 157°C [315°F]. At 170°C [338°F], the polypropylene samples lost all tension due to polymer melting while the polyester samples maintained a generally uniform tension.

4   Conclusions

The results of these tests indicate that the physical properties of both shrinkage and shrinkage force should be examined before specifying any paving fabric for hot asphalt overlay application. Observations from actual field installations have demonstrated that occasionally paving fabrics made from polypropylene actually cause cracks to form in the new overlay prior to compaction of the asphalt concrete. It is believed by the authors that these cracks are a direct result of high shrinkage forces contracting the fabric in areas where ends overlap and / or wrinkles are cut thereby causing the uncompacted asphalt concrete to move with the fabric.

5 References

1.   Button, J.W., et al, [1982] *Fabric Interlayer for Pavement Overlays*, Proceedings Second International Conference on Geotextiles, Las Vegas, U.S.A., Vol. II, Session 7B: Paved Roads, pp. 523 - 528.
2.   Bundesministerium Für Verkehr, Abteilung Stra8enbau. [1994] *ZTV Asphalt - StB 94*, Bonn, Germany, pp. 17 - 18.
3.   John, N.W.M. [1987] *Geotextiles*, Blackie and Sons Ltd., London, United Kingdom, pp. 115 -116.

Part five

Performance of maintenance techniques

Preoccupations of Iptana-Search Romania in pursuance of overlay systems behaviour

S.L. HARATAU, G. FODOR, C. CAPITANU and S. CIOCA
Iptana-Search, Bucuresti, Romania

Abstract
This paper presents the results of studies performed with Dynatest 8000 FWD to establish estimation criteria of deformability and load transfer at joints and transversal crackings correlated with the behavior of overlay systems. To characterize the asphalt mix, the indirect tension test was introduced. Determining of this characteristic on main bituminous mixes allows the presentation of its influence factors. The paper also presents a study case concerning utilization of Dynatest 8000 FWD in performance pursuance of a flexible composite pavement link strengthened with cement concrete as a reflective cracking retardation solution.
Keywords: Bituminous mixes, criteria of roads performance, deflection, FWD, load transfer at joints, tensile strength, traffic loading simulation.

1 Introduction

Social and economic changes in Romania have overtaken the network roads in an inadequate technical condition.
   The management operation program requires the existence of the essential elements of technical efficiency needed to make decisions. The endowment of IPTANA SEARCH with a Dynatest 8000 FWD equipment, allows the managers of the Romania roads network to have the necessary data to estimate both the technical condition and the pavement capacity to sustain the traffic loading.
   The processing of the test results performed with this equipment allows to establish the elastic stiffness of the pavement layers and of the subgrade, characteristics needed to establish the optimal pavement strengthening solution. To this end, the consideration of reflective cracking presents an extreme importance determined by roads maintenance historic. The economic constraints during the 1970 – 1989 period have determined the adoption of a

Reflective Cracking in Pavements. Edited by L. Francken, E. Beuving and A.A.A. Molenaar. © 1996 RILEM. Published by E & FN Spon, 2–6 Boundary Row, London, SE1 8HN. ISBN 0 419 22260 X.

strengthening solution of the flexible pavements, with reduced thicknesses of the bituminous layers (5 - 10 cm), regardless of the required computed thickness of those. These periodic maintenance works have determined that on a great part of road networks, the pavement to be made of bituminous layers with total thickness of 7.5 - 35 cm on granular subbases (generally well graded aggregate). In general, the surfacing of these roads presents damages like transversal and longitudinal cracks or crackings. Consequently, the designer is confronted with the difficult problem of establishing the optimal road rehabilitation solution, taking into consideration the reflective cracking.

2 Traffic loading simulation with Dynatest 8000 FWD.

The high variety of vehicle types, implicitly of the axle loads and of travel speeds have required the equalization of these in actions of a standard axle. The present tendency of increasing the permissible load on single axle has determined the adoption of the standard axle load of 115 kN, with the following characteristics:

● Tire contact pressure on pavement .... 0.685 MPa
● Contact area diameter .... 0.342 m

The standard axle load is simulated by FWD through adopting of the following test parameters [1]:

● Magnitude of load .... 250 kg
● Falling height .... 0.200 m
● Time application of the impulse .... 0.028 secs
● Load plate diameter .... 0.300 m

The time application of the impulse corresponds to a travel speed of 30 km/h . The configuration of the seven deflection transducers varies with pavement type and test purpose. Thus, this is specific to the determination of load transfer at joints of cement concrete pavements, according to figure 1.
The measured bowl deflections and the pavement structure constitute the input data for the ELMOD/ELCON program to determine the stiffness of the pavement layers and the subgrade, to establish the residual life and the needed thickness of the bituminous strengthening overlays. An assessment of the road performance can be made through the analysis of the folowing bowl deflection characteristics:

1.   Tests performed on flexible or rigid pavements, in the centre of the slabs: the central deflection (d_1), and the difference between the central deflection and the one measured at the radial distance of 900 mm with sensor 4, ($d_1 - d_4$).

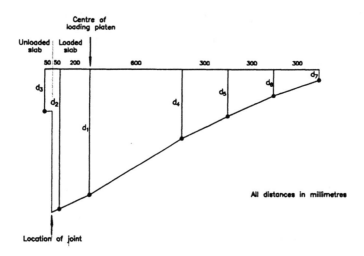

Fig. 1. The configuration of deflection transducers at joint

2.   Tests performed on rigid pavements at joints: the difference between the measured deflection on one side of the ,joint, on the loaded slab with sensor 2 and the one measured at the other side of the joint, on the adjacent slab, with sensor 3 (d_2 – d_3) and the ratio (d_3/d_2).

The values of these characteristics constitute estimate criteria of roads performance.

3 Assessment criteria of roads performance

3.1 Permissible deflection criterion
The ratio between the characteristic measured deflection on a homogeneous road link and permissible deflection represents one of the parameters which characterizes the technical roads condition, in the management operation program. While the permissible values of the deflections for deformability tests with the Benkelman Beam or the Lacroix deflectograph are established, the use of FWD imposed the specification of values specific to this methodology. So far, the attempt to correlate the deflection measured with Benkelman Beam and FWD central deflection (d_1) have not led to satisfactory results.

The tests performed with Dynatest 8000 FWD on about 350 km flexible roads allowed the setting of the permissible deflection d_1 for pavements made up from 3 – 20 cm bituminous layers and 20 – 50 cm granular layers. For this analysis were retained the test results performed on road links characterized by 8 – 12 years residual life, which do not require strengthening, or require only thicknesses of maximum 1.5 cm of bituminous overlays and of which the structural state is considered at limit. Thus were obtained significant correlations between the central deflection, the mean temperature of the bituminous layers and the traffic volume, as a linear regression:

$$d_1 = a_0 + a_1 \log \frac{\theta}{20} + a_2 \frac{1}{\log N_c} \tag{1}$$

where:

d_1 = central deflection, in μm
θ = the mean temperature of the bituminous layers, between 5°C and 30°C
N_c = average annual traffic for a design period of 10 years, between 10^4 and 10^5 standard axles loads on a lane.

The values of the constant term a_0 and of the regression coefficients a_1 and a_2 depend on the total bituminous layer thickness, according to table 1.

Table 1. Values of the regression parameters

Total thickness of the bituminous layers (cm)	a_0	a_1	a_2	r
3.0 – 7.5	– 197	487	3768	0.76
7.6 – 12.0	–1840	894	11847	0.88
12.1 – 20.0	– 736	696	6448	0.82

The correlation coefficient (r) with values ranged between 0.76 and 0.88 shows a significant correlation of the testing data.

For the mean temperature 20°C of the bituminous layers, the central deflection (d_1), varies between closed limits (529 μm ... 557 μm) for N_c = 10^5 115 s.a., and between large limits (745 μm ... 1122 μm) for N_c = 10^4 115 s.a. as results from figure 2.

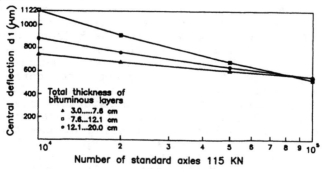

Fig. 2.   Deflection permissible variation with number of 115 kN standard axles

The estimate criteria for road bearing capacity, based on deflection (d_1), corresponding to temperature of 20°C and usual values of thickness bituminous layers (7.5 − 12 cm) and of the traffic volume, are given in table 2.

Table 2. Estimate criteria d_1 for bearing capacity

d_1 (μm)	< 500	500 - 650	651 - 800	>800
Rating	Very good	Good	Fair	Poor
Residual life (years)	10	5 - 10	2 - 5	< 2

3.2 Estimate criterion (d_1 − d_4) for stiffness of the granular layers

For some road links were obtained significant correlations between characteristic (d_1 − d_4) of the bowl deflection, total thickness of the granular layers and traffic volume, with the following linear regression form:

$$(d_1 - d_4) = a'_0 + a'_1 h_2 + a'_2 \frac{1}{\log N_c} \qquad (2)$$

where:

$d_1 - d_4 =$ difference between measured deflection by sensors 1 and 4, in μm

$h_2 =$ total thickenss of the granular layers, in cm

$N_c =$ the average annual traffic for a design period of 10 years, between 10^4 and 10^5 standard axle loads of 115 kN on a lane.

The values of the constant term a_0 and the regression coefficients a'_1 and a'_2 are given in the table 3.

Table 3. Values of the regression parameters

Total thickness of the bituminous layers (cm)	a_0'	a_1'	a_2'	r
3.0 - 7.5	- 501	- 6.7	5766	0.72
7.6 - 12.0	- 694	- 10.0	7603	0.76
12.1 - 20.0	- 513	- 0.2	4589	0.79

The correlation coefficient (r) has values between 0.72 and 0.79 and shows a significant correlation of the testing data.

It is considered that (d_1 − d_4) can characterize the stiffness, implicitly the base quality. Given the results obtained, the following estimate criteria of this bowl deflection characteristics were adopted, according to table 4.

Table 4. Estimate criteria ($d_1 - d_4$) of the subbase quality

$d_1 - d_4$ (μm)	< 400	400 - 700	> 700
Rating	Good	Fair	Bad

The relatively reduced length of the flexible composite and rigid pavements tested with Dynatest 8000 FWD does not allow the setting of the criteria permissible (d_1) and ($d_1 - d_4$) for these types of pavements.

3.3 Estimate criteria ($d_2 - d_3$) and d_3/d_2 for load transfer at joints

For these characteristics of the deflection bowl at the joints of cement concrete pavement, it have been adopted the criteria from table 5, based on data from speciality literature.

The reduced number of tests performed so far on the rigid roads does not allow to set others values.

Table 5. Estimate criteria of load transfer at joints

$d_2 - d_3$ (μm) d_2/d_3	< 25 > 0.75	25 - 80 > 0.75	81 - 150 > 0.50	>150 ≤ 0.50
Rating	Very good	Good	Satisfactory	Unsatisfactory

3.4 Homogenity criterion of bowl deflection characteristics

The homogenity of the bowl deflection characteristics along of a tested road is assessed by means of the variation coefficient C_v, computed on homogeneous road links, according to table 6.

Table 6. Estimate criteria of homogenity

C_v %	< 20	20 - 30	31 - 40	> 40
Rating	Good	Medium	Low	Unhomogenity

4 On the tensile strength of the bituminous mixes

The cracks progression is characteristic to rigid or flexible composite pavements with damages, strengthened with bituminous overlays. This process can appear in case of the flexible roads with periodic strengthening with bituminous overlays. Thus, the d_1 values computed with relation (1) are deformability thresholds under which the prevalent damage process is reflective cracking and not fatigue.

To retard the progression of cracks from old pavement to wearing course requires to adopt high thicknesses of new bituminous layers [2]. However when the design traffic does not require strengthening, this is a bad economic solution. The anti-cracking solutions involve high costs and are adopted for rehabilitation works of some important roads. The possible solution for the majority of national and local road networks is to improve the tensile strength of the bituminous mix of the road base and base course.

There were special preocupation to this end: working out of a simple test of tensile strength and investigation of influence factors of tensile behaviour of bituminous mix.

4.1 Test of tensile strength

The indirect tensile test was worked out in 1992 [3]. According to this methodology, the tensile strength is determined with the following relationship:

$$R_t = k\ R_{tB} \tag{3}$$

where:

R_t = the tensile strength (MPa)
R_{tB} = indirect tensile strength, (MPa) determined on Marshall sample
k = multiplication factor, depend on internal friction angle.

The internal friction angle is established at the same time with compression strength of bituminous mix on a three-cylindred overllapped Marshall sample, through analysis of the cracks on lateral surface.

Sistematic tests on different bituminous mixes used in Romania [4][5], allowed to establish some average values of factor k, which vary between 1.62 for the road bases and 1.75 for the base courses.

4.2 Influence factors of tensile strength of bituminous mix

The viscoelastic behaviour of the bituminous mix under load make that tensile strength depends on numerous parameters.

If the load and environment parameters are typical of test, a special interest presents the composition parameters. The following composition parameters have been taken into consideration: the source and type of bitumen, the type of filler, the petrographical nature of the rock of chippings, the composition of bituminous mix.

The study of the tensile strength of the bituminous mixes was made on mixtures with mechanical characteristics corresponding to the admissibility requirements in our country.

The indirect tensile strength decreases at the same time with the increase of the trial temperature, as is shown in figure 3 for semi-open mixtures with crushed stone and bitumen type D 50/80 (penetration 50 - 80 0.1 mm at 25°C).

It is interesting to note that from the viewpoint of indirect tensile strength, the optimum content of bitumen is different from that which give the greater compressive strength of the mixture.

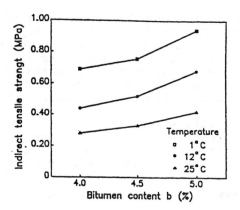

Fig. 3.   The indirect tensile strength variation with the
           content of bitumen

The studies performed emphasized that from the composition parameters, the best correlation was obtained between the indirect tensile strength, parameter $V_b/(V_b + V_g)$ and temperature $T(°C)$, where V_b is the volume of binder and V_g is the volume of air voids, in percentages, from the bituminous mix laboratory sample. Thus, correlations:

$$\log R_{tB} = a_0 + a_i\, V_b/(V_b + V_g) + a_2 T \qquad (4)$$

are considered representatives for the variation of the indirect tensile strength depending on the components of the bituminous mix. The correlation coefficients have values between 0.88 and 0.99. The values of the constant term a_0 and the regression coefficients a_1 and a_2 vary with type of bituminous mix, type and source of bitumen and range between:

$$a_0 = - 0.047 \ldots - 1.332$$
$$a_1 = 0.067 \ldots 1.910$$
$$a_2 = - 0.016 \ldots - 0.026$$

For a variation interval of 4 – 5% of the bitumen content, the indirect tensile strength is found to increase with $V_b/(V_b + V_g)$ for all types of studied mixtures.

It is mentioned that the indirect tensile strength variation is more sensitive at the variation of this parameters than at the change of mix density. Thus, it was not possible to obtain significant correlations $R_{tB}(\gamma_m)$, especially at high temperature.

A good correlation (the correlation coefficient has values between 0.77 and 0.82) was obtained between indirect tensile strength, the bitumen penetration at 25°C temperature (pen$_{25}$ 0.1 mm) and temperature $T°C$. Thus, the most significant relation is:

$$\log R_{tB} = a_m + a_n pen_{25} + a_p T, \tag{5}$$

where:

$a_m = 0.180 \ldots 0.403$
$a_n = -0.002 \ldots -0.003$
$a_p = -0.015 \ldots -0.026$

However, the low values of a_n show the reduced influence of the penetration on the indirect tensile strength. Other factors, with a reduced influence on the indirect tensile strength are grading of the total aggregate if their variation is within the standardised limits, the petrographical nature of chippings, content of crushed aggregate from mixtures for road bases, mineralogical nature of the filler and otheres.

Design of optimal bituminous mix composition requires consideration of maximum tensile strength criterion as well.

5 Reflective cracking retardation solutions

Utilization of some reflective cracking retardation solutions is a topical interest problem. Thus, different solutions have been conceived and tested and namely, inlaying of geosynthetical materials between the new wearing course and base course, reinforced bituminous treatment or even anti-cracking bituminous mixtures [6]. Unfortunatelly, the short operation period of test links did not allow getting of some conclusions.

Strengthening with concrete surface slabs of some damaged flexible composite roads had in view the retardation of reflective cracking. This is clearly from performed tests with Dynatest 8000 FWD. The results are presented in the following study case.

A highway link with flexible composite pavement performed in 1968 (15 cm bituminous mixture, 20 cement bound material and 20 cm well graded aggregates) was reinforced in 1992 with 21 cm cement concrete. The tests performed with Dynatest 8000 FWD before reinforcement show a correlation between the damaged area and the deflection bowl characteristics, as is illustrated in table 7. Tests performed in 1994 show a different load transfer at joints depending on the bearing capacity of the old road according to table 8. On the link 3, after two years of operation there are five transversal cracks.

Table 7. Statistic summary of deflection bowl characteristics before reinforcement

Link No	1		2		3	
Characteristics	d_1	$d_1 - d_4$	d_1	$d_1 - d_4$	d_1	$d_1 - d_4$
Mean	247 μm	177 μm	350 μm	220 μm	409 μm	284 μm
Standard deviation	66 μm	62 μm	84 μm	70 μm	96 μm	100 μm
Variance	27%	35%	24%	32%	23%	35%
Max. value 85% percentile	318 μm	244 μm	440 μm	295 μm	513 μm	392 μm
Damaged area %		6.5		45.3		60.3

Table 8. Statistic summary of the deflection bowl characteristic two years after reinforcement

Link No	1		2		3	
Characteristics	$d_2 - d_3$	d_3/d_2	$d_2 - d_3$	d_3/d_2	$d_2 - d_3$	d_3/d_2
Mean	10 μm	0.93	54 μm	0.71	83 μm	0.62
Standard deviation			37 mm	0.15	80 mm	0.23
Variance	70%	4.3%	68%	21%	96%	37%
Max. value 85 percentile	18 μm	-	95 μm	-	169 μm	-
Min. value 85 percentile	-	0.89	-	0.55	-	0.37
Load transfer	Very good		Satisfactory		Unsatisfactory	

Utilization of Dynatest 8000 FWD proved that this equipment can be an extremely useful nondestructive investigation instrument of the road bearing capacity. Thus, the data given to designer lead to optimal roads rehabilitation solution, depending on the prevalent distress mechanism of the strengthening overlays.

References

1. DYNATEST 8000 FWD, TEST SYSTEM
2. Goagolu, H., Marchand, J – P. and Mouratidis, A (1983) La méthode des éléments finis: application à la fissuration des chaussées et au calcul du temps de remontée des fissures. Bull. liaison Labo. P. et Ch – 125, mai – juin, pp. 76
3. Fodor, G. and Bradler, M. (1995) Asupra factorilor de influenta a rezistentei la intindere a mixturilor asfaltice. Proceedings of the workshop "Modern road pavements", Cluj – Napoca (Romania) pp. 35 – 41.
4. STAS 174 – 83 Imbracaminti bituminoase cilindrate executate la cald.
5. STAS 7970 – 83 Straturi de baza din mixturi bituminoase cilindrate executate la cald.
6. Belc, F. (1995) The prevention of the fissures at the mixted structure pavements. Drumuri si Poduri Romania, Nr. 25 – 28, pp. 10 – 13.

Experience with measures to prevent reflective cracking at road rehabilitation

H. PIBER
Amt der Kärntner Landesregierung, Bautechnik, Klagenfurt, Austria

Abstract
This contribution describes the rehabilitation of old cement concrete roads. Before deciding how to repair the roads, different solutions were tested on short test sections. Depending on bedding of slabs and heavy-duty traffic volume different types of unbounded and bituminous courses were used. The development of reflective cracks over the transverse joints was observed and recorded. The evaluations of the tests showed relevant information for the future work.
Keywords: Concrete pavement, deflection measurement, rehabilitation, test section

1 Introduction

In Carinthia, the southern region of Austria, about 70 km of the federal roads were constructed with concrete pavements in the fifties. The pavement consists of following two courses.

20 cm concrete pavement
<u>20 cm base course</u>
subgrade

The concrete pavement is carried out in two layers and it is not reinforced in the most cases. The wearing course was even built by using crushed diabase stone. Skid resistance and resistance to abrasion correspond even to the today's standard. The slab's size is 3.75 m x 7,50 m and they are connected with dowels.

The base course is built of unbounded material, a gravel sand mixture. The subgrade

Reflective Cracking in Pavements. Edited by L. Francken, E. Beuving and A.A.A. Molenaar. © 1996 RILEM. Published by E & FN Spon, 2–6 Boundary Row, London, SE1 8HN. ISBN 0 419 22260 X.

is carried out in a very simple and cheap way and drainage systems are hardly existing.

During time the longitudinal evenness became very inadequate. Evenness measurements made by a pump integrator showed values of about 70 to 90 inch/100 km. These values are twice the size of asphalt surfaces. The slabs started striking. The Carinthian road administration had the task to repair the existing pavement. The rehabilitation should be carried out by using asphalt in an economical way and the concrete slabs should not be smashed.

The first attempt was made 1982. A through road in a small village was repaired about 500 m in length. The annual daily traffic volume (ADTV) is 7000 vehicles per day in both directions with a ratio of heavy-duty traffic of 6 %.

The rehabilitation in this case was:

> 4 cm   wearing course of asphalt concrete
> 8 cm   asphalt base course
> _____
>   old concrete pavement.

The asphalt concrete of the wearing course was an AB 11 [2]. This is a mixture of a standard bitumen B 100 [1] and a crushed diabase stone. The maximum size of grain was 11 mm. The asphalt mixture of the base course was an BT-II-22 [3]. In this case a gravel-sand-material was also mixed with a standard bitumen B 100. The maximum size of the gravel was 22 mm.

The first attempt was not successful. This method could not prevent reflective cracking. After 13 years we noticed following distribution of cracks.

Table 1:   Development of cracking

	Number		Total length	
	(-)	(%)	(m)	(%)
Transverse joints of the old concrete pavement	124	100	434	100
reflective cracks over the whole width	27	22	94,5	22
reflective cracks over a part of the width	70	56	100,5	23
sum of reflective cracks	97	78	195	45

This unpleasant experience led to more detailed investigations of the existing concrete pavements.

2 Investigation of the bearing capacity

Our tests are concentrating on the vertical movement of the slabs. The main problem of reflective cracking is the vertical movement and not the strain depending on temperature. The test method is the same as we are using at the measurement of bearing capacity and is well known a deflection measurement [5]. The loading takes place by a truck with an 100 kN axle-load. The wheel load is 50 kN and the tyre pressure 0.7 N/mm². The vertical deflection is measured after the truck had left the point of measurement by an optical device. The frost-thaw-period is the right time for this type of test. In contrast to measurements on asphalt pavements the position of the truck plays an important roll. On a concrete slab the truck can be placed in different ways. We chose for a pilot test the following two positions.

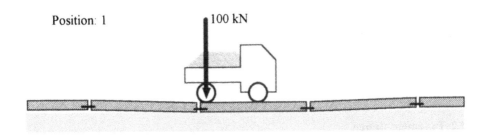

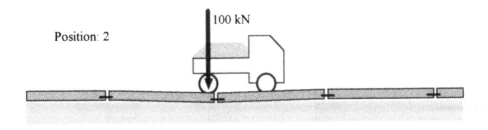

Fig. 1. Loading positions of the deflection measurement.

A short test section showed different results.

Position 1: The truck is placed on one slab. The back axle is near the edge of the slab. The critical deflection value was 0.69 mm.

Position 2: The truck is placed on two slabs. The back axle is placed near the edge of the first slab. In this case the critical deflection was 0.47 mm. Always the critical deflection value was evaluated from a statistical analysis. It is the sum of the main value plus three times the standard deviation. On account of this pilot test we choose the position 1 for all investigations.

3 Methods of rehabilitation and practical experiences

We decided to lay out four test sections on two federal roads. Test section A is a cross-town link of the B 106. The other three are rural roads of the B 106 (B, C) and the B 94 (D). At the cross town link the tolerance in level had to be very small and so only a thin course could be built over.

3.1 Basis data, traffic, deflection values

Table 2. Basis data of the test sections

Test section	Length of the section (m)	Annual daily traffic volume (Vehicle/day)	Percent of heavy-duty traffic (%)	Critical deflection value (mm)
A	1000	4500	7	0,45
B	300	4500	7	0,51
C	500	4500	7	0,51
D	4200	7000	12	0,61

3.2 Pavement design

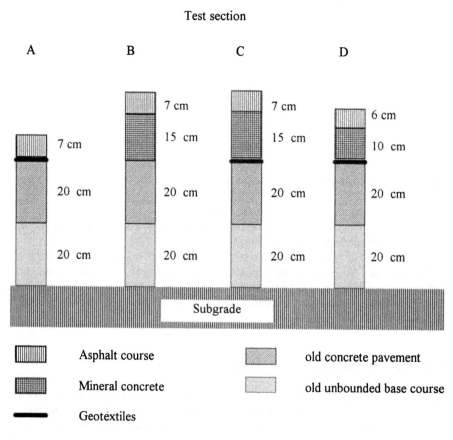

Fig. 2. Pavement design of the different test sections.

3.3   Material properties

3.3.1 Bituminous road base.

- *Section A, B, C:* The mixture type BTD 16 [3] was used to produce the bituminous road base This type has a dual purpose namely a road base and a wearing course. Therefore it is mixed by using crushed limestone and an original Venezuela-bitumen B 100. The maximum size of stones is 15 mm. The grading curve takes a continual course and the binder content is about 5.8 % by mass. The Marshall specimen consists 2.8 % by volume air voids of which 83 % are filled with bitumen. On-site the degree of compacting was measured with 99 %.
- *Section D:* In section D we used the mixture type AB 11 [2]. Usually this type serves to construct wearing courses. Crushed diabase stones and bitumen B 100 formed the two components of the mixture. The grading curve of the mineral aggregate ran constantly. The binder content was determined with 5.5 % by mass. The air void of the Marshall specimen amounted to 2.8 % by volume and the rate of compacting 98 %.

3.3.2 Road base
The road base of the section B, C, D was carried out by using a well graded aggregate of crushed stone [4]. This type of mineral concrete is mixed at mixing plants. The water content corresponds to the optimum water content determined by the proctor test. The road base is laid by asphalt paver and compacted by rollers. The required compacting was checked by a loading plate device. The ration E_{V2}/E_{V1} has to be smaller than 2.2 and E_{V2} more than 200 MN/m²

3.3.3 Geotextiles
On our test sections we tested two types of geotextiles.
- *Geotextiles 1:* (Test section A, C) This type is produced by a mechanical hardening of Polypropylene. Table 3 shows the technical data. When using this type a mass of bituminous pretreating of 1 kg/m² was necessary.
- *Geotextiles 2:* (Test section D) This geotextiles is produced by a thermal hardening of Polypropylene. In table 3 the technical data can be compared with geotextiles 1. The mass of a bituminous pretreating amount to maximum 0.5 kg/m².

Table 3. Technical data of geotextiles

Property Geotextiles	Thickness: (mm)	Weight (g/m²)	Tensile strength (kN/m)	Strain at a tensile strength of 100 % (%)
1	1,5	140	8,5	65
2	0,92	159	3,7	43

3.4 Development of reflective cracks

The following tables show the test results of the four sections.

Table 4 Development of cracking in section A
Construction year: 1987   Year of observation: 1996

	Number		Total length	
	(-)	(%)	(m)	(%)
Transverse joints of the old concrete pavement	268	100	1005	100
reflective cracks over the whole width	9	3	34	4
reflective cracks over a part of the width	29	11	33	3
sum of reflective cracks	38	14	67	7

Table 5 Development of cracking in section B
Construction year: 1987   Year of observation: 1996

	Number		Total length	
	(-)	(%)	(m)	(%)
Transverse joints of the old concrete pavement	82	100	308	100
reflective cracks over the whole width	4	5	15	5
reflective cracks over a part of the width	2	2	2	1
sum of reflective cracks	6	7	17	6

Table 6 Development of cracking in section C
Construction year: 1987   Year of observation: 1996

	Number		Total length	
	(-)	(%)	(m)	(%)
Transverse joints of the old concrete pavement	134	100	503	100
reflective cracks over the whole width	4	3	15	3
reflective cracks over a part of the width	8	6	11	2
sum of reflective cracks	12	9	26	5

Table 7 Development of cracking in section D
Construction year: 1989   Year of observation: 1994

	Number		Total length	
	(-)	(%)	(m)	(%)
Transverse joints of the old concrete pavement	1128	100	3948	100
reflective cracks over the whole width	15	2	63	2
reflective cracks over a part of the width	25	2	39	1
sum of reflective cracks	43	4	102	3

4 Conclusion

The tests show that concrete roads can be built over without smashing the concrete slabs. One requirement is a stable bedding of the slabs. Checks by using a deflection measurement, as described in chapter 2, should guarantee this point. Critical deflection values of fewer than approximately 0.7 mm are aimed at the frost-thaw-period. If the asphalt course is laid directly on the concrete pavement, a bitumen membrane should be used. We recommend a mechanical hardened geotextiles because the mass of bituminous pretreating is higher. The design of the asphalt course corresponds to the number of heavy traffic volume. For light traffic (less than 500 heavy vehicles per day in both direction) the asphalt pavement's thickness of about 8 cm can be accepted.

For roads with a middle traffic (500 - 1000 heavy vehicles per day in both directions) we suggest to construct a pavement by using asphalt and mineral concrete courses. The application of geotextiles is in this case no significant advantage. The thickness of the mineral concrete course should be between 10 to 15 cm. The total thickness of the asphalt courses is to design with 10 cm. In generally it is possible to prevent reflective cracking by these methods. It is also an economical opportunity of old concrete pavements' rehabilitation.

For this reason the Carinthian road administration decides to repair the remaining old concrete roads in this way.

5 References

1. Österreichisches Normeninstitut. (1985) *Erdölbitumen für Straßenbauzwecke* ÖNI, Wien. ÖNORM B 3610
2. Österreichische Forschungsgesellschaft. (1986) *Walzasphalt*. FVS, Wien. RVS 8.06.27
3. Österreichische Forschungsgesellschaft. (1986) *Bituminöse Tragschichten im Heißmischverfahren*. FVS, Wien. RVS 8.05.14
4. Österreichische Forschungsgesellschaft. (1986) *Mechanisch stabilisierte Tragschichten aus zentralgemischten Kantkörnungen*. FVS, Wien. RVS 8.512
5. Österreichische Forschungsgesellschaft. (1995) *Feldprüfungen, III Deflektionsmessung mit dem Benkelmanbalken, nach der optischen Methode und mit dem Deflektograf Lacroix*. FVS, Wien. RVS 11.066

On site behaviour of overlay systems for the prevention of reflective cracking

A. VANELSTRAETE and L. FRANCKEN
Belgian Road Research Centre, Brussels, Belgium

Abstract

This paper deals with the results of an evaluation of different sites in Belgium where interface systems were used for the prevention of reflective cracking. The selected projects, either surveyed at the time of realisation and/or evaluated afterwards, concern asphaltic overlays on cement concrete slabs, where steel reinforcement nettings, grids and nonwoven textiles impregnated with modified bitumen were used. For some projects, measurements were performed of the vertical movements at the edges of the cement concrete slabs.

The survey of the realisation of these projects led to recommendations concerning the laying procedures of the different types of interface systems. These were implemented in the new Flemish standard tender specifications. They are briefly described in this paper.

The results of the visual inspections of some projects are reported. They give insight in the reasons of failure of certain projects and the effectiveness of the used techniques.

Keywords: Reflective cracking, laying procedures, evaluation, cement concrete slabs

1 Introduction

Road works are nowadays increasingly carried out to rehabilitate the existing network. Repairing cracked road surfaces by merely applying an additional layer of asphalt is rarely a lasting solution. Deficiencies in the old pavement are very rapidly reflected at the surface as a result of the combined effects of thermally induced stresses and traffic loading.

Reflective Cracking in Pavements. Edited by L. Francken, E. Beuving and A.A. Molenaar. © 1996 RILEM. Published by E & FN Spon, 2–6 Boundary Row, London, SE1 8HN. ISBN 0 419 22260 X.

To extend the service lives of road structures, interface systems are now used for the prevention of reflective cracking.

During the last years, the Belgian Road Research Centre had the opportunity to assist at the realisation of several sites, where nonwovens, grids and steel reinforcement nettings were applied for the prevention of reflective cracking. This experience allowed the authors to get a good overview of the difficulties encountered during placement of different types of products and to give recommendations for the laying of these products in order to overcome these problems. These recommendations were implemented in the new Flemish tender specifications [1].

In addition, the Belgian Road Research Centre has performed evaluations of different projects with interface systems. In this paper results of evaluations of overlays on cement concrete slabs with steel reinforcement nettings and nonwovens will be given. The results of our evaluations of projects with nonwovens and woven textiles being evaluated previously by others [2, 3] are given in [4].

2 Laying of interface systems: an overview of encountered difficulties and recommendations to avoid problems

Careful installation of interface systems is the key to the good performance of these systems, with specific rules to respect for each type of interface product. The laying procedure of interface systems comprises the following consecutive stages: prior repairs, the application of a fixing layer, the application of the interlayer product itself, the placement of a protective layer in some cases and the application of the bituminous overlay. We will now describe encountered difficulties during placement and give recommendations concerning the different stages in order to avoid these problems.

Phase 1: Prior repairs
Filling existing potholes, repairing serious deficiencies and smoothing out major surface irregularities by the application of a levelling course make it possible to avoid cavities under the interface product, which may remain present after overlaying or result in locally inadequate compacted spots. They can give rise to potholes already very shortly after rehabilitation. Potholes and premature reflective cracking can also be the result of fixing interface products on loose or moving parts, e.g. severely broken cement concrete underlayers or concrete slabs showing high vertical movements at the crack edges. Stabilisation of the underlayer is necessary in such cases, e.g. stabilisation of concrete slabs by injection or by crack and seating techniques. Any loose surface material should also be removed.

Phase 2: Application of a fixing layer
It is now generally accepted that a good functioning of the interlayer requires a good bond to the underlying course, being applied homogeneously over the road surface. For nonwoven textiles this is achieved through a tack coat of emulsion or binder preferably admixed with polymers. The rate of spray is generally rather high: 1 to 1.5 kg/m^2, depending on the product. If an excess of bitumen is used,

problems might occur due to sticking of the bitumen to the tires of the vehicles, leading to tearing off the product. A too small quantity leads to a lack of adherence with the underlayer/overlay or to an insufficient impregnation of the product with binder, which prevents its good functioning. The exact quantity can be determined by the BRRC-testmethod given in [5].

Woven textiles and grids are fixed to the existing pavement surface by an emulsion tack coat, with a minimum rate of spread of 400 g/m² residual bitumen. Selfadhesive geogrids do not require a separate tack coat, provided the underlying surface is clean and dry at the time of installation. Steel reinforcing nettings are embedded in a slurry and locally fixed with nails. In order to achieve an overall bond of the interlayer product to the underlayer, it is recommended to spread out the fixing layer before unrolling the interface product, except for steel reinforcing nettings which are embedded in a slurry.

Phase 3: Application of the interlayer product

To prevent problems during overlying with asphalt, the interface product must be laid perfectly flat. Any folds, creases or wrinkles, which are unavoidable when the product is unrolled on bendy roads, must therefore be cut, after which the edges should be pressed down over each other and fixed to the underlying surface. Seams between adjacent rolls (placed side by side) are usually made with an overlap of 10 to 15 cm. Consecutive rolls in the longitudinal direction of the road are generally laid with an overlap of 10 to 15 cm for nonwoven textiles and 25 to 30 cm for grids. To prevent grids and steel reinforcing nettings from sliding under the asphalt paver, the beginning of any new roll should always be laid underneath the end of the previous one, in the direction of the paving. The ends should be nailed to the underlayer.

Fig. 1   Eliminating wrinkles from a glass fibre grid: cutting, pressing down the edges, and fixing onto the underlying surface (Photo CRR 3320/10A)

Phase 4: Application of a protective layer
Even if traffic is not allowed on grids as long they have not been overlaid, interlayers are often teared or damaged by the site traffic. It is therefore recommended to apply a protective layer. Grids are therefore covered with a surface dressing. In case of steel reinforcing nettings embedded in a slurry, the protection is already guaranteed by the slurry.

Application of the bituminous overlay
The bituminous overlay is applied in the conventional way. Nonwovens in polypropylene may deform or even melt if overlaid with mixes requiring laying temperatures above 160 °C. They are therefore not recommended in such cases.
A minimum overlay thickness is recommended: 3 cm on textiles, 4 cm on grids and steel reinforcing nettings.

3 Evaluation of sites

The Belgian Road Research Centre has performed evaluations of different projects with interface systems. The six projects described here concern overlays of cement concrete slabs, in most cases without foundation and with severe vertical movements ("rocking") at the edges of the slabs. Two different interface products have been placed: steel reinforcement nettings (either nailed or embedded in a slurry in order to fix it with the underlayer) and polyester nonwovens impregnated with bitumen. The results of other projects with nonwovens and wovens being evaluated previously by others [2, 3] are given in [4]. Table 1 gives for each of the six projects: year of renovation, size of the project, traffic level in number of vehicles per day for each direction, original road structure, type of interface, method of fixing with the underlayer, overlay type and thickness.

Table 2 gives an overview of the results of the evaluations, carried out during the periods 6/1994 - 10/1994 and 6/1995 - 10/1995. Additional information about the inspections is given below. For each project, the exact location of the reflective cracks and other type of damage was noted; this allows to follow up these projects in the future. The transverse cracks were divided into three categories, depending on their length: cracks smaller in length than 1/3, between 1/3 and 2/3, between 2/3 and 3/3 of the traffic lane width. Hence, apart from the total number of reflected cracks, the number of so-called "equivalent" reflected cracks can be deduced. This value takes into account that not all cracks have reached their full length. The number of equivalent reflected cracks is calculated as follows:

$$x_1.(1/6) + x_2.(3/6) + x_3.(5/6)$$

in which x_1, x_2, x_3 represent the number of cracks with length $\leq 1/3$; $> 1/3$ and $\leq 2/3$; $> 2/3$ and ≤ 1 of the traffic lane width, respectively. Knowing the average length of the concrete slabs, the number of original joints/cracks can be calculated and the percentage of (equivalent) cracks reflected can be deduced (see table 2).

Table 1: Description of the different sites

Place	Year Repair	Surface (m²)	Traffic level Type	Type of interlayer	Bitumen + quantity	Bituminous overlay
Tournai Mt.St-Aubert	1989	7200	<2000 medium heavy	steel reinforcing nettings	nailing	8 cm type IIIA(*) + surface dressing
N415 (Zwalm)	1991	13000	>2000 and <8000 medium heavy	steel reinforcing nettings	nailing	4cm type IIIC(*) + interface + 6cm type IIIA(*) + 4cm porous mix
N499 (Maldegem)	1991	13000	>2000 and <8000 medium heavy	steel reinforcing nettings	bitum. slurry with elastomers 24 - 26 kg/m²	4 cm porous mix
Aalter Nijverheidslaan	1992	4500	<2000 heavy	steel reinforcing nettings	bituminous slurry 24 - 26 kg/m²	5 cm type IIIA(*) + interface + 4cm type IV(**)
Genappe Rue Croisette	1993	1060	>2000 and <8000 medium heavy	nonwoven polyester	modified binder 1.5 kg/m²	4 cm ultrathin layer(***)
N442 Berlare	1995	20000	>2000 and <8000 heavy	(1)steel reinforcing nettings (2) no interface	bitum. slurry with elastomers 17 kg/m²	4 cm split-mastic-asphalt

(*) The type III-mix is a dense bituminous mix used for levelling courses and underlayers
(**) The type IV-mix is a dense mix for top layers
(***) The so-called ultrathin layer is a bituminous mix consisting of 70 - 80 % stones, 20 - 30 % sand, 10 - 14 % filler and 5 - 6 % binder (on 100 % aggregates) [6]

Table 2: Summary of the results of the inspections (for further details about the description of the sites: see table 1)

Project Year repair	Original structure	Type of interlayer	Overlay	Cracks reflected	Equivalent cracks reflected	Other damage
Tournai 1989	concrete slabs	steel reinforcement nettings, nailed	8 cm dense mix + surface dressing	1994: 1 on 154	1994: 1 on 154	1994: 1 pothole
N415 1991	concrete slabs	steel reinforcement nettings, nailed	10 cm dense + 4cm porous mix	1994: no	1994: no	1994: no
N499 1991	concrete slabs (length: 6 m)	steel reinforcement nettings, in slurry	4 cm porous mix	1994: 12 % 1995: 34 %	1994: 7 % 1995: 25 %	1994: no 1995: no
Aalter 1992	concrete slabs cracked and seated	steel reinforcement nettings, in slurry	9 cm dense mix	1994: no 1995: 1 crack	1994: no 1995: 1/2 crack	1994: no 1995: no
Genappe 1993	concrete slabs (length: 5.8 m)	nonwoven polyester, 1.5 kg/m^2 modif. binder	4 cm ultrathin layer	1995: 23 %	1995: 17 %	1995: no
Berlare 1995	concrete slabs partly cracked and seated	(1)steel reinforcement nettings, in slurry (*) (2)no interface (*)	4 cm SMA	1995: (1) no 1995: (2) no	1995: (1) no 1995: (2) no	1995: (1) no 1995: (2) no

(*) There are six test sections; they are described in detail in the text

Project Tournai (Mont St-Aubert)
This overlay with steel reinforcement nettings on heavily cracked cement concrete slabs is in a very good state. Only one reflection crack was observed five years after repair. There is one pothole, appearing just above a broken corner in the original concrete slab. As already mentioned in 2, this can be due to an improper fixing of the interface to the underlayer, to the instability of the underlying broken corner, or to cavities or insufficiently compacted spots under the interface as a result of differences in level between the (parts of) slabs.

Project N499 (Maldegem)
This overlay with steel reinforcement nettings on heavily cracked cement concrete slabs shows transverse reflective cracks: 12 % cracks/joints reappeared after three years of repair, 34 % have reflected after four years of repair. The cracks are fine and do not lead to discomfort for the user yet. Some of the cracks are double cracks, spaced about 2 - 3 cm from each other. The high percentage of reflective cracking is due to the limited 4 cm overlay thickness, combined with large vertical movements ("rocking") at the edges of the concrete slabs. Measurements of these movements have been performed with the faultimeter (see fig. 2); they were in average of the order of 1 mm before overlaying. Better results undoubtedly could be obtained by cracking and seating the slabs before overlaying and/or by increasing the overlay thickness. The high vertical movements at the crack edges also explain the appearance of double cracks at some places. In 1994, measurements of the vertical movements at the edges of the transverse cracks in the overlay have been performed: the movements are of course smaller than before overlaying, about 0.1 mm instead of 1 mm.

Fig. 2 Measuring slab rocking on a concrete pavement by means of a faultimeter (Photo CRR S/5409)

Project Aalter (Nijverheidslaan)
No reflective cracking was observed in 1994, one reflective crack was seen in 1995. In contrast to the project N499, the slabs have been cracked and seated before overlaying and a 5 cm (levelling) course was placed in addition to the 4 cm thick top layer. These two facts undoubtedly lead to these high differences in reflective cracking between these projects.

Project Genappe
For this project where impregnated nonwovens in polyester were applied, 23 % of the joints/cracks have reflected two years after repair. The cracks are fine and do not lead to discomfort for the users yet. Better results undoubtedly could be obtained by cracking and seating the slabs before overlying and/or by increasing the overlay thickness.

Project Berlare
Prior to repair, measurements of the vertical movements at the edges of the cement concrete slabs have been performed. Based on these results, six test sections have been determined. At the time of rehabilitation (5/95), they have been overlaid with the same type (Split-mastic-asphalt) and thickness (4 cm) of overlay, but they were realised with/without interface, with/without cracking and seating the concrete slabs and with different vertical movements at the edges of the concrete slabs (see table 3).

Table 3: Description of the six test sections for the project Berlare: with/without interface system, with/without crack and seating, with different vertical movements at the crack edges

Section	Interlayer System	Crack and seat	Vertical movements at crack edges
1	Steel reinforcement nettings	Yes	-
2	Steel reinforcement nettings	No	± 0.5 mm
3	Steel reinforcement nettings	No	± 1.0 mm
4	No	Yes	-
5	No	No	± 0.5 mm
6	No	No	± 1.0 mm

No reflective cracking has been observed so far (12/95). These test sections have to be followed in the future in order to compare these techniques and to determine limits for the vertical movements at the crack edges above which crack and seating is recommended.

4 Conclusions

During the last years, the Belgian Road Research Centre assisted at the realisation of several sites, where interface systems were applied for the prevention of reflective cracking. This experience allowed the authors to get a good overview of the difficulties encountered during placement of the different types of products and to give recommendations for the laying of these products in order to overcome these problems. These recommendations were implemented in the new Flemish tender specifications.

The authors can conclude that careful installation of interface systems is predominant for a good performance of these systems, with specific rules to respect for each type of interface product. The laying procedure of interface systems comprises the following consecutive stages: prior repairs, the application of a fixing layer (for grids), the application of the interlayer product itself, the placement of a protective layer and the application of the bituminous overlay. Very important rules to respect are:

- Concerning prior repairs, it is absolutely necessary to avoid cavities under the interface product. These may remain present after overlaying or result in locally inadequate compacted spots. They can give rise to potholes already very shortly after rehabilitation. Interface systems may also not be placed on unstable or moving (parts of) underlayers.
- For the fixing layer:
  In order to achieve an overall and homogenous fixing of the interlayer product to the underlayer, it is recommended to spread out the fixing layer before unrolling the interface product, except for steel reinforcing nettings which are embedded in a slurry being applied after placement of the interlayer product.
- For the application of the interlayer product:
  To prevent problems during overlying, any folds, creases or wrinkles, which are unavoidable when the product is unrolled on bendy roads, must therefore be cut, after which the edges should be pressed down over each other and fixed.
- Grids are often partly teared off or damaged by the site traffic. It is therefore recommended to apply a surface dressing as protective layer.
- It is recommended that the overlay thickness is minimum 3 cm on textiles and 4 cm on grids and steel reinforcing nettings.

The present paper also describes the results of an evaluation of six overlays on cement concrete slabs, where steel reinforcement nettings and nonwovens in polyester were used for the prevention of reflective cracking. In addition to the recommendations to be respected during the different stages of placement, we come to the following conclusions:

- The overlay thickness remains one of the predominant factors for what concerns reflective cracking, even with the use of an interface system. The highest percentage of cracks being reflected was observed in case of 4 cm thick overlays on cement concrete slabs, whereas no reflective cracking appeared on a 14 cm thick overlay.

- There are limits concerning the vertical movements at the crack edges of cement concrete slabs. Cement concrete slabs showing high vertical movements have to be cracked and seated before placing the overlay system.
  In 1995, six test sections were built with/without interface, with/without crack and seating, with different vertical movements at the crack edges, all with the same overlay type and thickness.
  No damage is observed so far. Future evaluations of these sections should help us to determine these limits.
- The projects with steel reinforcing nettings where all rules mentioned above have been respected, performed very well after three to six years of repair.

Acknowledgments

The authors gratefully thank Mr. P. Vanelven and Mr. P. Peaureaux (Belgian Road Research Centre) for their collaboration to the evaluation of the sites. They also thank Y. Vancraeynest and J. Veys of N.V. Bekaert, G. Baert (Mobilmat) and Y. Decoene (Screg Belgium) for providing additional information about the sites. The Belgian Road Research Centre thanks the IRSIA ("Institut pour l'Encouragement de la Recherche dans l'Industrie et l'Agriculture") for their financial support in this research project.

References

1. Flemish Standard Tender Specifications 250, Ministry of the Flemish Community, Department of Environment and Infrastructure, Administration of Road Infrastructure and Traffic, 1996.
2. P. Silence and J. Estival, "Applications de Tissés de polyester pour lutter contre la remontée des fissures", Proceedings of the 1st RILEM-Conference on Reflective Cracking in Pavements, pp. 281 - 287, 1987.
3. Y. Decoene, "Belgian apllications of geotextiles to avoid reflective cracking in pavements", Proceedings of the 2nd RILEM-Conference on Reflective Cracking in Pavements, pp. 391 - 397, 1993.
4. A. Vanelstraete and Y. Decoene, "Behaviour of Belgian applications of geotextiles to avoid reflective cracking in pavements", Proceedings of the 3rd RILEM-Conference on Reflective Cracking in Pavements, Maastricht, 1996.
5. Belgian Road Research Centre, Testmethod "Mesure de la quantité de liant retenu dans une membrane géotextile non tissée pour interface anti-fissure".
6. J.-P. Serfass, P. Bense, J. Bonnot, J. Samanos, "A new type of ultrathin friction course", Transportation Reserach Board, Washington, January 1991.

Asphalt reinforcement in practice

M. HUHNHOLZ
Retired Director of the Federal Roads Authority, Itzehoe Department, Germany

Abstract
Cracks, gaping joints and the so-called alligator cracking skin of bituminous layers have always been of concern to the roads engineer. In this paper the development of an asphalt reinforcement is shown by means of personal practical experience of the author with different extensive test sections -beginning in 1976- and with the most different materials. Summarizing conclusions for the practice and principles for the installation, which have been worked out during the years, are given.
Keywords: Large-scale projects, Non-wovens, Orientated geogrids, Woven geogrids.

1 Introduction

Cracks, gaping joints and the so-called alligator cracking skin of bituminous layers have always been of concern to the roads engineer because the structure shows damages which, under the load of traffic, may soon result in the structure being unusable.

Questions concerning the reasons for the development of cracks are several:
- horizontal movements due to changes in temperature in concrete slabs,
- vertical movements,
- horizontal and vertical movements of the substructure and/or the sub-soil and
- may be caused by the construction, or construction details.

In the past, the problem was solved with a corresponding strengthening of the pavement. This led to a thickening of the road construction and in case of a

Reflective Cracking in Pavements. Edited by L. Francken, E. Beuving and A.A.A. Molenaar. © 1996 RILEM. Published by E & FN Spon, 2–6 Boundary Row, London, SE1 8HN. ISBN 0 419 22260 X.

substructure with a low bearing capacity to new settlements with all the resulting subsequent measures and therefore to an unnecessary increase in costs.

2 Application examples

2.1 Development of geotextile fabrics
With the production of geosynthetics at the end of the sixties, the use of these materials in earthworks started on a large scale. Non-wovens and fabrics are used for substructures with a low bearing capacity to achieve an improvement of the bearing capacity, even if only for a short period.

After the geosynthetics proved their effectiveness as separation layers, filters and reinforcement in road constructions, the experience of these materials was applied to bituminous road construction.

2.2 Types of geosynthetics
Geosynthetics in road construction are defined as geotextiles, geogrids and geomembranes (Fig. 1).

In the practical examples, the applications of non-wovens, woven geogrids and orientated geogrids in asphalt constructions will be clarified.

2.3 Non-wovens and woven geogrids
Starting point for the application of these materials was a cracked bituminous wearing course of an asphalt road L 276 in Kudensee in an area with an approximately 16 m thick substructure with a low bearing capacity. The cracks ran

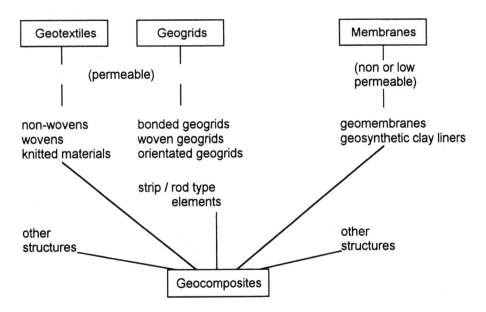

Fig. 1. Types of geosynthetics.

in longitudinal direction over the complete width of the road. There were also partial transverse cracks.

Due to changes of the ground water and an increasing traffic load, the road pavement did not stand the influence of these forces. The reasons for the cracks were horizontal and vertical movements. To increase the intervals for necessary maintenance works in road pavements in 1976, three different geotextile types were used in several sections of this highway underneath the new binders and wearing courses.

- A needle punched non-woven (raw material: polyethylene) - approximately 250 g/m².
- A thermally bonded non-woven (raw material: polyethylene) - approximately 220 g/m².
- A grid fabric of polymer covered polyethylene-threads - screen opening: 30 mm.

The old road surface was provided with a bitumen emulsion and the mentioned geotextiles were installed on it. After a further brushing of the non-woven, but not of the grid fabric, an asphalt binder 0/16 mm with a thickness of 3.5 cm followed by an asphalt wearing course 0/11 mm with a thickness of approximately 3.0 cm was installed on these layers.

Problems occurred when asphalt placement vehicles passed over the non-wovens. The non-wovens impregnated with bitumen, partially stuck to the wheels of the lorries and were pulled up. The renewed arrangement of the non-wovens was very difficult and time consuming. Folds in the non-wovens were unavoidable. They were covered by the asphalt binder. The installation of the binder course in the area of the installed grid fabrics, however, was not as difficult, although some lifting up of the layer by the lorry wheels was observed. The final asphalt concrete wearing course was installed without any problems.

Approximately 1.5 years after the installation of the asphalt concrete layers, in total 6 to 7 cm thick, cracks occurred again in the area of the non-wovens under the load of moderate traffic, in the area of the woven grid the first cracks occurred after approximately 3 years.

Result:
- A reduction in the rate of the crack growth was achieved but in view of the previous costs the result was not very satisfactory.

2.4 Orientated geogrids

With the development of orientated geogrids, produced by biaxial stretching of polypropylene sheets punched with a regular pattern of holes, since the end of the eighties, a building material was offered which was resistant to deformation. Under certain conditions these orientated geogrids withstand the installation temperatures of the asphalt binder and the asphalt concrete and therefore justify their use in asphalt pavements.

The reinforcement of the bituminous layer is achieved as horizontal forces are absorbed by the geogrids.

This task can, however, only be fulfilled if an effective and material-adequate transfer of force into the reinforcement material is possible. A grain skeleton of double-broken chippings 8/11 mm would contribute to this.

The junctions of the geogrid have the effect of anchors in the double-broken chippings in which the geogrid was fastened and fixed. The highest shear stresses occur 8 to 10 cm underneath the top edge asphalt cover layers (stress distribution in the semi-infinite body according to Boussinesq).

With the installation of a grid underneath the bituminous layer (with a thickness of at least 8 cm) of an existing road, the development of cracks would be delayed for several years. The first application was carried out in 1989 in Schleswig-Holstein.

The approach road to the motorway A 23 near Itzehoe, Germany, was selected which has a heavy traffic load (cement bunker vehicles). This road is based on a sub-soil partially consisting of clay and in a longer section ($\approx$ 1.5 km) consisting of peat to a depth of 7 m.

In spite of multiple overlays in recent decades, cracks and deformations of the asphalt concrete layers have so far been unavoidable. Cracks and deformations occurred again 2 years at the latest, after the corresponding repairs and strengthening of the bituminous pavement. In the meantime the asphalt pavement reached a thickness of up to 45 cm.

An orientated geogrid, Tensar® AR 1, was installed.

The following principles have to be considered during the installation. The surface should, if possible, be even. In case of greater unevenness, for example at single deep-seated points caused by spallings, a bituminous regulation layer should be used.

In the present case the existing asphalt pavement was removed 10 to 12 cm deep so as not to change the weight relation on the substructure with a low bearing capacity. Only a few spots were repaired with asphalt mixed material 0/5 mm. After this the surface was even, except some grooves of a few millimetres, the provided transverse inclination was achieved.

On the prepared substructure 3 m wide geogrids were placed. The beginning of the roll was fixed on the bituminous layer with a tape band and steel nails. The geogrid was then stretched and according to the requirement fixed with small steel nails at $\geq$ 5 cm intervals of approximately 20 m on the subjacent asphalt layer.

The individual grid rolls were connected with metal clips. The lateral overlapping is minimally 2 and maximally 3 grid openings (Fig. 2). In case of a radius of curve $\leq$ 600 m the grids were installed by sections and adapted to the radius of curve.

After the installation of the stiff geogrids, a tack coat for the assurance of the force transmission was placed. A bitumen emulsion U 70 K with 1.6 kg/m² was sprayed. In case of greater surface area due to the milling grooves, the amount is increased to 1.9 kg/m².

After that a layer of double-broken chippings, 8/11 mm with an amount of 10 kg/m² was installed on the surface (Fig. 3). This bitumen coat was pressed with a light roller to achieve a good bedding of the grain particles.

The geogrid was covered with double-broken chippings, 8/11 mm, so that the grid stood out lightly against the substructure. Loose fine gravel particles were brushed off. Then the asphalt binder course, 0/16 mm, and the asphalt concrete wearing

Fig. 2. Connection of two grids by overlapping and clipping.

Fig. 3. Placing of chips 8/11 mm.

course, 0/11 mm, were installed each with a thickness of 4 cm. The prescribed compaction (97 %) was completely achieved.

The on-site vehicles and the resident traffic passing over the geogrids and the tack coat did not cause any problems or damage.

Results:
- After seven years under heavy traffic load the approach road still does not show any cracks. A considerable increase in the service life has been achieved.
- An essential criterion for the successful use of this method was the durable bond between the layers. Two years after the installation, shear tests at drill cores ($\varnothing$ = 160 mm) revealed that the compaction: binder course/geogrid + tack coat/ base, matched or exceeded the required minimum value in 8 cm depth of 6.0 kN at all the drill cores.

3 Further reflections concerning the use of geogrids

The success of the installation of orientated geogrids onto the pavement of the approach road was not completely satisfactory. How can the bond of the different layers be improved? There have not been a lot of possibilities.

Two modification were chosen for the next project to achieve an improvement of the bond:
- The fine gravel 8/11 mm of the chip seal was delivered and installed coated with binder film.
- A polymer-modified bitumen should be used instead of an emulsion.

A precondition was that the temperature sensivity of the geogrid was improved so that there would be no considerable decrease in strength of the geogrid due to the temporary influence of the high temperature of the binder and the asphalt binder course during the installation.

These preconditions could be fulfilled.

4 Large-scale project site B 5: Wilster - Brunsbüttel

In the years 1978 to 1984, the Federal Highway B5 from Wilster to Brunsbüttel was built with a length of 18 km crossing the Nordost-Kanal. The highway embankment was built on clay and peat layers with a thickness of 14 to 19 m (average thickness of the sand layer was approximately 3.5 m).

In the first phase of the installation, a wearing course was not installed as decades of secondary settlements were expected. After 6 years under traffic load, innumerable longitudinal and transverse cracks were detected, due to the movements of the substructure and the subsoil. A new design of the bituminous pavement according to the latest knowledge (considering the settling embankment) revealed the necessity of 2 additional binder courses and the missing asphalt concrete wearing course.

As every new load generates new settlements the additional weight has to be kept as low as possible. This can be achieved by using the orientated geogrid. If it is used the binder course with a thickness of 4 cm would not be necessary. Another advantage of the geogrid is that a considerably finer crack distribution can be expected.

Fig. 4. Good bond between geogrid and substructure.

The works went out to tender in Spring 1992. The geogrid was installed as described before on 130,000 m².

A polymer-modified bitumen PMB 65 of 1.7 to 1.9 kg/m² was applied to the base and grid to get the adhesion of the grid with the subjacent bituminous layer. Fine gravel 8/11 mm coated with a binder was then spread without delay. These stages were carried out without any problems (Fig. 4).

The installation of the fine gravel was carried out without any difficulty. The installed binder and wearing course bonded well with the corresponding base layer. The dry weather in Summer 1992 favoured the complete project. To date the complete highway does not have any cracks.

Examination of the compaction between the layers at drill cores with a diameter of 15 cm revealed excellent results. The measured maximum shear forces in the geogrid plane lie just under those without an installed grid. The compaction was considerably improved.

In September 1992 measurements of the asphalt pavement, with a FWD at 9 points of measurement were carried out. A subjacent measurement at the end of September 1994 revealed that the results at 5 points of measurement stayed the same or even slightly improved. This meant that the measured slight deflection corresponded with a high stiffness.

The values of the bearing capacity stayed constant and the bonding of the asphalt construction as well as the longitudinal gradient did not change.

The pavement of the Federal Highway B5 has been optimally installed with regard to the substructure and the traffic load.

The highway will be continuously observed during the next few years.

5 Consequences

Due to this successful application of the orientated geogrids for road constructions, strengthening of the asphalt pavements on substructures with a low bearing capacity, the orientated geogrid was successfully used in similar cases of asphalt pavements. It is especially suitable for the support of town or village roads in case the vertical alignment cannot be changed due to the approach to crossings and entries to adjacent houses and properties.

These projects require sufficient covering of the geogrid with asphalt binder and asphalt concrete ($\geq$ 8.0 cm).

As a summary, it has to be stated that during the application of orientated geogrids the following basic preconditions have to be achieved:

- The base for the installation of grids has to be even, dry and clean.
- Potholes, gaping joints have to be filled in advance.
- The groove depth in case of milling should be < 4 mm.
- The geogrid has to be fixed well on the base layer, it has to be installed and stretched without any folds.
- The installed geogrid should not be trafficked by vehicles.
- For a better tack coat, fine gravel 8/11 mm with an amount of 12-14 kg/m² shall be installed only.
- The fine gravel should be bituminously coated.
- The installation of the binder course, minimum 4 cm, preferably 6 cm thick, should follow the installation of the geogrid as soon as possible.
- The total thickness of the asphalt layers on top of the geogrid should be **at least** 8 cm, preferably 10 cm.

If all these instructions are adapted, the installation of the orientated geogrid will be successful.

Behaviour of Belgian applications of geotextiles to avoid reflective cracking in pavements

A. VANELSTRAETE
Belgian Road Research Centre, Brussels, Belgium
Y. DECOENE
SCREG Belgium, Brussels, Belgium

Abstract

During the previous conferences on "Reflective cracking in pavements" (Liège, 1989 & 1993), papers were presented about thirteen Belgian experiments with geotextiles to avoid reflective cracking in pavements, mostly in overlays on cement concrete pavements. Geotextiles impregnated with polymer-bitumen were covered with a porous or dense asphalt wearing course. Many types of geotextiles were used: nonwoven polyester (manufactured on the jobsite or in a factory), nonwoven polypropylene, woven polyester. Different polymer-bitumens were chosen for the impregnation of the geotextile: bitumen with new elastomers, bitumen with recycled elastomers, emulsion of bitumen with new elastomers. This paper deals with the results of an evaluation of these thirteen test sites.
Keywords: Reflective cracking, geotextile, woven, nonwoven, cement concrete

1 Introduction

At the first RILEM-Conference on "Reflective Cracking in Pavements" in 1989, P. Silence et al. presented a paper in which three experimental sites with woven geotextiles, carried out between 1985 and 1987, were described and evaluated [1]. Four years later, for the second RILEM-Conference on Reflective Cracking of 1993, Y. Decoene reported the results of a second evaluation of these sites, together with the results of ten other projects, carried out by SCREG (Belgium) between 1989 and 1993, where different types of woven and non-woven geotextiles were used for the prevention of reflective cracking [2]. The present paper describes the results of a detailed inspection of these thirteen roads, being exposed now to two to ten years of traffic.

Reflective Cracking in Pavements. Edited by L. Francken, E. Beuving and A.A.A. Molenaar. © 1996 RILEM. Published by E & FN Spon, 2–6 Boundary Row, London, SE1 8HN. ISBN 0 419 22260 X.

2 Description of the sites and reparation techniques

The location of the thirteen sites is given in fig. 1.

Nine of the thirteen projects concern overlays on concrete slabs, in most cases without foundation and with severe vertical movements ("rocking") at the edges of the slabs. The slabs have not been cracked and seated, except for the project at Tourpes. Two sites are overlays on heavily damaged continuous reinforced concrete [3] and two projects concern overlays on flexible pavements. Different interface products have been placed: a woven in polyester, a nonwoven in polyester, a nonwoven in polypropylene and a nonwoven in polyester manufactured on site [4]. Different types of binders have been used for the impregnation of the nonwoven and/or the adhesion with the support: new elastomers, recycled elastomers and an emulsion of bitumen with new elastomers (natural latex) [4]. As wearing course different thicknesses and types of mixes have been applied (see table 1). The porous asphalt, indicated with (*) in table 1 is the so-called Fixtone double coated method. Type III-mix is a dense bituminous mix used for levelling courses and underlayers. The ultrathin layer (indicated with ** in table 1) is a bituminous mix consisting of 70 - 80 % stones, 20 - 30 % sand, 4 - 6 % filler and 5 - 6 % binder (on 100 % aggregates) [5]. The mix with high mastic content (indicated with *** in table 1) is a French Stone-mastic-asphalt mix and is composed of 67 - 74 % stones, 16 - 23 % sand, 10 - 14 % filler, 6.8 - 7.6 % binder (on 100 % aggregates) and stabilised with fibres [6].

Table 1 gives for each project: year of renovation, size of the project, traffic level (in number of vehicles per day, for each direction), original road structure, type of geotextile, type and quantity of bitumen used for adhesion and impregnation of the geotextile, overlay type and thickness.

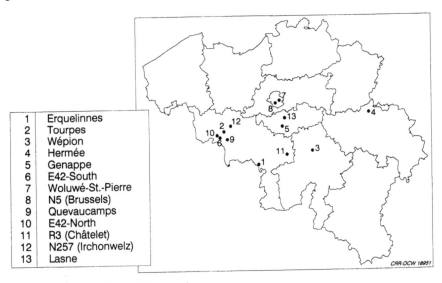

1	Erquelinnes
2	Tourpes
3	Wépion
4	Hermée
5	Genappe
6	E42-South
7	Woluwé-St.-Pierre
8	N5 (Brussels)
9	Quevaucamps
10	E42-North
11	R3 (Châtelet)
12	N257 (Irchonwelz)
13	Lasne

Fig. 1 Location of the different sites.

Table 1: Description of the different sites [(*), (**), (***) for details: see text]

Project and no.	Year repair	Surface (m²)	Traffic level type	Original structure	Type of interlayer	Bitumen + quantity	Bituminous overlay
1:Erquelinnes	1985	2500	<2000,med. heavy	concrete slabs	woven polyester	new elast.,1.5 kg/m²	4 cm porous mix(*)
2:Tourpes	1987	19200	< 2000, heavy	concrete slabs	woven and nonwoven polyester	new elast. 0.6 kg/m²	3cm IIIC+interface + 5cm IIIB + 4cm porous mix (*)
3:Wépion	1987	20600	<2000, light	flexible pav.	woven polyester	new elast.,0.5 kg/m²	3 cm porous mix(*)
4:Hermée	1989	1300	<2000, light	concrete slabs	nonwoven polyester	new elast.,1 kg/m²	4 cm porous mix
5:Genappe	1990	1060	< 2000, heavy	concrete slabs	woven poyester	recycl. elast.,1.5kg/m²	4 cm porous mix
6:E42-South	1990	75190	>8000, heavy	cont.reinf. concrete	polyester manuf. on site	recycl. elast., 3 kg/m²	4 cm porous mix.
7:Woluwe S.-P.	1990	1820	<2000, med. heavy	concrete slabs	nonwoven polyprop.	recycl. elast.,1.5kg/m²	4 cm porous mix
8:N5 (Brussels)	1990	1120	>8000, heavy	concrete slabs	nonwoven polyester	recycl. elast.,1.5kg/m²	4 cm porous mix
9:Quevaucamps	1991	2580	< 2000, light	concrete slabs	(1) nonwoven polyester (2) no interface	new elast.,1kg/m² -	1.5 cm ultrathin layer (**)
10:E42-North	1991	141200	>8000, heavy	cont. reinf. concrete	polyester manuf. on site	recycl. elast., 3 kg/m²	5 cm porous mix
11:R3 (Châtelet)	1991	22000	>8000, heavy	flexible pav.	polyester manuf. on site	recycl. elast.,2.5kg/m²	2.5 cm porous mix
12:N527 (Irchonwelz)	1991	3600	<2000, med. heavy	concrete slabs	nonwoven polypropylene	new elast.,1kg/m²	3 cm mix with high mastic content(***)
13:Lasne	(1)1992 (2)1993	13500 13000	>2000 and <8000 light	concrete slabs	nonwoven polyester manuf. on site	new elast. (latex) 1 kg/m²	(1) 3cm porous mix (2) 3cm mix with high mastic content (***)

3 Evaluation of the sites

During the previous evaluation no reflective cracking was observed, except for the projects at Erquelinnes (36 cracks) and at Genappe (2 cracks).

Table 2 gives an overview of the results of the present evaluation, carried out during the period 12/1995 - 1/1996. Additional information about the inspections is given below. For each project, the exact location of the reflective cracks and other type of damage was noted; this allows to follow up these projects in the future. The transverse cracks were divided into three categories, depending on their length: cracks smaller in length than 1/3, between 1/3 and 2/3, between 2/3 and 3/3 of the traffic lane width. Hence, apart from the total number of cracks reflected, the number of so-called "equivalent" reflected cracks can be deduced. This value takes into account that not all cracks have reached their full length. The number of equivalent cracks reflected is calculated as follows:

$$x_1 \cdot (1/6) + x_2 \cdot (3/6) + x_3 \cdot (5/6)$$

in which x_1, x_2, x_3 represent the number of cracks with length $\leq 1/3$; $> 1/3$ and $\leq 2/3$; $> 2/3$ and ≤ 1 of the traffic lane width, respectively. Knowing the average length of the concrete slabs, the number of original joints/cracks can be calculated and the percentage of (equivalent) reflected cracks can be deduced (see table 2). We note that many transverse cracks are double cracks (see fig. 2), 2 - 3 cm spaced from each other. They are most probably the result of high vertical movements ("rocking") of the slabs at the joints. In the calculations of the number of (equivalent) cracks, a double crack is counted as one single crack.

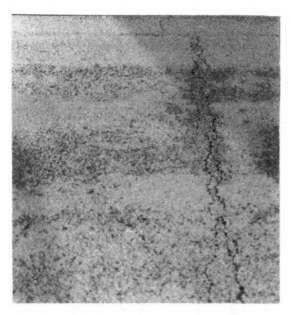

Fig. 2 Example of a double crack in the overlay above a joint between two cement concrete slabs with high vertical movements (project Quevaucamps) (Photo CRR 3525/13A)

Table 2: Summary of the results of the inspections [(*) for details: see text]

Project and no.	Year repair	Original structure	Type of interlayer	Overlay	Reflected cracks	Equivalent reflected cracks	Other damage
1:Erquelinnes	1985	concrete slabs length: 3.3m , 10m(*)	woven polyester	4 cm porous mix	55%(l=3.3m) 18 %(l=10m)	43%(l=3.3m) 14 %(l=10m)	local spalling at borders
2:Tourpes	1987	concrete slabs	woven, nonwoven polyester	8cm dense, 4cm porous mix	no	no	local subsidence
3:Wépion	1987	bituminous mix	woven polyester	3 cm porous mix	no	no	longit.cracks, at sides: subsidence, spalling
4:Hermée	1989	concrete slabs (l=5m)	nonwoven polyester	4 cm porous mix	53 %	28 %	-
5:Genappe	1990	concrete slabs (l=11.5m)	woven polyester	4 cm porous mix	55 %	41 %	fatigue cracking
6:E42-South	1990	cont.reinf.concrete	polyester manuf. on site	4 cm porous mix	no	no	14 potholes over 16 km
7:Woluwe S.P.	1990	concr. slabs (l=12m)	nonwoven polyprop.	4 cm porous mix	31 %	19 %	-
8:N5 Brussels	1990	concr. slabs (l=8.5m)	nonwoven polyester	4 cm porous mix	65 %	48 %	1 pothole, local spalling
9:Quevaucamps	1991	concrete slabs (length: 8.5 m)	(1) nonwoven polyester (2) no interface	1.5 cm ultrathin layer	(1) 73 % (2) 100 %	(1) 49 % (2) 82 %	- -
10:E42-North	1991	cont.reinf.concrete	polyester manuf. on site	5 cm porous mix	no	no	85 potholes over 16 km
11:R3 (Chât.)	1991	flexible pav.	polyester manuf. on site	2.5 cm porous mix	no	no	fatigue cracking
12:Irchonwelz	1991	concrete slabs (length: 9 m)	nonwoven polypropylene	3 cm mix with high mastic content	35 %	21 %	2 potholes
13:Lasne	(1)1992 (2)1993	concrete slabs (length: 12 m)	polyester manuf. on site	(1)3 cm porous mix (2)3 cm mix with high mastic content	(1) 100 % (2) 4 %	(1) 63 % (2) 3 %	(1) no (2) no

Project No.1: Erquelinnes: Rue des Bonniers et Rue du Conroye
In addition to the high appearance of reflective cracking (Fig. 3), there is a lack of an efficient drainage system with consequently lots of spalling at the sides of the road. For the "Rue des Bonniers" a difference in reflective cracking has been observed between both directions (see table 3):

Table 3: Appearance of transverse cracks for project No. 1 (Erquelinnes)

	Rue des Bonniers		Rue Couroy		All
	Direction N40	Direction Erquelinnes	Direction N40	Direction Erquelinnes	
% of reflected cracks (3.3 m between joints/cracks)	77	51	46	46	55
% of reflected cracks (10 m between joints/cracks)	62	38	37	36	43

The original concrete slabs were in a very bad state: many slabs were broken. During the evaluation, we noticed that more reflective cracking appeared at joints between slabs, than at cracks in slabs. This is also clear from tables 2 and 3, where the percentage of reflected cracks is determined in two ways:
- with respect to the number of joints and cracks in the slabs. In this case the average length between two joints or/and cracks is 3.3m.
- with respect to the number of joints (cracks not included) between the slabs. The average length is then 10 m.

Fig. 3 General view of the project at Erquelinnes: a lot of reflective cracking, spalling at the sides of the road due to inefficient drainage (Photo CRR 3525/22A)

Project No. 2: Tourpes
No reflective cracking has been observed. This is mainly the result of the large overlay thickness, being 12 cm in total, and the fact that the slabs have been cracked and seated before rehabilitation. Over a few hundred meters of this 2.5 km long project, subsidence has been observed locally, most probably due to an insufficient bearing capacity. At a few places, the porous asphalt shows ravelling.

Project No.3: Wépion (Rue Lecomte and Rue Suary)
This project concerns an overlay of a heavily cracked flexible pavement with potholes. No reflective cracking has been observed (fig. 4). At the sides of the road, local subsidence with spalling has been observed, most probably due to a lack of drainage system. The interface system is clearly visible at some of these places. There are many longitudinal reflective cracks.

Project No. 5: Genappe
A lot of reflective cracking is observed (see table 2). Most of the cracks are double, indicating large vertical movements at the edges of the slabs. Part of the project shows fatigue cracking.

Project No. 7: Woluwe S.-P.
At many places, even if no reflective cracking was observed, the underlying structure of the cement concrete slabs can be clearly noticed, due to differences in level. The beginning of the appearance of a longitudinal crack is observed at the limit of adjacent slabs over the almost entire length of the project.

Fig. 4 General view of the project at Wépion: no reflective cracking, local spalling at the sides of the road due to a lack of drainage system (Photo CRR 3526/8)

Project No. 8: N5 Brussels
Besides reflective cracking with many double transverse cracks, local spalling is observed at a few places in the middle of the road, at the longitudinal joint in the overlay.

Project No. 9: Quevaucamps
This project was partly carried out without interface product and partly with a nonwoven polyester. As can be seen in table 2, a clear difference was found between the project with and without interface: for the project without interface a larger number of reflected cracks was observed: 100 % compared to 72 % (with interface). Cracks are generally longer: 82 % equivalent cracks compared to 49 % with interface. Furthermore they are larger in width than the cracks in the part with interface product. Although a high level of reflective cracking was obtained, clear evidence was thus found for a positive effect of the use of nonwovens.

Project No. 10: E42 -North
There are many potholes, mainly in the right wheel track of the right lane, they are the reflection of heavily damaged spots in the continuous reinforced concrete near the joint with the emergency lane in bituminous concrete.

Project No. 11: R3 (Châtelet)
Some parts of this viaduct show cracking on the right lane. These cracks only appear in the 2.5 cm thick top layer of porous asphalt and are not related to reflective cracking. They are most probably due to fatigue. We note that a porous mix is more subjective to fatigue cracking than a dense bituminous mix, because of its high void content.

Project No. 13: Lasne
One part of the project dates from 1992, the other is from 1993. The same interlayer system and thickness of overlay were applied, but the type of overlay with a 70/100 pure bitumen was different: porous asphalt for the site of 1992 and a mix with high mastic content [6] for the site of 1993. A very high difference in reflective cracking was observed: all cracks/joints reappeared in the case of the 3 cm porous asphalt, only 4 % reappeared for the 3 cm overlay with high mastic content. Although both projects are not fully comparable, because of their difference in year of repair, the large difference in reflective cracking indicates the advantage of the use of mixes with high mastic content for the prevention of reflective cracking. For the project of 1993, we note that even if no reflective cracking was observed, the underlying structure of the cement concrete slabs can be clearly noticed, due to differences in level.

4 Conclusions

The present paper describes the results of a detailed inspection of thirteen roads, where different types of woven and non-woven geotextiles were used for the prevention of reflective cracking. These projects have now been exposed to two to ten years of traffic and have been described and evaluated previously in the framework of the first and second RILEM-conferences on Reflective Cracking in Pavements. Nine projects concern overlays on concrete slabs, in most cases without foundation and with severe vertical movements ("rocking") at the edges of the slabs. Two sites are overlays on severely cracked continuous reinforced concrete and two projects concern overlays on flexible pavements. Different types of binders have been used for the impregnation of the nonwoven and/or the adhesion with the support and different thicknesses and types of bituminous mixes have been used as overlay (see table 1).

One of the conclusions is that the percentage of reflected cracks depends highly on the nature and state of the underlayer:

- Very few or no reflective cracking is observed on overlays with woven or nonwoven interface products on existing flexible pavements.
- No reflective cracking is observed on overlays on continuous reinforced concrete. However, in cases of very severe local damage of the underlying concrete structure, potholes reappear at the surface of the new road structure, even with the use of a nonwoven manufactured on site.
- Reflective cracking largely appears on overlays of cement concrete slabs, not being cracked and seated at the time of repair, even with the use of an interface system. Percentages of cracks reflected after three to eight years of repair range from 31 to 100 %, depending on the project; percentages of equivalent cracks reflected range from 19 % to 63 %. Reflective cracking mainly occurs at the joints and to a less extent at the cracks in the cement concrete slabs. One joint often reflects in two adjacent cracks at the new road surface. They indicate high vertical movements at the edges of the cement concrete slabs.

From the evaluation of the projects it follows clearly that the different elements of an overlay system (the interface system and the overlay) are of importance:

- The use of an nonwoven impregnated with modified bitumen offers a positive effect on reflective cracking as compared to an overlay system without interface (project Quevaucamps). Less cracks have reflected: 73 % with interface and 100 % without interface for a 1.5 cm thick overlay on cement concrete slabs, four years after repair. They are generally smaller in length (49 % equivalent cracks reflected with interface, 82 % without interface) and in width.
- The overlay thickness remains one of the predominant factors as what concerns reflective cracking, even with the use of an interface system. The highest percentage of cracks being reflected was observed in the case of an 1.5 cm ultrathin overlay (project Quevaucamps), whereas no reflective cracking occurred in the case of a 12 cm thick overlay (project Tourpes).
- These evaluations indicate that the use of bituminous mixes with a high mastic content is advantageous for the prevention of reflective cracking (project Lasne). For the section with porous asphalt all cracks have reflected three years after repair ; 4 % of the joints/cracks have reflected for the mix with the high mastic content two years after repair.

Besides reflective cracking, other types of damage have been observed: local spalling at the borders due to an insufficient drainage and subsidence due to a lack in bearing capacity.

Despite the positive effect of woven and nonwoven geotextiles, important reflective cracking has been observed in the case of overlays on cement concrete slabs with large vertical movements at the edges of the slabs.

In such cases, it is advisable to crack and seat the concrete slabs before overlaying, in order to limit the movements of the underlayer. In order to guarantee that cracks do not reappear shortly after rehabilitation, a sufficient overlay thickness is required. Therefore, the Screg-system being a 1.5 - 2 cm coated sand asphalt, rich in elastomeric binder (resistant to deformations of the underlayer), combined with a 3 - 4 cm wearing course with a high mastic content [7], might be promising to prevent reflective cracking in cases where no reinforcement is necessary or where no thicker overlay is possible.

Acknowledgments

The authors gratefully thank Mr. P. Vanelven (Belgian Road Research Centre) and Mr. K. Vandecasteele (Screg Belgium) for their collaboration to the evaluation of the sites. The Belgian Road Research Centre thanks the IRSIA ("Institut pour l'Encouragement de la Recherche dans l'Industrie et l'Agriculture") for their financial support in this research project.

References

1. P. Silence and J. Estival, "Applications de tissés de polyester pour lutter contre la remontée des fissures", Proceedings of the 1st RILEM-Conference on Reflective Cracking in Pavements, pp. 281-287, 1987.
2. Y. Decoene, "Belgian apllications of geotextiles to avoid reflective cracking in pavements", Proceedings of the 2nd RILEM-Conference on Reflective Cracking in Pavements, pp. 391-397, 1993.
3. R. Dumont and Y. Decoene, "The application of a geotextile manufactured on site on the Belgian motorway Mons-Tournai", Proceedings of the 2nd RILEM-Conference on Reflective Cracking in Pavements, pp. 384-390, 1993.
4. J. Samanos, H. Tessonneau, "New system for preventing reflective cracking: membrane using reinforcement manufactured on site (MURMOS)", Proceedings of the 2nd RILEM-Conference on Reflective Cracking in Pavements, pp. 307-315, 1993.
5. J.-P. Serfass, P. Bense, J. Bonnot, J. Samanos, "A new type of ultrathin friction course", Transportation Research Board, No. 1304, pp. 66-72, January 1991.
6. G. Lelardeux, "Une expérience intéressante de renforcement de chaussées", Revue Générale des Routes et Aérodromes, No. 563, April 1980.
7. J.-P. Serfass, J. Samanos, "Anti-reflective cracking two-layer asphalt systems: Assessment and comparisons", Revue Générale des Routes et Aérodromes, No. 724, pp. 42-43, December 1994.

Geotextile reinforced bituminous surfacing

H. VAN DEUREN
VicRoads, Metropolitan South East Region, Melbourne, Australia
J. ESNOUF
VicRoads, Northern Region, Bendigo, Australia

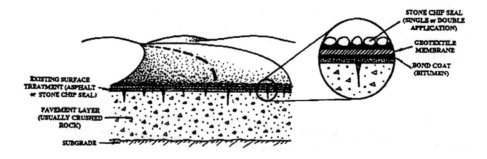

Abstract
This paper presents VicRoads' experience since 1979 with geotextile reinforced sprayed bitumen chip seals for rehabilitating pavements to delay and/or inhibit the propagation of cracking, and as an initial treatment on pavements and bicycle paths. The paper discusses techniques, materials, equipment, typical applications and case studies.
Keywords: Cracked pavement, fatigued asphalt, geotextiles, reinforced chip seal.

1 Introduction

VicRoads has developed a sprayed bitumen chip seal reinforced with a geotextile membrane which provides a safe, low cost surfacing treatment for cracked road pavements.

VicRoads is the State Road Authority of Victoria located in the south eastern corner of Australia and has been successfully providing sprayed bitumen chip seal surfaces on its road network for more than 50 years. However, with higher traffic volumes and increased vehicular loadings in an ever tightening financial environment it was found that conventional sprayed chip seal treatments were not always able to cope, particularly over cracked surfaces.

Geotextile reinforced chip seals have been applied to a large variety of road surfaces ranging from unsealed low trafficked pavements consisting of high clay content materials to fatigued asphalt pavements on major metropolitan freeways.

The first use of a geotextile chip seal in Victoria was in 1979 when the geotextile was placed by hand over very small areas of distressed pavement. With a number of

Reflective Cracking in Pavements. Edited by L. Francken, E. Beuving and A.A.A. Molenaar. © 1996 RILEM. Published by E & FN Spon, 2–6 Boundary Row, London, SE1 8HN. ISBN 0 419 22260 X.

improvements particularly in handling and placing geotextile VicRoads is now able to place six lane kilometres or more per day on highways with high volume traffic.

Currently at least 100 lane kilometres per year of geotextile reinforced chip seals are being placed in Victoria, and the usage of this treatment is rapidly increasing.

Geotextile seals provide low cost effective surfacing treatments over extensive cracked pavements and give a safe, waterproof surface to roads that normally would require reconstruction at much higher costs.

2 Description

Geotextile reinforced chip seals consist of a paving geotextile impregnated with bitumen and covered with either a single or double application bitumen chip seal. They provide a waterproof membrane capable of inhibiting cracking in an existing pavement. Geotextile chip seals employ the tensile strength of the geotextile fabric and the elastic recovery of bitumen to bridge cracks and allow for some vertical movement to ensure that no moisture penetrates the pavement.

Geotextile seals have been used extensively on old cracked flexible pavements where the cracking may be due to:

- surfacing age and oxidation of the bitumen;
- environmental conditions (thermal or swell/shrinkage);
- poor pavement design (asphalt fatigue);
- inferior material quality of either the surfacing or the pavement itself.

Geotextile seals have also been used on pavements constructed with a base course material stabilised with cement where the cracking is usually due to shrinkage, and/or fatigue; and as initial treatments on high clay content roads.

2.1 Typical uses for geotextile seals
There are many applications for the use of geotextile seals to extend the life of the pavement or defer rehabilitation/reconstruction including:

- Old/cracked pavements;
- Fatigued Asphalt;
- Cement treated base courses;
- Shoulders/bicycle paths.

2.2 Old/cracked pavements
Cracking can occur in flexible unbound granular or old Macadam pavements as a result of oxidisation of the bitumen leaving the asphalt or sprayed bituminous surface brittle and unable to absorb or cope with deflections caused by traffic loading. The cracks allow water to penetrate the pavement and reduce the sub-grade strength by wetting it, which results in pavement failures.

Geotextile seals are used to provide a waterproof membrane to inhibit reflective cracking and prevent penetration of moisture and thus defer pavement reconstruction.

Polymer Modified Binder and Bituminous Scrap Rubber sprayed seals have had limited success in inhibiting the cracking. Cracks have often reappeared within 2-3 years after the modified or scrap rubber seals has been placed.

2.3 Fatigued Asphalt

Fatigued asphalt is a common problem in older type flexible pavements where inadequate depth of base course does not provide the necessary stiffness to support the asphalt. Even though the pavement is structurally sound, fatigue cracking of the asphalt may occur and these cracks will allow moisture to enter the pavement eventually causing potholes and failures.

A geotextile seal provides a waterproof membrane that prevents moisture from entering the pavement and weakening the base. The oldest geotextile seal applied onto a fatigued asphalt pavement in Victoria was laid in 1990 and shows no sign of the underlying cracks reflecting to the surface.

2.4 Cement treated base courses

Pavements constructed with a cement treated or stabilised crushed rock base course have proven to be a viable and economical alternative to conventional full depth granular or asphalt pavements. However, these pavements may display regular transverse and/or block cracking which when combined with wet weather and high traffic volumes, may lead to pavement distress.

Geotextile seals provide a uniform, waterproof membrane over the pavement that allows cracks to be spanned, prevents moisture from entering the pavement and leeching of fines from the base course.

2.5 Shoulders/bicycle paths

Geotextile seals can be a very economical alternative to constructing and/or strengthening shoulders and bicycle paths. Due to the low traffic loading and volumes the existing base could be (re)shaped without adding any material to strengthen the shoulder or bicycle path. Geotextile with a sprayed bituminous stone chip surface is then applied to provide a satisfactory waterproof pavement surface.

3 Types of Geotextiles

Non-woven, needle-punched geotextiles are used because they have uniform elongation, good resistance to tearing and excellent bitumen/ fabric adhesion. The most common types of geotextiles used for bituminous sprayed seals are discussed below:

3.1 Polyester

Polyesters are very suitable because their melting point is typically 250 °C, they absorb only small amounts of water, are not sensitive to ultra violet light and are suitable for all applications including polymer modified binders. The polyester is porous so it retains sufficient bitumen to perform the waterproofing function and is not affected by hydrocarbons in the bitumen.

The polyester geotextile used has the following properties:

- Unit mass - 140 g/m²
- Thickness - 0.6 mm (20 Kpa)
- Tensile strength - 10 kN/m (longitudinal)
- (Two directional tension) - 10 kN/m (transverse)
- Elongation - 27-30% (lonitudinal)
                  - 27-30% (transverse)
- Melting point - 260 °C

3.2 Polypropylene

Polypropylene may be used where the bitumen temperatures do not approach or exceed 175 °C which is the softening/melting point of polypropylene. Where they have been used they have achieved good results providing care has been taken to ensure that bitumen is applied below 175 °C. Polypropylene is more sensitive to ultra-violet light than polyester.

4 Placing geotextile reinforced chip seals

4.1 Surface preparation

The pavement should be basically sound and self draining with any visually distressed areas and those that are severely out of shape being removed and replaced with sound material in accordance with normal working practices.

Surface irregularities should be corrected and pavement rutting regulated. Existing cracks greater than 5 mm should be filled and sealed to support the geotextile and prevent it from being drained of bitumen in the vicinity of cracks.

4.2 Bond coat

The bond coat is the bitumen that enables the geotextile to adhere to the road surface.

For resurfacing the recommended bond coat binder is usually Class 170 bitumen preferably without thinners (usually kerosene). Bitumen emulsions could also be used but have not been used by VicRoads at this stage.

4.3 Laying of geotextile

Immediately after the bond coat is sprayed, the geotextile is spread using a special laying machine. The geotextile should be held close to the ground to prevent billowing and with sufficient tension to minimise wrinkling. Adjoining or adjacent runs should be overlapped by at least 100 mm, with the overlapped joint receiving additional bitumen. Overlapping joins are preferred to butted types because they are more effective in providing a continuous membrane.

4.4 Binder

The binder is the bitumen sprayed onto the geotextile to hold the stone chips in place.

A suitable bituminous binder is Class 170 bitumen but Polymer Modified and Granulated Scrap Rubber Binders can also be used to maximise stone retention and binder elasticity. If polymer modified binders are used then Polypropylene geotextile should be avoided because the higher spraying temperature of the binder would cause the geotextile to melt. Also some modified binders may occasionally present a problem with absorption into the geotextile as they have a tendency to form a skin.

An allowance of 0.9 l/m² should be made when determining the application rate for the chip seal for absorption of bitumen into the geotextile (with a grade of 140 g/m²).

4.5 Stone chips

The stone chips used should be clean and pre-coated with a bituminous based product at the quarry or on-site. For single-coat seals a 10 or 14 mm size stone is normally used. For two-coat seals it is normal to use 14 and 7 mm stone sizes and sometimes a 10 and 5 mm combination.

5 Discussion

5.1 Placing geotextile seals

When applying geotextile stone chip seals consideration should be given to the following:

1. Single application stone chip seal should not be placed where there is:

- high volumes of turning traffic;
- high percentage of large commercial vehicles especially on climbing lanes;
- stop/start traffic such as on approaches to signalised intersections.

2. Double application stone chip seals have a distinct advantage over (1) above because they are able to absorb the additional stresses introduced by turning cars and commercial vehicles.
3. Geotextile seals obviously perform best on straight sections of road but can also be laid on large radius curves. This may result in the formation of minor creases in the geotextile which do not diminish the effectiveness of this treatment. They can be placed on tight bends/curves if the geotextile is cut and wedges removed.
4. Removal of the geotextile as part of future rehabilitation or reconstruction work can be done by cold planing. Paving fabrics are recyclable in both hot and cold stabilisation processes with the geotextile providing some tensile reinforcement.

5.2 Additional surfacings

The main objective of geotextile seals is to waterproof a cracked pavement to prevent further deterioration and defer or remove the need for major rehabilitation or reconstruction. However, an exposed stone chip surface may not be appropriate at some locations because of noise or riding qualities. VicRoads has used a number of additional surfacing treatments over geotextile chip seals.

5.3 Asphalt
Dense graded, Stone Mastic or Polymer Modified Asphalts are suitable for use over geotextile chip seals.

A typical treatment would consist of a 10 mm stone chip geotextile seal to which 30-40 mm minimum depth Size 10 or 14 Asphalt is applied.

5.4 Open graded friction course (OGFC) asphalt
OGFC Asphalt provides a very quiet surface which improves skid resistance, minimises water spray and improves riding qualities.

A typical treatment would consist of a 10 mm stone chip geotextile seal to which a 25 mm minimum Size 10 OGFC Asphalt layer is applied.

5.5 Paver laid chip seal/ultra thin asphalt
This treatment has similar properties to OGFC Asphalt.

A typical treatment consists of a 10 mm chip stone geotextile sprayed bitumen seal to which a 12-15 mm depth size 10 Paver Laid Chip Seal/Ultra Thin Asphalt overlay is applied.

5.6 Future pavement maintenance
It is anticipated once a geotextile seal has been placed, the section of road will be able to be maintained using normal resurfacing practices such as another sprayed chip seal, an asphalt overlay, Paver Laid Chip Seal/Ultra Thin Asphalt or Open Graded Friction Course Asphalt.

6 Case Studies

6.1 Reflective cracking at Maroondah Highway between Stirling Road and Oban Road
This section of highway consists of a three lane pavement in each direction located in an urban area with traffic volumes of 50,000 vehicles per day with 6% commercial vehicles.

The pavement was constructed using cement-treated crushed rock (4-6% cement) as a base course with a 75 mm depth asphalt overlay. Due to the heavy traffic loading and stiffness of the cement treated crushed rock base course, cracks developed that reflected to the asphalt surfacing. A chip seal incorporating 20% by volume of granulated scrap rubber in the bitumen had been applied over the pavement 4 years earlier but the cracks had reflected through this also and extensive leeching of fines from the base course was evident on the surface.

Deflection testing carried out indicated the highway pavement was structurally sound and still had sufficient strength for the type and volumes of traffic travelling on this road.

It was recommended a geotextile double application stone chip seal be applied as an alternative to reconstruction.

Minor patching and regulation was carried out before the geotextile reinforced two-coat chip seal was applied in 1989 using the following materials and standards:

- Bond coat - Class 170 bitumen with 4% of thinners sprayed at 0.6 l/m²;
- Geotextile type - Polypropylene with a minimum fabric grade of 140 g/m², spread with a special fabric applicator and given two passes with a multi-wheeled roller;
- First application of chip seal - Class 170 bitumen with 5% of granulated scrap rubber and 4% of thinners sprayed at 1.5 l/m² and covered with 14 mm stone chips spread at 90 m²/m³ and rolled with multi-wheeled rollers before opening to traffic;
- Second application of chip seal - after the 14 mm stone chip seal was given a light sweep, Class 170 bitumen with 8% of thinners was sprayed at an application rate of 0.5 l/m² and covered with 7 mm stone chips spread at 150 m²/m³.

Because a Polypropylene geotextile fabric was used the spraying temperature of the bitumen was reduced to less than 175 °C.

The bitumen was modified with 5% by volume using granulated scrap rubber prior to spraying to improve retention of the stone.

Six years after the laying of the geotextile reinforced double application stone chip seal there is no evidence of the underlying cracked pavement or leeching of fines from the cement stabilised crushed rock base course. It is proposed to resurface this section of highway with a Ultra Thin Asphalt overlay in approximately 3-4 years time to further extend the life of the pavement by another ten years.

6.2 Fatigued Asphalt at Burwood Highway from Sophia Grove to Wattle Avenue

This section of highway consists of a 14 metres wide pavement in a two way traffic situation through the shopping centre of a township. The traffic volume is approximately 20,000 vehicles per day of which 8% is commercial vehicles.

The pavement at this location has a very flexible base which caused the very old, oxidised asphalt surfacing to develop extensive fatigue cracks.

Following deflection testing it was recommended 150 mm depth of pavement be removed and replaced with asphalt to achieve a 20 year design life at a cost of Aus$40.00/m². However, it was decided to apply a double application geotextile reinforced stone chip seal at a cost of Aus$6.00/m² to extend the life of the pavement by 10 years.

Minor patching and regulation was carried out before a geotextile reinforced double application chip seal was applied to the pavement in 1991 using the following materials and standards:

- Bond coat - Class 170 bitumen with 4% of thinners sprayed at 0.6 l/m²;
- Geotextile type - Polyester with a minimum fabric grade of 140 g/m², spread with a special fabric applicator and given two passes with a multi-wheeled roller;

- First application of chip seal - Class 170 bitumen with 4% of thinners sprayed at 1.5 l/m² and covered with 14 mm stone chips spread at 90 m²/m³ and rolled with multi-wheeled rollers before opening to traffic;
- Second application of chip seal - after the 14 mm stone chip seal was given a light sweep, Class 170 bitumen with 8% of thinners was sprayed at an application rate of 0.5 l/m² and covered with 7 mm stone chips spread at 150 m²/m³.

Since the geotextile reinforced chip seal was placed on the highway 5 years ago there is no evidence of the underlying cracks reflecting to the surface nor has any pavement maintenance been required, while surface texture has remained good. There is some flushing of bitumen on the approach to a signalised pedestrian crossing but the treatment is achieving its design purpose.

Future pavement resurfacing will consist of placing an Ultra Thin Asphalt overlay with a normal dense graded asphalt overlay on the approaches to the signalised pedestrian crossing in approximately 4-5 years time to further extend the life of the pavement by another ten years.

6.3 Old, cracked stone chip seal pavements at Hogan Street (Main Street), Tatura Township

This section of road consists of a 14 metres wide pavement with two way traffic and is the main street through the shopping centre of a rural township. The traffic volume is approximately 3,000 vehicles per day of which 20% is commercial vehicles, mainly large milk tankers.

The existing road was designed and constructed as a flexible unbound granular pavement and had extensive cracking due the brittleness of the surfacing caused by oxidisation of the bitumen in the stone chip seal. Because of this, the road surface was unable to absorb or cope with the deflections introduced by the heavy traffic. It was decided to apply a geotextile reinforced chip seal to maintain the existing pavement in a trafficable and maintenance free condition for five years until funds could be provided for full reconstruction and to allow sufficient time to prepare a detailed design for the pavement and service relocation works to proceed.

A single-coat 14 mm stone chip sprayed bitumen surface was applied in April 1989, as follows:

- Bond coat - Class 170 bitumen with 4% of thinners sprayed at 0.9 l/m²;
- Geotextile type - Polyester with a minimum fabric grade of 140 g/m², spread with a special fabric applicator and given two passes with a multi-wheeled roller;
- Chip Seal - Class 170 bitumen with 4% of thinners sprayed at 1.4 l/m² and covered with 14 mm stone chips spread at 85 m²/m³ and rolled with multi- wheeled rollers before opening to traffic.

In the five years the geotextile reinforced chip seal was in place the treatment was successful with little maintenance having to be carried out on this section of road. The road was fully reconstructed during Nov/Dec 1994 with a geotextile seal having provided the distressed pavement with an extra five years of serviceable life.

7 Conclusion

As can be seen from the case studies, geotextile reinforced chip seals have been very successfully applied in Victoria, Australia, over badly cracked pavements.

The types of pavements have ranged from cement stabilised or high clay content pavements in low trafficked areas as initial treatments through to high volume freeways with fatigued asphalt. Unfortunately, space has prevented other cases being discussed in this paper.

A geotextile reinforced chip seal can be a very successful treatment on its own or as the first stage rehabilitation of a distressed pavement without the need for full reconstruction with associated high costs and inconvenience to the public.

Geotextile reinforced chip seals have been rapidly growing in popularity in Australia as they have proven their effectiveness and versatility in inhibiting reflective cracking in pavements. Whilst they may not be a miracle solution to cracked pavements they have proven to be very effective and have deferred the need for expensive reconstructions of many sections of road.

Performance of the crack and seat method for inhibiting reflection cracking

J.F. POTTER
Transport Research Laboratory, United Kingdom

J. MERCER
Road Engineering and Environment Division, Highways Agency,
Department of Transport, United Kingdom

Abstract

Trials of the crack and seat method of maintaining jointed concrete roads and flexible composite roads have been undertaken at four sites in the UK. The technique requires that fine, full-depth vertical cracks are induced in the exposed concrete which is then firmly seated into the foundation using a heavy roller before overlaying with bituminous material. The induced cracks should enable the expansion and contraction of the cement-bound layer to be distributed more evenly which should minimise the occurrence of transverse reflection cracks in the surface of the bituminous surfacing.

The trials were designed to investigate the effects of crack spacing, pattern of cracking and overlay thickness on the performance of the technique. This is being assessed by monitoring development of reflective cracking, change in transient deflection, longitudinal profile and wheelpath rutting. After about four years in service, the test sections incorporating crack and seat are showing very little deterioration in comparison to the substantial number of transverse reflection cracks which have developed in the equivalent control sections.

A methodology for designing the thickness of bituminous overlay is being developed based on the effective stiffness modulus of the cracked and seated concrete.

Keywords: Reflection cracking, crack and seat, bituminous overlay, design, performance.

1 Introduction

In the UK, the performance of unreinforced jointed concrete roads usually depends on the condition of the load transfer dowel bars between slabs. Corrosion of the bars inhibits thermal movement which can cause spalling at the joints and lead to mid-bay cracking. Poor load transfer at the joints increases the stresses on the foundation under the action of traffic and this can lead to settlement of the slabs and pumping of fines from the underlying materials. To improve the serviceability, it is common practice to overlay the pavement with bituminous material after carrying out selected repairs to the concrete. Although the overlay strengthens the pavement, it is usually carried out to improve the riding quality of the road. However, thermal movements within the concrete frequently caused transverse reflection cracks to develop in the new surfacing above the joints. To minimise the cracking, the present overlay recommendations are to apply at least 180mm of bituminous material. In addition, geogrids or stress absorbing membranes are sometimes incorporated at joints and cracks to help reduce the occurrence of reflection cracking.

Flexible composite roads also suffer from transverse reflection cracks at the road surface which develop above shrinkage cracks in the underlying cement-bound layer. Typically these reflection cracks are spaced between 6m and 10m apart.

Since 1991, the technique of crack and seat, in which fine cracks are created in unreinforced concrete slabs or exposed lean concrete roadbases before overlaying, has been trialed in the UK.

2 Crack and seat methodology

2.1 Theoretical considerations
The approach used for the trials in the UK was to induce fine transverse cracks in the concrete in order to create more locations where thermal contraction could occur whilst at the same time retaining satisfactory load transfer characteristics at the cracks. As the crack spacing in the concrete base is reduced, the horizontal strains resulting from thermal movements should be distributed more evenly throughout the pavement and are therefore less likely to cause transverse reflection cracks in the surface of the overlay. Provided that the cracks induced in the concrete are fine, load transfer between the newly formed slabs should be satisfactory due to good aggregate interlock. This is the basic principle of de-stressing used in the crack and seat method, which produces shorter slabs whilst retaining satisfactory structural integrity.

Bituminous overlays are usually applied to jointed concrete roads to restore riding quality and/or skid resistance rather than to improve the structural capacity. In most instances the structural capacity of a jointed unreinforced concrete road is greater than that required for the weight of traffic being carried. For flexible composite construction, the requirement to cover transverse reflection cracks is a frequent reason for overlaying or resurfacing. When the concrete or cement-bound roadbase is cracked and seated, its loadspreading ability is reduced due to a reduction in the effective stiffness modulus of the cemented layer. The amount by which the modulus is reduced depends on the degree of interlock at the cracks and on the crack spacing. The finer the cracks and the wider the spacing, the smaller the reduction in stiffness modulus.

Consequently, it is necessary to establish the balance between the crack spacing, the reduction in stiffness modulus, the overlay thickness and the occurrence of reflection cracks. It is necessary to take the reduction in stiffness modulus into account when designing an overlay on a cracked and seated pavement.

2.2 Practical considerations

For cracking and seating on jointed concrete construction it is not necessary to repair defective joints and cracks, prior to overlay, in the manner needed for a conventional overlay. This means that maintenance costs should be reduced and also that the complete maintenance operation should be carried out more quickly, not only because concrete repairs are not required, but also because no curing time is then necessary. Effectively, the process changes the jointed concrete pavement into a strong cement-bound roadbase with fine, closely-spaced cracks. Strong concrete does not usually suffer from degradation at the cracks, caused by mechanical abrasion under traffic, or from the opening and closing of the crack with changes in temperature.

For roads with cement-bound roadbases, reducing the crack spacing mechanically requires the removal of the upper bituminous roadbase and surface layers before the cracking operation can be carried out.

It is important that the cracking operation produces only very fine cracks in order to provide the good aggregate interlock needed for load transfer. However, it is also essential that the cracks penetrate to the full depth of the concrete layer to allow small thermal movements to take place at each crack. In order to see the cracks following the cracking operation, it is usually necessary to wet the surface of the concrete and observe the cracks as it dries. Cores should be taken though a selection of the cracks to check that the cracks penetrate the full depth. It is important to prevent the formation of longitudinal cracks which can contribute to a reduction in load-spreading ability of the overlaid road.

At the start of any scheme it is necessary to adjust the cracking machine accurately to achieve the required results. The amount of force that the machine needs to apply depends on the strength and the thickness of the concrete layer. On both types of construction, the cracked surface is seated using a pneumatic tyred, heavy roller prior to application of the new bituminous overlay. This operation is carried out to ensure that any voids under the slabs are filled prior to overlaying.

Two types of equipment have been used, and are available in the UK for the cracking operation - those using a guillotine action and those using a whiphammer action. The guillotine machines are capable of generating fine transverse cracks but in normal operation, the whiphammer produces a diamond shaped crack pattern. With each machine, it is recommended that adjustments are made to produce the required fineness and depth of cracking, without causing undue spalling on a length of concrete prior to treating the trial areas. The force applied by the guillotine blade or by the hammer is adjusted according to the strength and thickness of the concrete. It is important that the induced cracks are vertical through the concrete and not inclined at an angle, because inclined cracks are likely to inhibit the thermal movements due to aggregate interlock enforced by the weight of the overlying pavement structure.

Crack and seat is not recommended for reinforced jointed concrete pavements because it would be necessary to break the concrete completely from the reinforcement in order to influence the expansion and contraction properties of the slabs.

3 Road trials

Four road trials have been constructed between 1991 and 1993; two on jointed unreinforced concrete roads (M5 and A14) and two on roads with lean concrete roadbases (A30 and A40). The M5 motorway has 3 lanes and a hard shoulder in each direction and the other sites are situated on two-lane dual carriageways.

The trials were designed to evaluate the effects of overlay thickness, crack spacing and the pattern of cracking, created by different types of cracking machine, on the in-service performance of the roads. The performance of the test sections is being assessed by direct comparison with control sections of conventional overlay. During the construction of the trials, measurements were made of the condition of the road before, during and after the various maintenance operations. Table 1 gives details of the design and construction of the trials.

Table 1. Design and construction of crack and seat trials

Location of trial	Date built	Original construction	Factors investigated	Number of test sections	Overlay thickness (mm)
M5 Taunton	May 1992	230mm JCP on 225mm granular	crack spacing (0.5m,1m, 2m)	8	100, 150
A14 Quy	July 1993	265mm JCP on 185mm granular	cracking machines	9	100, 150, 180
A30 Exeter	Feb. 1991	115mm HRA on 190mm lean concrete on 320mm granular	crack spacing (1m, 2m)	8	115, 175
A40 Whitchurch	March 1992	175mm HRA on 175mm lean concrete on 200mm granular	overlay thickness	6	175, 355

JCP - jointed unreinforced concrete pavement HRA - hot rolled asphalt roadbase and surfacing

Initial assessments at the trial sites were made to determine the condition of the road in order to produce a detailed design for the trials and to provide an objective rating, which enables a quantified comparison to be made of the effectiveness of the crack and seat treatment in relation to the load transfer efficiencies and the severity of the original cracks. All the measurements have been recorded on computer spreadsheets to carry out systematic analyses as condition measurements become available.

4 In-service performance

The in-service performance of the road trials is being assessed by measuring the condition of the test sections in the spring and in the autumn. The measurements being made are as follows:

* visual assessment of the development of reflection cracking (growth and severity)
* Falling Weight Deflectometer (FWD) deflections above joints and shrinkage cracks to assess load transfer behaviour
* FWD deflections in mid-slab locations on the overlaid jointed concrete pavements to investigate structural performance
* longitudinal surface profile using the TRL High-speed Survey Vehicle (HSV)
* depth of wheelpath ruts using the HSV

4.1 Development of cracking

Detailed visual inspections were carried out under road closure conditions. At the trials with jointed concrete pavements, the location of the transverse joints between slabs was marked at the edge of lane 1 immediately after being overlaid to identify the location of the original joints. On the flexible composite trials, the locations of the original reflection cracks in the surfacing were recorded by position along the road so that any subsequent reflection cracking in the new surfacing could be related to the original cracking by direct measurement.

The occurrence of any reflection cracking in the new overlays was recorded; the location was identified, the length of cracking was measured and its severity estimated by experienced inspectors. The location and severity of other surface defects were also recorded and the weather conditions at the time of the inspection were noted. The severity of cracking determined by visual inspection was recorded as falling into one of three severity gradings. These gradings were also used to assess the visual severity of the original cracking. The visual severity ratings are as follows:

A - wide (greater than or equal to 2mm), often with spalling or bifurcation

B - easily visible (less than 2mm wide)

C - fine cracking (including that seen only when the road is drying)

At the trial on M5 near Taunton, no visual deterioration was observed in any test section during the first year in service. However, 18 months after the road was re-opened to traffic, eleven cracks were observed in lane 1 above the joints in the untreated control section with the 100mm overlay. The cracks were fine or very fine and were located above joints that had poor load transfer ability before the treatment. At subsequent inspections the number and length of cracks have steadily increased as quantified by the development of cracking shown in Figure 1. The results are expressed as a percentage of the number of cracks compared with the number of original joints in the test sections and also in terms of the percentage length of cracking that has occurred compared with the total length of transverse joints. All of the reflection cracks were categorised as fine.

In the crack and seat section with 100mm of overlay, after three years in service, three transverse cracks have been noted above joints in the section with 0.5m crack spacing and two in the section with 1m crack spacing. This shows that the cracking operation does not necessarily "lock up" the original joints. No reflection cracks have occurred above any cracks induced in the slabs by the crack and seat technique.

At the trial on the A14 near Quy, the development of cracking on the control section is following the same pattern as observed on the M5 even though all of the joints had good load transfer characteristics; these results are also given in Figure 1. No cracks have been observed in the overlays on any of the crack and seat treated sections.

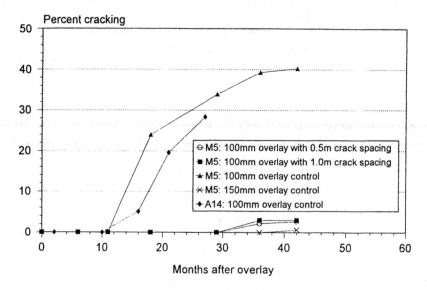

Fig 1a: Development of cracking by length

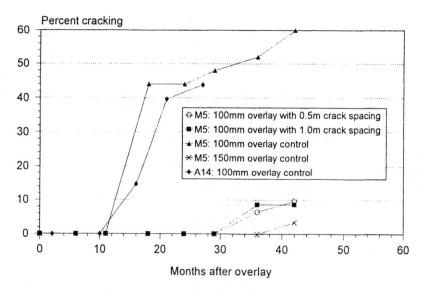

Fig 1b: Development of cracking by number of cracks

At the trial on the A30 near Exeter, the cracking has only occurred in the control section with the thinnest (115mm) bituminous layer. Most of the cracks are in the same location as cracks that were generally classed as severe prior to the maintenance works. On the A40 near Whitchurch, no transverse cracks have been observed after nearly four years in service.

4.2 Transient deflection

FWD deflections were measured annually at the sites on M5 and A14. Measurements were made with the loading plate close to the underlying joints to determine load transfer efficiencies as shown in Figure 2a and also within each original slab with the geophones arranged in positions to enable the effective stiffness moduli to be back-calculated (Figure 2b).

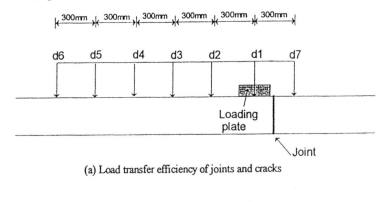

(a) Load transfer efficiency of joints and cracks

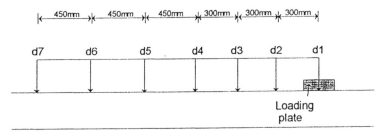

(b) Mid-slab location for modulus testing

Figure 2. FWD geophone test configurations

Figure 3 shows some typical load transfer efficiencies measured on the control section with 100mm thickness of overlay on the M5 at Taunton. It can be seen that before overlay the load transfer efficiency of the four joints varied from poor to good. After overlaying, the load transfer for each joint was excellent showing that the bituminous material was making a significant contribution to the load transfer efficiency.

Figure 3 shows that joint number 86, with good initial load transfer ratio, has retained a ratio of unity and no reflection crack has occurred. Above joint 85, a crack was observed at the inspection 18 months after overlaying. This crack extended across the full width of the lane and the load transfer efficiency has steadily reduced since being overlaid. Joint number 82 had the lowest load transfer ratio prior to overlay. At the inspection after 18 months, a crack was observed which extended over less than one half of the width of the lane. This crack is now starting to grow and it will be interesting to observe how the load transfer efficiency changes as the crack develops.

On the crack and seat sections, the restoration of load transfer efficiency was similar to the control sections for the equivalent range of initial value. The load transfer efficiencies across the original joints remain close to unity after 42 months in service.

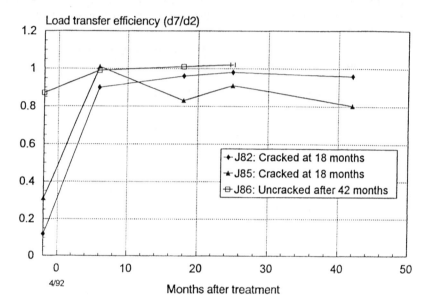

Figure 3: Load transfer efficiency of selected cracks on 100mm control M5 Taunton

The deflections measured in the overlaid concrete slabs have remained constant with time indicating no deterioration in the structural condition of the pavements.

4.3 Longitudinal profile
The variance in longitudinal profile was measured on the original constructions, then after cracking and seating and again after the overlays were applied. Subsequently, measurements are made annually to investigate whether the development of unevenness in the crack and seat sections is different to that in the control sections.

On the M5 at Taunton, during the 7 months before the road was treated, a general slight deterioration in the longitudinal profile was measured on all test sections. After cracking and seating, the existing profile did not change significantly which implies that there was no voiding underneath the concrete. The bituminous overlay improved the longitudinal profile substantially and, since the road was re-opened to traffic in June 1992, the evenness of all test sections at the shorter wavelengths has improved slightly whilst there has been no change at the longer wavelengths. The change in longitudinal profile of the crack and seat sections is similar to that for the untreated control sections. At the other sites, there has been no difference between the change in profile on the crack and seat sections and the control sections.

4.4 Wheelpath rutting

Although the development of rutting in the wheelpaths of the bituminous overlays should not be influenced by the crack and seat treatment, a bituminous surface that deforms is less likely to crack. Therefore, it is considered important to check that the development of rutting at the individual sites is similar on the various test sections. Although ruts are beginning to develop following the hot summer of 1995, the development at individual sites has been generally at a similar rate on the crack and seat test sections and the control sections.

5 Structural design

Although the principal objective of the crack and seat treatment is to reduce the occurrence of reflection cracking in the bituminous overlay, it is important that the design of the overlay is based on the anticipated traffic loading that the structure is expected to carry. From a structural viewpoint, the required thickness of overlay will depend on the loadspreading ability of the existing pavement structure after the crack and seat treatment. With a knowledge of the thicknesses and the stiffness moduli of the treated cemented layers and the foundation layers, the critical stresses and strains in the pavement structure can be calculated for various thicknesses of overlay assuming typical stiffness moduli of the bituminous materials. The required overlay thickness can then be determined on the basis of ensuring that the traffic-induced tensile strains at the bottom of the bituminous overlay and the vertical strain in the subgrade, using the criteria given by Powell et al [1], are maintained within the acceptable limits.

5.1 Effective stiffness modulus

After cracking, the stiffness moduli of the concrete layer or slabs for the different crack spacings were determined by back-calculation of the deflections measured by the FWD. Although FWD deflections can be used to backcalculate stiffness moduli with reasonable accuracy, the calculations assume that the deflection bowl is measured on a uniform structure. In practice, cracks introduce discontinuities into the pavement and, therefore, the results of the backcalculations produce values of an "effective" stiffness modulus for the cracked cement-bound layer. At individual sites, the moduli of the slabs before cracking were similar and after cracking the moduli typically reduced by between 50 and 70 per cent depending on the crack spacing. A greater reduction in modulus was observed as the crack spacing was reduced.

Careful application of the crack and seat operation ensures that good aggregate interlock is maintained so that the concrete will react to loading in a reasonably consistent manner and the effective stiffness modulus should also be reasonably consistent. However, it is important that when making measurements with the FWD to ensure that geophones are not touching the road surface directly on a crack.

5.2 Design of overlay thickness

A design exercise was carried out based on the back-calculated moduli for the cracked concrete derived from the trials and typical moduli for the bituminous overlay, the sub-base and the subgrade. The TRL computer program AXSTRESS was used to calculate the values of the tensile strains at the bottom of the overlay and the vertical strains at the top of the subgrade and then the equations used in UK analytical design were used to predict pavement life.

The predicted traffic-carrying capacity of the test sections at Taunton, for example, based on conventional fatigue and deformation characteristics, were all in excess of 100 million equivalent 8.2Mg axles. The predicted lives increased with increase in crack spacing and with thickness of overlay.

A draft specification for the crack and seat operation has been developed based on the experience gained in trials.

6 Conclusions

1.   The crack and seat technique has been used successfully in road trials on jointed concrete roads and on exposed lean concrete roadbases. On jointed concrete roads the technique is quicker and less expensive to carry out than the conventional treatment of repairing defective joints and overlaying with 180mm of bituminous material.

2.   Early indications of the process inhibiting the occurrence of transverse reflection cracks are encouraging. After four and one half years in-service, on flexible composite construction with 115mm of bituminous surfacing no cracks are visible in the crack and seat treated section, whilst nearly 40 per cent of the transverse cracks have appeared in the control section. Similarly, on unreinforced jointed concrete pavements, after three and one half years in service, cracks have appeared in the 100mm overlay control section above about 60 per cent of the joints and not on the corresponding crack and seat treated section with 2m crack spacing. On the sections with 0.5m and 1m crack spacing short fine cracks have occurred above less than 10 per cent of the joints.

3.   The longitudinal profile on the overlaid crack and seat treated sections is similar to that on the control sections.

4.   A structural design methodology is being developed based on the reduction in stiffness modulus of the concrete after cracking and seating. A draft specification for crack and seat has also been developed for use in the UK based on experience gained from the trials.

7 Acknowledgements

The work described in this report forms part of the Highways Agency's research programme carried out by the Transport Research Laboratory and is published by permission of the Chief Executives of the Highways Agency DOT and TRL.

8 References

1.   Powell, W.D., Potter, J.F., Mayhew, H.C. and Nunn, M.E. (1984) The structural design of bituminous roads. TRRL Laboratory Report LR 1132. The Department of Transport. Crowthorne.

Reflective cracking in Swedish semi rigid pavements

J. SILFWERBRAND
Royal Institute of Technology, Stockholm, Sweden
B. CARLSSON
Swedish Road and Transport Research Institute, Linköping, Sweden

Abstract
The Swedish use of semi rigid pavements has been more frequent during the latest decade. The use of cement bound roadbases would, however, increase rapidly if the problem with reflective cracking in the asphalt layers could be solved definitively. However, several measures to prevent and/or postpone reflective cracking exist. In Sweden, some promising test sections have been built during the last decade.

Test sections have been carried out on three sites, in Linköping in 1985 and in 1989 and in Kristinehamn in 1986. Test variables are: asphalt concrete thickness, type of asphalt concrete including polymer modified binders, and presence of geogrids and geotextiles. Minor cracking and some damages exist, but the overall performance is promising. In the paper, the influences of the different variables on the cracking behaviour will be reported and discussed.
Keywords: asphalt pavements, cement bound roadbases, field tests, reflective cracking

1 Introduction

Cement bound roadbases were introduced in Swedish test roads during the 1950s, see Höbeda [1]. A 7 km long road - the first major project of this kind - was constructed in Vårgårda close to Göteborg in 1969. Three layers of asphalt with a total thickness of about 120 mm were placed on top of a 175 mm thick cement bound roadbase. Severe transverse cracking was, however, observed few years later. Water has penetrated into the unbound layers through the wide cracks. Frost heave was observed at the cracks. The damage progress is typical for the ultimate effects of reflective cracking, see, e.g., Foulkes [2].

To prevent or at least postpone the reflective cracking, the Swedish National Road Administration prescribed rather thick asphalt layers on cement bound roadbases, considerably thicker than corresponding asphalt thickness on asphalt bound roadbases, [3]. The prescribed thickness made, however, semi rigid pavements uneconomical. In

Reflective Cracking in Pavements. Edited by L. Francken, E. Beuving and A.A.A. Molenaar. © 1996 RILEM. Published by E & FN Spon, 2–6 Boundary Row, London, SE1 8HN. ISBN 0 419 22260 X.

order to reduce the asphalt thickness, field tests have been carried out with other methods to reduce the risk of reflective cracking.

Test sections have been constructed in Linköping in 1985 and in 1989 and in Kristinehamn in 1986. The thickness of the asphalt layers has been varied. Polymer modified binders in the asphalt layer, geogrids, and geotextiles have been used in some test sections. The field tests will be described and discussed in the following two sections.

The successful results of these test sections made it possible for the Swedish National Road Administration to reduce the prescribed asphalt thickness in the new code, [4].

2 Description of field tests

2.1 Linköping 1985
The field tests consist of five test sections labelled 0, 1, 2, 3, and 4. The cross section consists of a 35 mm thick asphalt wearing course, an asphalt base course of varying thickness, a 150 mm thick cement bound roadbase, and a gravel layer of varying thickness (Fig. 1). Test section 0 is made with an untreated gravel roadbase instead of the cement bound base. All test sections are about 100 m long. They were constructed in 1985. The road was designed for 500 to 1500 heavy vehicles daily. Traffic measurements showed 200 heavy vehicles in 1986 and 700 in 1995. The predominant subgrade consists of clayey moraine, clayey mo, and clay.

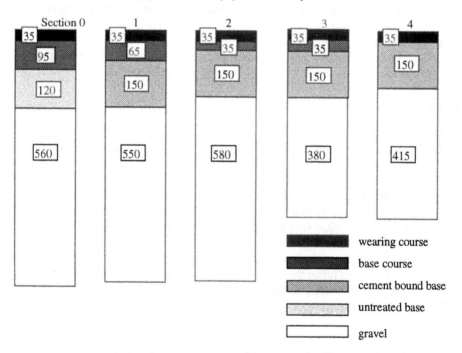

Fig. 1. Cross sections for test sections Nos. 0-4 in Linköping 1985. Measurements in mm.

The cement bound roadbase was made of a mixture containing 110 kg/m^3 cement, 1400 kg/m^3 sand (0-8 mm), 185 kg/m^3 gravel (8-16 mm), 585 kg/m^3 gravel (16-32 mm), and 95 kg/m^3 water. The compressive strength of cores from the roadbase was 8.5 MPa.

The asphalt base course layer consisted of bitumen stabilised gravel with a nominal binder content of 4.5 % and a nominal void content of 6 %. The wearing course consisted of asphalt concrete with bitumen penetration 85, a nominal binder content of 6.5 %, and a nominal void content of 3 %.

Further information is given by Petersson, Carlsson, and Ydrevik [5].

Linköping is situated about 200 km south-west of Stockholm. Sweden is divided into six climate zones. Linköping belongs to the second warmest zone. The winter (average daily temperature below 0°C) is limited to 80 days.

2.2 Linköping 1989

The 1989 field tests are situated close to the previous field tests in Linköping described above. The new field tests consist of eight test sections labelled 1-8 (Table 1). The cross section consists of a 40 mm thick asphalt wearing course, a 25-40 mm thick asphalt base course, a 160 mm thick cement bound roadbase, and a 425 mm thick gravel layer (Fig. 2). Different measures to prevent reflective cracking were studied: polymer modified binders, geogrids, geotextiles, admixture of fibres in the cement bound layer, and admixture of granulate from old car tyres in the wearing course. The test sections are 75 to 90 m long. They were constructed in 1989. The road was designed for 250-1000 heavy vehicles daily. In 1990, the number of heavy vehicles was about 350 and in 1995 about 400. The subgrade consists of a 3 to 6 m thick clay layer resting on a mo layer with a thickness of about 3 m.

The cement bound roadbase was made of a mixture containing 100 kg/m^3 cement, 1430 kg/m^3 sand (0-8 mm), 220 kg/m^3 gravel (8-16 mm), 550 kg/m^3 gravel (16-32 mm), and 110 kg/m^3 water. The compressive strength of cores from the roadbase was 11 MPa at 28 days. To the cement bound roadbase of test section No. 8, polypropylene fibres were added. The fibres had a length of 18 mm and the fibre content was 900 g/m^3.

Seven base course layers consisted of asphalt concrete with bitumen penetration 180, a maximum aggregate size of 12 mm, a binder content of 6.2 %, and a void content of 2.8 %. The base course of test section No. 6 consisted of bitumen stabilised gravel with a maximum aggregate size of 25 mm, a binder content of 4.2 %, and a void content of 6.7 %.

Six wearing courses consisted of asphalt concrete with bitumen penetration 85, a maximum aggregate size of 16 mm, a binder content of 6 to 6.5 %, and a void content of approximately 2 %. The wearing course of test section No. 1 consisted of a rubber modified asphalt material (RUBIT) with a maximum aggregate size of 12 mm (60 % aggregate above 8 mm), a binder content of 8.3 %, a void content of 2 %, and an admixture of granulated old car tyres. The wearing course of test section No. 2 consisted of asphalt concrete with a modified binder. The binder had a polymer content of 6 %.

The geogrid (REHAU ARMAPAL 6030) was made of polyester with a mesh width of 30 mm. The geotextile (FIBERTEX F-2B) was made by polypropylene and had a nominal thickness of 0.95 mm. Both geogrid and geotextile were glued to the base course layer with a bituminous emulsion. The emulsion was also applied to the geogrid or geotextile prior to wearing course placement.

Further information is given by Hultquist and Carlsson [6].

Table 1. Test sections at the field tests in Linköping 1989.

Test Section No.	Type of base course	Base course thickness (mm)	Measure to prevent reflective cracking
1	Asphalt concrete (bitumen penetration 180)	25	Granulate from old tyres added to the wearing course
2	Asphalt concrete (180)	25	Polymer modified binders used in the wearing course
3	Asphalt concrete (180)	25	Geogrid placed between wearing and base courses
4	Asphalt concrete (180)	25	Geotextile placed between wearing and base courses
5	Asphalt concrete (180)	25	Nothing (reference section)
6	Bitumen stabilised gravel	40	Nothing (reference section)
7	Asphalt concrete (180)	25	Nothing (reference section, placed later than the other sections)
8	Asphalt concrete (180)	25	Polypropylene fibres added to the cement bound roadbase

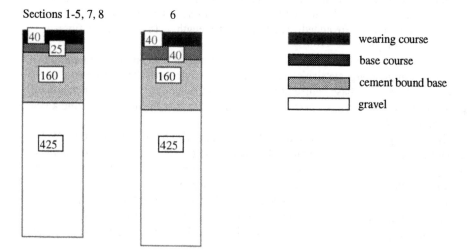

Fig. 2. Cross sections for test sections Nos. 1-8 in Linköping 1989. Measurements in mm.

2.3 Kristinehamn 1986

The field tests in Kristinehamn consist of four test sections labelled 0, 1, 2, and 3. The cross section consists of a 35 mm thick wearing courses of different bituminous materials, an asphalt base course of different type and varying thickness, a 160 mm thick cement bound roadbase, and a gravel layer of varying thickness (Fig. 3). The length of the test sections varies between 100 and 200 m. They were constructed in 1986. The road was designed for 500 to 1500 heavy vehicles daily. In 1995, the number of heavy vehicles was about 350 on test sections Nos. 0-2 and about 500 on test section No. 3. The subgrade consists of a 5-12 m thick clay layer resting on moraine.

The cement bound roadbase was made of a mixture containing 100 kg/m^3 cement, 185 kg/m^3 sand (0-8 mm, including water), 2078 kg/m^3 gravel (0-32 mm, including water), and 30 kg/m^3 water.

The base course of test section No. 3 consists of asphalt concrete with bitumen penetration 180. The base courses of the other test sections consists of bitumen stabilised gravel. The wearing course of test No. 0 consists of RUBIT containing an admixture of granulated tyres. The other wearing courses consist of asphalt concrete with bitumen penetration 85. All the bituminous materials have been described in the previous section.

Further information is given by Petersson & Karlsson [7].

Kristinehamn is situated at the large Swedish lake Vänern between Stockholm and Göteborg. Kristinehamn belongs to the same climate zone as Linköping, so the climate factors at the three field tests in Linköping and Kristinehamn are roughly the same.

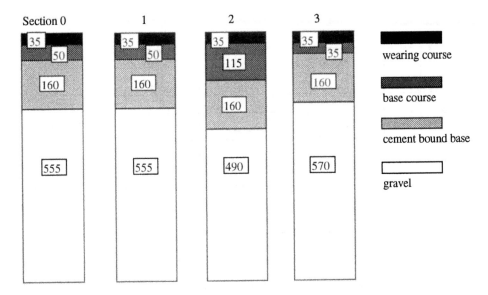

Fig. 3. Cross sections for test sections Nos. 0-3 in Kristinehamn 1986. Measurements in mm.

3 Evaluation

The field tests have been evaluated through measurements with falling weight deflectometer, evenness measurements, and crack surveys, see Hultquist and Carlsson [8]. In this paper, the discussion is limited to the crack surveys. For comparisons, the amount of cracking has to be defined. Here, a crack percentage has been defined by the following equation:

$$\text{crack percentage} = \frac{\text{number of square meters containing transverse cracks}}{\text{total number of square meters}} \cdot 100$$

In practice, each test section has been divided into 1 m² squares. Subsequently, the number of squares with visible transverse cracks have been counted and compared with the total number of squares of the test section. To avoid disturbances from edge effects, the first and latest 5 m of each test sections have been omitted in the evaluation. The crack surveys have been carried out once or twice each year. Hence, it has been possible to follow the development of reflective cracking. The results from the three field tests are presented and discussed in the following sections.

3.1 Linköping 1985
Crack surveys have been carried out between 1986 and 1992 (Fig. 4). In 1993, the test sections were overlaid with a 40 mm thick bituminous surfacing in order to improve the evenness. Only a few tiny transverse cracks existed on the new surfacing in November 1995.

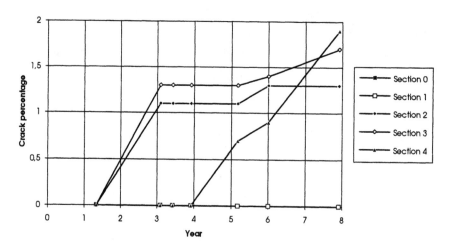

Fig. 4. Percentage of transverse cracks for test sections Nos. 0-4 in Linköping 1985. "Year=0" is defined as January 1st, 1985. The measuring series was discontinued in 1993 due to placement of a new overlay.

The overall performance was excellent. The crack percentage was limited to 2. No transverse cracks were observed in test sections Nos. 0 and 1 (containing thick asphalt base courses). Also test section No. 4 (without asphalt base course) was uncracked during the first four years, but since 1990 (year 6), the reflective cracking increased rapidly on this test section. No difference in crack development rate between summer and winter was observed.

3.2 Linköping 1989
Crack surveys have been carried out since 1990 (Fig. 5).

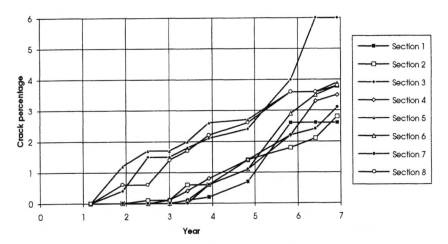

Fig. 5. Percentage of transverse cracks for test sections Nos. 1-8 in Linköping 1989. "Year=0" is defined as January 1st, 1989.

With exception for test section No. 3, also these test sections have performed well with crack percentages less than 4. During the first years, transverse cracking was most frequent on test sections Nos. 3, 5, and 8, but later the crack performances have been more even.

In November 1995, the worst performance was observed on test section No. 3 (containing geogrid). Beside the transverse cracking, also some surface damages could be found on this test section. The unsatisfactory performance is probably caused by an incorrect placement of the geogrid. Best performance was observed on test sections Nos. 1 and 2 (containing granulate from old tyres and polymer modified binder in wearing course, respectively). The crack development rate seems to be the same during summers and winters.

3.3 Kristinehamn 1986
Crack surveys have been carried out between 1988 and 1995. The amount of transverse cracking is larger than corresponding cracking on the field tests in Linköping (Fig. 6), but far from alarming. No simple explanation has been found. Differences in pavement width, differences in subgrade conditions, and different types of cement in the cement bound base might be influencing factors.

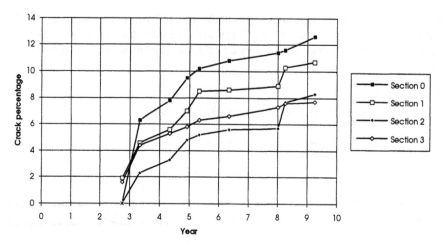

Fig. 6. Percentage of transverse cracks for test sections Nos. 0-3 in Kristinehamn 1986. "Year=0" is defined as January 1st, 1986.

Test sections Nos. 2 (containing a thick asphalt base course) and 3 (containing a base course of asphalt concrete with bitumen penetration 180) showed the best performance. The worst performance was observed on test section No. 0 (containing a wearing course with granulated tyres). No obvious difference in crack development rate between summer and winter was observed.

4 Concluding remarks

Based on the field tests, the following conclusions can be drawn:

i) An asphalt surfacing consisting of a 40 mm thick wearing course and a 40 mm thick asphalt base course layer resting on a cement bound base seems to perform well during a period of at least 5 to 8 years. Transverse crack percentage varied between 3 and 7 for different field tests.

ii) Geotextiles and geogrids do not seem to have any beneficial effect on the transverse crack performance. The bad performance may have been caused by incorrect placement of the geogrid. Besides, the advantages with geotextiles and geogrids may be more apparent when longer sections are evaluated. (Here, each test section was less than 100 m).

iii) Use of polymer modified binders and admixture of granulated tyres in the wearing course may have a beneficial effect on the transverse crack performance.

iv) Homogeneous asphalt surfacings or surfacings consisting of two layers with similar properties seem to be less prone to reflective cracking than surfacings consisting of two materials with strongly different properties. Base courses of asphalt concrete were better than those of bitumen stabilised gravel. A surfacing consisting of solely a wearing course performed excellently during four years. This conclusion agrees well with theoretical calculations according to Silfwerbrand [9].

v) No difference in crack development rate was observed between winter and summer. It means that Foulkes' hypothesis [2] stating that reflective cracking is more likely to occur during cold winter days in northern countries has not been proven. According to this hypothesis, a reflective crack starts from the top surface of the asphalt surfacing as a result of the tensile yield crack mechanism. This mechanism is most important during cold winter days.

5 References

1. Höbeda, P. (1985) Cement stabilisation of road bases - a survey of experiences in Sweden. *Bulletin*, No. 423, Swedish Road and Transport Research Institute, Linköping, Sweden, 50 pp. (In Swedish).
2. Foulkes, M.D. (1988) Assessment of asphalt materials to relieve reflection cracking of highway surfacings. *Thesis*, Plymouth Polytechnic, Plymouth, UK, 238 pp.
3. Swedish National Road Administration (1984) Swedish Pavement Code. *Bulletin* No. TU 154, Borlänge, Sweden. (In Swedish).
4. Swedish National Road Administration (1994) Swedish Pavement Code. Part 1, Common assumptions, and part 3, Pavement design. *Bulletins* Nos. 1994:21 and 1994:23, Borlänge, Sweden. (In Swedish).
5. Petersson, Ö, Carlsson, B., & Ydrevik K. (1987) Test road Lambohovsleden, Linköping - construction report. *Bulletin*, No. 524, Swedish Road and Transport Research Institute, Linköping, Sweden, 21 pp. (In Swedish).
6. Hultquist, B.-Å., & Carlsson, B. (1990) Reinforced and modified asphalt layers on cement bound bases - test sections on Lambohovsleden part 3, Linköping 1989 - construction report. *Bulletin*, No. 632, Swedish Road and Transport Research Institute, Linköping, Sweden, 23 pp. (In Swedish).
7. Peterson, Ö., Karlsson, B. (1987) Test road East Ringvägen, Kristinehamn - construction report. *Bulletin*, No. 521, Swedish Road and Transport Research Institute, Linköping, Sweden, 20 pp. (In Swedish).
8. Carlsson, B. (1993) The 1992 year evaluation of test roads with cement bound bases. *Report*, No. V175, Swedish Road and Transport Research Institute, Linköping, Sweden, 23 pp. (In Swedish).
9. Silfwerbrand, J. (1993) Reflective cracking in asphalt pavements on cement bound road bases under Swedish conditions. *Proceedings*, 2nd International RILEM Conference on Reflective Cracking in Pavements, Liege, Belgium, March 10-12, 1993, E & FN Spon, London, pp. 228-236.

Paving fabrics, how to increase the benefits

E. RAMBERG STEEN
Fibertex A/S, Technical Textiles Division, Aalborg, Denmark

Abstract
Paving fabrics, used in road maintenance, inserted between the existing, cracked road surface and the new asphalt wearing course, saturated with bitumen, act as a stress-absorbing, water-proofing interlayer membrane system prolonging the lifetime of the entire road construction if:

• the material (bitumen/paving fabric/asphalt material) are suitable

• the working procedure is qualified

• the installation of the paving fabric is done properly

This paper will focus on why the contractor, both before and while doing a road maintenance work, has to pay attention to a lot of important differences in each single component; the condition of the existing road surface, the bitumen used as tack-coat for the paving fabric, the paving fabric itself, the asphalt material, the weather conditions, the available equipment and the possibilities for solid workmanship of the local shifts etc. That the contractor is to combine all these components during the installation and completion highlightens the necessity of having qualified working procedures, and focusing on what happens, if the paving fabrics are not properly installed.
Keywords: Asphalt material, bituminous tack-coat, paving fabric, reflective cracking, road maintenance, stress-absorbing water-proofing interlayer membrane system, working procedure.

Reflective Cracking in Pavements. Edited by L. Francken, E. Beuving and A.A.A. Molenaar. © 1996 RILEM. Published by E & FN Spon, 2–6 Boundary Row, London, SE1 8HN. ISBN 0 419 22260 X.

1 Introduction

When reasphalting old worn down and cracked road surfaces, the cracks rapidly penetrate the new asphalt layer. In order to delay or counteract this reflective cracking, incorporation of paving fabrics is widely used in many different asphalt pavement construction types, including surface dressings, cold mixed asphalt materials or hot mixed asphalt materials, paving courses all varying from thin survival layers to heavy reinforcement layers. The actual site conditions are seldom the same, so whether the main purpose for using the paving fabric is to keep up the bearing capacity by preventing surface water from softening the subsoil or it is to reduce reflective cracking by stress relieving various expansions and contractions, it is for the contractor necessary to have a sufficiently detailed but still practical working procedure, covering the specific information about weather condition, type of bitumen, paving fabric, asphalt construction type etc.

In this paper, the main topic will be the use of paving fabrics in hot mixed asphalt materials, however, when it is relevant, there will be added remarks about other construction types, materials etc.

Figure 1 illustrates the most commonly used construction procedure: Unrolling a paving fabric onto a bituminous tack-coat. The heat from the succeeding hot asphalt material laid on the unrolled paving fabric will soften the underlying bitumen in order for this to penetrate and saturate the paving fabric, bonding the new asphalt layer to the old existing road surface and at the same time creating a stress-absorbing water-proofing membrane in between.

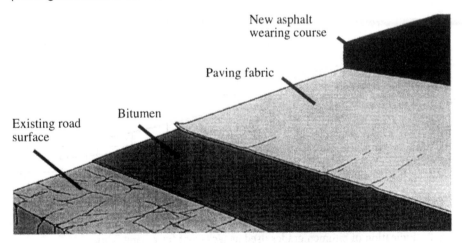

Fig. 1. An illustration of a general road renovation using paving fabric.

2 Preparation of existing road surfaces

The road surface shall present a clean uniform appearance, possibly after mending and levelling large holes, cracked crocodile areas or fissures. In general there shall be no pretreatment to cracks unless these exceed 5 mm in width.

3 The Material

3.1 The bituminous tack-coat

The most important part of a paving fabric application is the uniformity and the consistency of the bituminous tack-coat prior to unrolling the paving fabric. The bitumen must be sprayed in a uniform pattern at the prescribed rate (see equation 1 in section 3.3) and have a consistency according to the actual site conditions.

If, prior to carrying out a paving fabric installation job, the contractor was given unlimited freedom to decide the quality of bitumen; values like grade of penetration, time of workability, character of adherence etc. it would have been far more easy, but such freedom is not always present. Moreover, the contractor has the suitability of his equipment to take into consideration. Normally, it will only be possible for the contractor or the awarding authority on specially chosen big enterprises to decide exactly which type of bitumen should be used.

Old practices or good experiences with one type of bitumen for one type of project, is not just transferable to another type of project - like installing of paving fabric. It is a necessity that the contractor understand the truth of the matter. Often, the contractor does not know much of which type of bitumen he is using - and even less, when additives are used. Such condition makes the concept difficult, which is why types of useful bitumen for paving fabrics should be lined up. Only pure bitumens, without any form of solvents and with a grade of penetration according to the actual weather conditions, should be taken into consideration. See fig. 2.

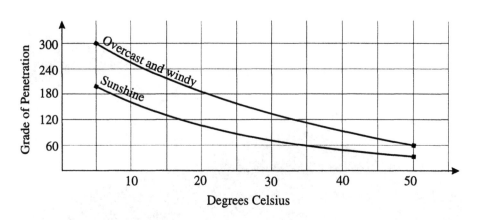

Fig. 2. Indication of bitumen grades used as tack-coat for paving fabrics.

At traditional paving projects, the paving contractors normally have good experiences using *softening additives*. Additives, which keeps the sprayed bitumen workable through a longer period of time. However, this is not the case when dealing with paving fabrics. The bitumen will far too soon find its way through the paving fabric, where it might be picked up onto vehicle tires, which could damage the fabric. *Polymer modified bitumens*, on the contrary, are normally advantageous to use, because the tack-coat then acts more sticky and viscous.

3.2 The equipment for spraying bituminous tack-coat

The proper operation of the spraying truck at the times required is essential for an organised paving fabric application. The installation procedure has to be adjusted, not only to the available or chosen bituminous tack-coat, but also to the actual equipment of the contractor. If the available equipment is only suitable for bitumen emulsions limited to keep the tack-coat heated up to maximum 80° C - or, the volume of the truck tank is less than needed for carrying out a practical day's work, the installation procedure has to differ from the procedure of matching a hot melted bitumen and a tank with plenty of store capacity.

The width of the spray pattern must match the width of paving fabric plus 5 cm on each side, and the spray-bar must be set and checked before spraying commences. The spread rate and uniformity shall be constantly monitored by the supervisor and occasionally checked by laboratory testing.

The driver shall be trained in applying a uniform rate and trial runs shall be carried out to determine the speed of the truck required to give the specified spread rate. Furthermore, the driver shall be able to carry out an acceptable alignment. Not as shown on the picture fig. 3 below.

Fig. 3. Incorrect alignment of the bituminous tack-coat.

To secure a correct alignment of the sprayed bituminous tack-coat, good visual angles for the truck driver is a necessity. Installing of a simple steel bar in front of the cabin, similar to the system used in asphalt paving machines, often solves direction problems for the truck driver. It makes it easier only to spray the bituminous tack-coat on exact, predetermined areas.

Spraying shall start and finish at predetermined points with as sharp a cut off as possible. At the end of the sprayed lane, after the spray-bar has been shut off, the driver shall accelerate away by at least 10 metres to avoid drips on the asphalt surface, and if not returning to spray again, shall park in a location off the asphalt where the bitumen will not be picked up by other vehicles. The driver shall avoid leaving bitumen puddles at all times.

3.3 The paving fabric

The purpose of using a paving fabric is to create a membrane effect, not to establish reinforcement [1]. Therefore the paving fabric must be flexible and easy to handle. Not too stiff with too much tensile strength counteracting the paving fabric to be unrolled in slight curves or on uneven or milled existing road surfaces.

However, in order for the fabric to live up to the expectations of it's containing the minimum quantity of bitumen to act as a satisfactory interlayer membrane, the paving fabric has to be needle-punched with a weight of minimum 140-150 g/m².

U.S.Department of Transportation, NHI, stated in their *Task Force 25*, that it has to be taken into consideration that different types of needle-punched paving fabrics demand different quantities of bitumen [2]. The quantity of bitumen has to be specified to the actual type of paving fabric. At the rate of 0,9 to 1,35 litres/m² or as recommended by the paving fabric manufacturer. NOTE: When using bitumen emulsions, the application rate must be increased to offset the water contents of the emulsion.

Over the last 25 years, one of the pioneers in using paving fabrics for retarding reflective cracking in pavements, the California Department of Transportation, CALTRANS, has made many tests, resulting in many reports. E.g. *Evaluation of paving fabric test installations in California* [3]. They have made the majority of the research with interlayers of paving fabric, and state in their *Paving Fabric Task Force* [4], that the needle-punched paving fabric shall be treated by heat or other processes causing the fibres on one side only to become bonded together, forming a glazed delamination free surface. The treated side turned upwards to increase the resistance against passing of the paving machinery's wheels or tracks.

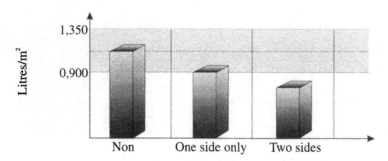

Treatments of 140-150 g/m² fabrics

Fig. 4. Retention properties of various non-woven paving fabrics, Q_{fabric}.

The specified rate must be sufficient to satisfy the bitumen retention properties of the paving fabric and bond the paving fabric and the new asphalt overlay to the old existing road surface. It is obvious, that a non-treated paving fabric demands more bitumen for creating the same interlayer membrane, as a paving fabric treated on one side only, and that a paving fabric treated on both sides, is not capable of containing the minimum quantity of required retention bitumen. See fig. 4. above.

In order to account for the variables in pavement structure, an additional quantity of retention bitumen, Q_c should be specified. An open and eroded road surface requires more additional bituminous tack-coat than a dense and smooth surface. Based on experience, a quantity between 50-150 g/m² is recommended by E. R. Steen [5]. The minimum on top of a dense surface and the maximum on top of an open surface. Depending on the quality of a new asphalt levelling course a large part of the tack-coat might be absorbed into this 'fresh" pavement, which means that the quantity of the tack-coat has to be increased, possibly up to around 200 g/m². This quantity will also be the case after milling, depending on the milling machine. Accounting all conditions of the existing road surface, the quantity, Q_c, should be estimated on site. See fig. 5.

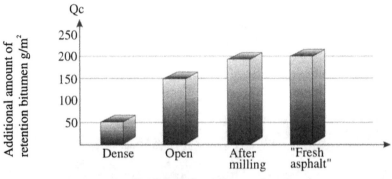

Fig. 5. Indication of the additional bituminous tack coat, Q_c

The whole necessary quantity of sprayed bituminous tack-coat, $Q_{necessary}$, before unrolling the paving fabric is then, in compliance with the two last-mentioned figures, to be calculated according to the following equation:

$$Q_{necessary} = Q_{fabric} + Q_c \qquad (1)$$

3.4 Unrolling the paving fabric

In order to maximise the speed of unrolling paving fabrics, sufficient rolls for one day's work should be placed ahead of the unrolling equipment or in some other way be within reach to avoid unnecessary hard work for the crew.

Usually, the paving fabric is unrolled by a crew of three using a small tractor or front-end loader rigged with special, mechanical equipment for handling rolls of fabric. The paving fabric can be placed manually (i.e. without using a tractor), this is, however, quite slow and cumbersome.

The unrolling of the paving fabric can take place when the hot bitumen has been sprayed and sufficiently cooled, or - when using bitumen emulsion - the bitumen emulsion has been allowed to set.

When hot bitumen is used as tack-coat, the weather temperature should be above +5° C, and if bitumen emulsion is used, the weather temperature should be above +10° C. Otherwise, the surface of the bitumen will not be sticky enough to ensure the bonding of the paving fabric during unrolling.

In order to ensure proper installation of the paving fabric, wrinkles should be avoided by keeping the paving fabric tight during unrolling. Folds, resulting in a triple thickness of fabric, must be cut with a knife, laid flat and if necessary treated with additional bitumen.

When stretching of the paving fabric during the unrolling by using special, mechanical equipment, slight curves can be performed. Sharp curves on the opposite, should be made by cutting sections of fabric, and making a small number of transverse joints to enable a smooth transition round the curve.

The paving fabric shall be unrolled with the treated side upwards and shall be seated with brooms after placing.

The paving fabric should be joined together by overlaps of between 10-15 cm. Not less than 5 cm and not more than 30 cm. Transverse overlaps should be made in the direction of paving to prevent edge pick-up by the asphalt paving machine. Besides, overlaps not already tacked should be added the necessary quantity of bituminous tack-coat in between the two layers of paving fabric. See fig. 6.

Fig. 6. Transverse overlaps made in the direction of the unrolling of paving fabric.

To minimise unnecessary driving on the unrolled paving fabric, this should not be placed more than about 200-300 meter ahead of the asphalt paving operations unless for a practical reason. Moreover, no more paving fabric should be placed than what can be covered the same day.

The paving fabric has to be kept dry. If the unrolled paving fabric contains water, the heat from the hot asphalt material cannot soften the bituminous tack-coat to such an extent that it can force out sufficient water for complete saturation of the paving fabric in order to create the expected membrane.

3.5 The hot mixed asphalt material

Hot mixed asphalt material is not just hot mixed asphalt material. It can act in totally different ways, depending on the grain mix design, the grade and quantity of bitumen, the kind of filler used, the aggregate and additives. Each sort of hot mixed asphalt material has its own special advantages for each distinct application.

It is important for the finished construction that the hot asphalt material contains enough retention heat when paved on top of the unrolled fabric, to soften the underlying bitumen enough to ensure the maximum saturation of the fabric. Standard hot mixed asphalt materials have plant temperatures varying between 140-170° C. These are temperatures suitable for paving fabrics made of polypropylene, which ensures a flexible behaviour in contact with the hot asphalt materials and, moreover, ensures that milled asphalt materials containing paving fabrics are recyclable [6]. If the temperature of the asphalt material for some reason should exceed 170° C, a small quantity of the asphalt material should be sprinkled manually before paving to protect the unrolled paving fabric. As a rule of thumb about 1 kg/m² should be applied for each degree Centigrade that exceeds 170° C .

The thickness of the asphalt overlay on top of the paving fabric should be between 2½-5 times larger than the diameter of the biggest grain in the asphalt material, which is a purely technical parameter for compacting the asphalt layer. Furthermore, the thickness of the asphalt overlay, regardless of the grain sizes, should as a minimum be more than 3 cm on top of old asphalt pavements, and more than 4 cm on top of concrete slabs. See fig. 7.

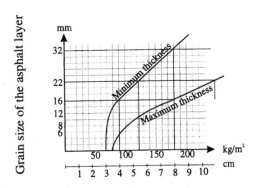

Thickness of the asphalt overlay

Fig. 7. Thickness of hot mixed asphalt layers on top of paving fabrics.

If carrying out the process in cold weather, more thickness of asphalt overlay should be added, due to the need for enough retention heat. Moreover, to assist in fabric saturation, the initial rolling of the asphalt layer should follow the paving machine closely.

3.6 The operation of paving machinery in connection with paving fabrics

The paving fabric will often be unrolled and the asphalt material distributed at the same time as allowing necessary travel of traffic. It can seems a little complicated, but good planning often solves the problem. Take care, however, that the trafficking does not destroyed the unrolled paving fabric.

If crawlers or tyres sticking and pick-up occurs, a small quantity of the asphalt material should be sprinkled manually on top of the paving fabric in the wheel paths of the vehicles.

The truck drivers are not allowed to use the brakes in the front of the asphalt paving machine. Furthermore, truck drivers should avoid unnecessary driving in fat spots or at longitudinal overlaps and avoid changing their direction in a violent manner.

4 The working procedure

When paying attention to differences in site conditions, weather conditions, material characteristics and local possibilities for solid workmanship, and not forgetting the divergence in local understanding of the need for a paving fabric interlayer, a suitable working procedure including a quality control of the whole installation process should be drawed up, and shall as a minimum cover:

- the preparation of the road surfaces to receive paving fabrics
- determine the grade and quantity of bituminous tack-coat
- the operation of the bitumen spray truck prior to paving fabric application
- determine the type of paving fabric
- the handling and unrolling of paving fabric
- determine the grade and thickness of the asphalt overlay
- the operation of the paving machinery in connection with paving fabrics
- determine the efforts of the quality control
- determine remedial repairs

5 Typical causes for bad installation

Wrinkles and overlaps in the paving fabric can cause cracks in the new asphalt overlay if not properly handled during the construction process, as well as insufficient bituminous tack-coat or too thin asphalt overlay normally results in slippage problems.

Bitumen emulsions work well in the membrane system, but runoff problems might occur in the application of emulsions on sloped and crowned roadways making the application rate difficult to control. Furthermore, if the fabric is placed before all the water has evaporated from the emulsion, slippage problems and eventual overlay failures might occur after a short period of time.

Using too soft a grade of bituminous tack-coat normally causes troubles for the paving equipment. If this happens the only thing to do is to sprinkle with some asphalt material in advance to avoid that the tyres stick to the fabric.

Fig. 8. Too soft a grade of bituminous tack-coat causes troubles.

Fig. 9. Too thin a overlay of asphalt might result in cracks.

6 Conclusion

To efficiently combine the materials suitable for the local weather conditions and equipment, qualified working procedures, including quality control and description for remedial repairs, should form the basis for increasing the benefits of using an interlayer of paving fabric for road maintenance projects.

It should never be allowed to use just any available product of random project conditions. The right materials for each actual project should be critically selected each time. Furthermore, following a qualified working procedure, the contractor should be able to combine the use of different types of bituminous tack-coats, paving fabrics, asphalt overlays etc. to get an overall package solution.

7 References

1. Vicelja, Joseph L. (1989) *Pavement Fabric Interlayers*, Presented at University of Liége Conference, Reflective Cracking in Pavements.
2. U.S.Department of Transportation - NHI (1992) *Task Force 25*, Publication No. FHWA-HI-9001, Revised April 1992.
3. CALTRANS (1990) *Evaluation of paving fabric test installations in California*, Technical Report No: FHWA/CA/TL-90/02.
4. CALTRANS (1991) *Paving Fabric Task Force*.
5. Steen, E.Ramberg (1994) *Paving Fabrics, Increasing the benefits by proper installation work*, presented in Singapore at the Fifth International Conference on Geotextiles, Geomembranes and Related Products.
6. CALTRANS (1986) *Recycling Asphalt Concrete (Experimental Construction), including information of recycling pavement containing various fabrics*, Technical Report No. FHWA-CA-TL-86/11.

Investigation on reflective cracking in semi-rigid pavements in Northern Poland

J. JUDYCKI and J. ALENOWICZ
Technical University of Gdansk, Highway Engineering Department,
Gdansk, Poland

Abstract
The paper presents results of an investigation on reflective cracking carried out in the Technical University of Gdansk. Three research projects, sponsored by District Directorate of Public Roads, are presented. The first project has been carried out since autumn 1988 on an 18 km long section of National Road No 7, which cracked transversally after cold winters of mid-eighties. Probable reasons for the cracking were investigated and repair solutions were proposed. In the second project a PCC pavement of National Route 22, which required a new overlay due to surface defects after over 40-year service was observed. Data on development of reflective cracks is given. Finally, a research, which started in 1994 on a 0.5 km long experimental section of semi-rigid pavement, is presented. Results of investigation on load transfer across cracks in cement stabilized base layer are given and discussed.
Keywords: Pavement, semi-rigid, reflective cracking, load transfer, Poland

1 Introduction

Poland is a country where use of cement treated bases is frequent and has a long tradition. One of the main reasons is the lack of rock deposits in the central and northern parts of the country. On semi-rigid pavements, reflective cracking of asphalt wearing course is normally observed, and occurs with different frequency (from few cracks per 1 km up to more than 10 per 100 m).

The problem of reflective crackings became an important issue in the Gdansk region after severe winters, which occurred in mid-eighties. Reflection crackings appeared on many sections of road pavements, even those which had been some years in service and did not show cracks before.

The Highway Engineering Department of the Technical University of Gdansk has been investigating the problem of reflective cracking for more than ten years now [1,2].

Reflective Cracking in Pavements. Edited by L. Francken, E. Beuving and A.A.A. Molenaar. © 1996 RILEM. Published by E & FN Spon, 2–6 Boundary Row, London, SE1 8HN. ISBN 0 419 22260 X.

Some aspects were investigated in cooperation with the Road and Bridge Research Institute in Warsaw [3,4] and the University of Oulu (Finland) [5,6]. Field investigation on some pavement sections was sponsored by the Distict Directorate of Public Roads in Gdansk. Three of such projects are briefly presented and discussed in the paper.

2 Research project on National Road No 7 Gdansk - Warsaw

2.1 General
The research project has been carried out since autumn 1988 on an 18 km long section of National Road No 7, from the border of Gdansk to the Kiezmark bridge, over the Vistula river. The road crosses Zulawy - a flat area in the Vistula estuary, partly located in a depression from 0.5 to 1.5 m below the sea level. The road was built on an embankment 1.5 - 2.5 m high. Climate in the area is moderate with sea influence.

The asphalt layer of the road cracked transversally after cold winters in mid-eighties. Crack intensity, as evaluated in 1988 and in 1996, was from zero to ten per 100 m. The aim of the research was to find out reasons for the crackings and to propose methods of repair.

The road was reconstructed and upgraded in 1966-70 and in 1980-ties. Before the first reconstruction the pavement was 5.5-6.0 m wide and its structure was the following: 20 cm drainage layer, 20 cm base course made either of crushed aggregate or crushed cobbles, and a thin asphalt layer. During the first reconstruction the pavement was levelled, widened and strengthened with asphalt concrete (up to 8 cm) or lean cement concrete (from 8 to 20 cm) with asphalt concrete surfacing. During the second reconstruction 2 m wide asphalt shoulders were added and a new asphalt wearing course was laid. Eventually, on some sections a semi-rigid pavement was formed with lean concrete 8 - 20 cm thick and asphalt layers 9 - 20 cm thick.

2.2 Scope of research
The research started with visual evaluation of pavement and registration of cracks on the whole 18 km long section. Cracking index was calculated from the formula:

$$CI = N + 0.5 \times N1 + 0.25 \times N2 \tag{1}$$

where:
CI - cracking index - number of transverse cracks per 100 m,
N - number of cracks on the full width of the pavement,
N1 - number of cracks on 1/2 width of the pavement,
N2 - number of cracks on 1/4 width of the pavement.

Nine typical sections, each 500 m long, were selected for detailed investigations. The sections were designated as:
- uncracked   -   cracking index below 0.5 (3 sections),
- medium cracked   -   cracking index from 1 to 4 (2 sections),
- intensively cracked -   cracking index above 4 (4 sections).

On the selected sections, coring of the pavement and boring of subgrade soil were performed. Specimens of the pavement and subgrade soil were tested in laboratory. For asphalt mixes, laboratory testing included evaluation of their composition, penetration and softening point of recovered bitumen, and indirect tensile strength of the mix at - 10ºC. For lean concrete compression strength was tested. For subgrade soils granulometric composition, natural moisture content and capillarity were tested.

On selected sections visual observations, and measurements of vertical movement of cracks were performed every month since December 1988 until June 1989. Visual observations were performed periodically until 1996.

Field and laboratory testing was performed to enable evaluation of possible reasons for cracking and to suggest suitable repair methods.

2.3 Crack intensity
Crack intensity is presented in Table 1.

Table 1. Crack intensity on the National Road No.7 Gdansk - Warsaw (1988)

Station Km - Km	Cracks per 100 m	Cracks per 1 000 m	Station Km - Km	Cracks per 100 m	Cracks per 1 000 m
7 - 8	0 - 4.5	11.0	16 - 17	2 - 6	37.0
8 - 9	0 - 3.5	18.5	17 - 18	0 - 4.5	25.5
9 - 10	0 - 3	15.0	18 - 19	0 - 3	18.0
10 - 11	0 - 4	14.5	19 - 20	0 - 2	6.5
11 - 12	0 - 8	23.5	20 - 21	0 - 1	4.0
12 - 13	1- 5	30.5	21 - 22	0 - 2.5	12.5
13 - 14	0 - 10	38.5	22 - 23	0 - 3	7.5
14 - 15	0 - 5	10.0	23 - 24	0 - 2	10.0
15 - 16	0 - 2	4.0	24 - 25	0 - 5	11.0

2.4 Analysis of testing results
Possible effects of the following factors on transverse cracking were analyzed: thickness of asphalt layers, thickness of lean concrete base, compressive strength of lean concrete base, indirect tensile strength of asphalt concrete, composition of asphalt concrete, properties of recovered bitumens, subgrade soil and subsurface water conditions.

2.4.1 Thicknesses of asphalt layers and lean concrete base
The effect of these two factors is presented in Fig. 1 and 2. On uncracked sections, average thickness of asphalt layers was greater than 16 cm and average thickness of lean concrete was lower than 10 cm. However, intensive cracking (CI = 10) occurred also on one section where average thickness of asphalt layers was rather high and equal to 16 cm. In general, there was a tendency that cracking intensity was lower for greater thickness of asphalt layers and for lower thickness of lean concrete base, but relationships were not very clear. Lower number of reflective cracks on sections with thin lean concrete base could be explained by lower tension forces applied to asphalt layers during thermal shrinkage of cracked base. It is worth noticing that there were no transverse cracks on sections with no lean concrete base. It indicates that crackings were reflective in nature.

2.4.2 Strength of lean concrete base
Fig. 3 presents relationship between compressive strength of lean concrete and cracking intensity of asphalt layers. There was rather clear tendency that the number of transverse cracks increased for greater compressive strength of lean concrete base.

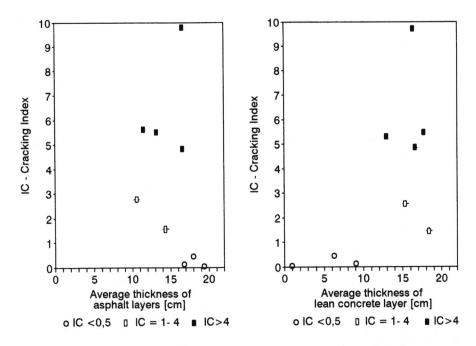

Fig.1 Cracking Index vs asphalt layers thickness

Fig. 2 Cracking Index vs lean concrete layer thickness

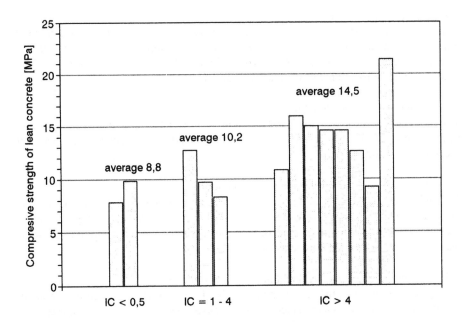

Fig. 3 Compressive strength of lean concrete vs Cracking Index

Most probably stronger base had cracked due to initial and/or thermal shrinkage with greater spacing which resulted in wider crack opening due to thermal shrinkage during cooling. Subsequently, more intensive tension in asphalt layers over cracks in the base was induced.

2.4.3 Indirect tensile strength of asphalt concrete

Table 2 presents the relationship between asphalt mix variables and intensity of cracking, including indirect tensile strength and deformation at -10°C. It can be noticed that tensile properties of asphalt concrete at low temperature are similar for all sections cracked and uncracked.

Table 2. Effect of wearing course asphalt mix variables on intensity of cracks

Asphalt mix variables	Sections		
	Uncracked	Medium cracked	Intensively cracked
	CI ≤ 0.5	CI = 1 - 4	CI ≥ 4
Indirect tensile strength at - 10°C, (MPa)	3.37, 3.14, 3.22	2.90, 2.90	2.77, 3.06, 2.48, 2.90
Deformation during indirect tensile test, (%)	0.87, 0.93, 1.09	0.78, 1.14	1.04, 1.03, 0.78, 0.79
Coarse aggregate content, (%)	46, 48, 47	44, 47	51, 56, 46, 51
Filler content,(%)	10, 10, 11	10, 9	9, 10, 11,8
Bitumen content, (%)	6.9, 6.5, 7.2	6.7, 7.4	6.4, 6.5, 6.9, 6.8
Penetration of extracted bitumen (0.1 mm)	32, 29, 37	27, 35	33, 29, 31, 28
Softening Point of extracted bitumen, (°C)	60, 57, 61	67, 57	60, 61, 60, 59

2.4.4 Asphalt concrete composition

Composition of asphalt concrete on all sections was similar and no conclusion could be drawn as to its effect on cracking intensity (Table 2). Bitumen content in the wearing course is rather high as it used to be typical for the Polish practice in the past. New Polish specifications assume lower bitumen content.

2.4.5 Properties of extracted bitumen

The original bitumen was D70 (penetration 70). It aged and hardened, reaching penetration of 27-37. No difference in bitumen properties on cracked and uncracked sections was found.

2.4.6 Subgrade soil and subsurface water conditions

The embankments were formed of clayey soils. A sandy drainage layer was constructed below the pavement structure. Water table is kept in the area by a pumping system at a level 0.5 - 1.0 m below the ground surface. No difference in the subgrade conditions was found on cracked and uncracked sections.

2.4.7 Deterioration of the pavement in the vicinity of cracks

Coring through most of the cracks indicated that lean concrete corroded due to action of salt used for winter maintenance. In some cases there was no solid concrete base under the crack in asphalt layer. The lean concrete had deteriorated into fine aggregate or small concrete blocks up to 0.2-0.3 m from a crack line.

2.4.8 Proposed repair technologies

It was proposed to measure deflections at cracks to assess load transfer across them and to find which ones required deep repair, including base layer. The following solutions were proposed as alternatives:

– Cracks with good load transfer: filling with bitumen, preferably - modified, "bridging" (covering of 10-20 cm wide strip along a crack with mastic asphalt 1-2 mm thick and application of fine aggregate) or 1 m wide and 8 cm deep milling of asphalt layers along cracks, installation of a geosynthetic and construction of a new asphalt layer.

– Cracks with poor load transfer: replacement of the pavement structure in a strip 1 m wide along the transverse crack and improvement of subgrade if necessary.

2.4.9 Pavement condition in January 1996

The pavement was not repaired during the 1988-1996 period. Inspection of the road in January 1996 indicated that in the vicinity of most of the reflective cracks pavement severely deteriorates with time. From single transverse cracks which had been observed eight years ago double and triple transverse cracks or crack network were formed. Moreover, fatigue crackings were developed under wheel paths starting from the reflective cracks. The fatigue cracking developed along the wheel path in a ladder or in a network pattern. It indicates that fatigue cracking occurred in the lean concrete base and in the asphalt layers, starting from transverse reflective cracks. It could be noticed that new transverse cracks did not develop during the last 8 years on all test sections. It could indicate that a kind of an equilibrium state in the thermal shrinkage of pavement layers had been achieved before 1988.

3 "Zblewo" research project

3.1 General

The "Zblewo" research project started in 1989. The concrete pavement of National Route 22 after over 40-year service required reconstruction. The pavement, which was 9,0 m wide, consisted of 4,5 m wide and 10,0 m long slabs with two concrete shoulders 0,5 m wide. The slabs were 0,22 m thick, the joints not doweled. Bearing capacity of the pavement was estimated as sufficient. The reconstruction works were deemed to be necessary due to spalling over the surface of the slabs and in the area of joints. Therefore the road administration decided that only a thin asphalt overlay should be applied as the maintenance measure.

Reflective crackings in the new overlay over existing joints and cracks were considered one of the main problems to be solved. The road authorities decided that a 1,5 km section, which showed the most severe surface defects would be used to test different technological solutions. The Technical University of Gdansk was asked to study the case and prepare a design for a test section.

3.2 Scope of the test

The trial included three sections with different technologies. Two solutions, which were regarded as the most promising and a control section were selected for construction:

- Section 1 - (control) unmodified asphaltic concrete, no geotextile over joints,
- Section 2 - asphaltic concrete modified with latex, no geotextile over joints,
- Section 3 - asphaltic concrete modified with latex, geotextile over joints.

The overlay which was 5 cm thick was placed on a levelling layer with a 3 cm average thickness. Both layers were made of 0/12.8 mm dense graded asphaltic concrete. Bitumen content was 6.0% in the case of conventional asphaltic concrete and 6.3% in the case of the mix modified with carboxylated butadiene-styrene SBR latex (5% of dry substance in the binder). Both mixes had Marshall stability exceeding 6.75 kN, which was the requirement according to the Polish Standard.

A nonwoven geotextile made of polypropylene was used over joints to retard reflective crackings in the overlay. The geotextile properties were the following: thickness - 2.5 mm, mass per unit area - 300 g/m^2, tensile strength - 8 kN/m, elongation at break - 140% (5cm wide strip index test).

The geotextile was installed over joints in the concrete pavement in 0.5 wide strips. The pavement surface was sprayed with hot modified asphalt and the strips were immediately unrolled. The geotextile was fully saturated with the binder.

3.3 Results

The effectiveness of the solutions used in the trial differed. The percentage of cracks, which reflected through the overlay on the control section and sections 2 and 3 are given in Table 3.

Table 3 Percentage of cracks reflected through the overlay

Test section	After 1 winter	After 2 winters	After 6 winters
Section 1 - control	50%	80%	95%
Section 2 - latex, no geotextile	40%	60%	80%
Section 3 - latex, geotextile	20%	40%	60%

It has been observed that introduction of the carboxylated SBR latex-modified binder in a relatively thin overlay gave only little improvement in its resistance to reflective cracking. Further introduction of a geotextile interlayer resulted in a more significant improvement in the overlay performance. This result could suggest that "dividing" effect of the binder-saturated geotextile interlayer was in the considered case more important than improvement of fracture properties of asphalt mix.

An interesting observation could be made due to use by the contractor of a thicker geotextile than specified. The geotextile, which was used by mistake over few pavement joints had mass per unit area around $450g/m^2$ and thickness 5 mm. Cracks on both edges of the geotextile strip appeared in the overlay after the first winter. Most probably the geotextile did not fully saturate with bitumen and it formed a weak layer in the pavement structure.

4 Interchange "Wysoka" research project

4.1 General

The research project has been executed since autumn 1994. On the interchange "Wysoka" near Gdansk a 440-meter long section of cement stabilized base course cracked severely immediately after construction. It was assessed that the cracks were caused by primary shrinkage due to extremely high temperature during construction of the base. Reflective crackings were expected as a future problem because the designed thickness of asphalt layers was 12 cm.

The pavement structure was the following:
- wearing course: 5 cm asphaltic concrete,
- binder course: 7 cm asphaltic concrete,
- base course: 22 cm cement stabilized natural aggregate,
- subbase: 50 cm sand,
The road administration decided to use the section for a field trial on reflection cracking. The Department of Highway Engineering of the Technical University of Gdansk was asked to study the problem and propose anti-reflective cracking measures.

4.2 Temperature during construction

The base course of the road in question was constructed in summer 1994. In July and in the first decade of August 1994 the summer was exceptionally hot. During the day air temperature was in the range from +30°C to +35°C. The sections of the cement stabilized base constructed at such high temperatures cracked severely due to primary shrinkage at intervals 3 - 10 m, immediately after construction. When air temperature felt below +25°C cracks of the base layer were rather rare.

4.3 Deflection measurements

Deflections were measured with use of a Benkelman beam on the cement stabilized base before laying of asphalt layers. Measurements were carried out at cracks and in the middle points between cracks. At cracks, deflections were measured using two Benkelman beams, which allowed simultaneous measurement of deflection on both edges of a crack. Deflections were measured under 50 kN load applied by a twin wheel.

The evaluation of the crack influence on bearing capacity was based on the following data:
- coefficient of load transfer efficiency across a crack,
- difference between deflections of the loaded and non-loaded edge of a crack,
- ratio of deflections at a crack edge and in the middle point between cracks.

The results of deflection measurements are given in Table 4. Load transfer across a crack was assessed with a load transfer efficiency coefficient, calculated from the following formula:

$$LTE = 2y_2/(y_1 + y_2) \tag{2}$$

where:
LTE - load transfer efficiency coefficient,
y_1 - elastic deflection of the loaded edge,
y_2 - elastic deflection of the non-loaded edge.

The load transfer is full when LTE = 1 i.e. ($y_1 = y_2$) and no load transfer occurs when LTE = 0 ($y_2 = 0$).

Difference of deflections at crack edges was 0.13 mm in average, which was 60% of the average deflection of the loaded edge.

The mean value of deflections measured on non cracked base, between cracks, was 0.07 mm and was over three times lower than the mean value at crack edges which was 0.22 mm. The ratio of deflections equal to three is greater than obtained from the Westergaard theory. Underlaying pavement layers were constructed according to specifications and properly compacted, so it were excluded as a factor which could influence the results.

Table 4. Results of deflection measurements carried out in October 1994

Data	Deflection of the loaded edge y_1 [mm]	Deflection of the non-loaded edge y_2 [mm]	Load Transfer Efficiency coefficient LTE	Difference of deflections $y_1 - y_2$ [mm]
No of measurements	18	18	18	18
Mean value	0.22	0.09	0.52	0.13
Maximum value	0.44	0.21	1.00	0.24
Minimum value	0.07	0.00	0.00	0.00
Standard deviation	0.09	0.06	0.28	0.08

The following main conclusions have been drawn from the results of deflection measurements:

1. Load transfer across cracks newly developed in cement bound layers can differ significantly and can be lower than 0.5. In the described trial LTE values lower than 0.5 were measured in 50% of cracks and LTE values lower than 0.66 were measured in 80% of cracks.
2. It can not be assumed that new cracks in cement stabilized bases assure good interlock and load tranfer across them.
3. Even new cracks form points of significantly lower bearing capacity than anticipated for uncracked pavement structure.

4.4 Anti-cracking measures

The following anti-cracking measures were used on the trial section with shrinkage cracks: (a) SAMI, (b) a geogrid, (c) a nonwoven geotextile on elastomer-modified bitumen. Additionally on a base section which was constructed during more favourable weather (20-25°C, cloudy) cutting of notches in the base at 2.5, 5.0 and 7.5 m spacing was done prior to laying of asphaltic concrete layers. Asphalt layers were constructed in summer 1995 and the road was opened to traffic in October 1995.

The section is under observation since autumn 1994. The first crack, probably of thermal origin occurred in the asphalt wearing course after over three weeks of severe frost in December 1995. Three subsequent cracks developed until the beginning of February 1996. The frost in January 1996 was exceptional, with temperatures below minus10°C to minus 20°C practically throughout all the month.

5 References

1. Judycki, J. (1984) Crackings in asphalt layers constructed on concrete slabs, *Drogownictwo*, Vol. 39, No 9, pp. 293-298.

2. Alenowicz ,J. (1995) Measures of limiting crack propagation in asphalt layers, Proceedings of *International Seminar "Durable and Safe Road Pavements*, Kielce, Poland (in Polish).
3. Judycki, J., and Alenowicz, J. (1991) *Mechanism of transverse crackings in pavements*, unpublished report for Road and Bridge Research Institute in Warsaw (in Polish).
4. Judycki, J., and Alenowicz, J. (1991) *Use of interlayers against reflective crackings*, unpublished report for Road and Bridge Research Institute in Warsaw (in Polish).
5. Alenowicz, J., Kekalainen, R. and Ehrola E., (1990) *Minimizing reflection and frost heave crackings in flexible and semi-rigid road pavements.*, University of Oulu, Finland, Publications of Road and Transport Laboratory, No 5.
6. Judycki J. (1991) *Structural characterization of road base materials treated with hydraulic binders*, University of Oulu, Finland, Publications of Road and Transport Laboratory, No 12.

The treatment of reflective cracking with modified asphalt and reinforcement

D.J. O'FARRELL

Civil Engineering Laboratory, Construction Services Department, Cumbria County Council, Penrith, United Kingdom

Abstract

The M6 motorway in the southern part of Cumbria was constructed with a flexible composite pavement and opened to traffic in 1970. By the late 1970's transverse cracking was evident in the surfacing and resurfacing works were undertaken. This work was limited in its effectiveness and cracks reappeared within 2 years. In 1987 further investigative works were undertaken which established that the pavement was predominantly structurally sound with a considerable residual life which could be realised if an effective method of crack sealing/inhibiting could be devised.

This paper describes remedial works which have been undertaken since 1990 using polymer modified bituminous materials and fabric or grid reinforcement in various combinations. A total of 8 remedial contracts have now been completed and a review of the effectiveness of the different alternatives has been undertaken following a visual inspection carried out in February 1995.

The results of this inspection are reported, together with some conclusions drawn from the observations.

Keywords: Fabric reinforcement, flexible composite pavement, grid reinforcement, lean concrete roadbase, polymer modified asphalt, reflective cracking.

1 Introduction

The M6 motorway is the main west coast route between England and Scotland running from the M1 motorway to the east of Coventry to Carlisle. The route through the south part of Cumbria runs from the coastal plain with an elevation of 45 metres above sea level at junction 36 to a rocky moorland plateau with an elevation of some 300 metres at junction 37. Within the 12 km section between

Reflective Cracking in Pavements. Edited by L. Francken, E. Beuving and A.A.A. Molenaar. © 1996 RILEM. Published by E & FN Spon, 2–6 Boundary Row, London, SE1 8HN. ISBN 0 419 22260 X.

junctions 36 and 37 there is a constant gradient of 2.6% for 5.7 km followed by a 5 km gradient of 3%.

The motorway is a dual carriageway construction of 3 lanes and a hard shoulder in each direction. In order to maintain the constant gradient for northbound climbing traffic the carriageways are split and stepped one above the other in places. This particular section was opened to traffic in October 1970 and was designed in accordance with national standards of the time for a traffic life of 20 million standard axles (msa) to failure. By January 1995 the structure had been subjected to approximately 40 msa. The Annual Average Daily Traffic flow is in the order of 30,000 vehicles some 25% of which are Heavy Goods Vehicles.

The construction is of flexible composite type with the following materials:

Thickness (mm)	Description
40	14 mm Hot Rolled Asphalt wearing course
60	20 mm Hot Rolled Asphalt basecourse
80	28 mm Hot Rolled Asphalt roadbase
180	Lean concrete roadbase
150	Granular subbase
300	Granular capping material

The lean concrete was manufactured in a site batching plant using locally available Silurian Gritstone as a coarse aggregate. Detailed pavement investigations undertaken on this section of the M6 in the latter part of 1994 revealed concrete strengths ranging from 9 MPa to 34 MPa with values typically in the order of 20 MPa.

2 Reflective cracking

By 1975 transverse cracking was apparent in sections of lane 1 in both carriageways and remedial works were undertaken to seal these using a hot poured bituminous sealant. This was of limited effect and during 1977/78 some 4.1 km of wearing course was removed and replaced with conventional Marshall design asphalt incorporating 50 pen bitumen binder. Transverse cracks reappeared by 1980 and further treatment with sealants was carried out.

The frequency of the transverse cracking increased and in early 1988 a detailed pavement evaluation was undertaken to identify the most appropriate treatment for the affected areas. The investigation and subsequent analysis indicated that at the time the pavement was structurally quite sound but that the reflective cracking would continue to cause problems and eventual structural deterioration unless treated. Two options for treatment were suggested both of which retained the lean concrete roadbase and involved replacing the bituminous layers to a depth of 180 mm with either

40 mm HRA wearing course with elastomer additive
140 mm HRA basecourse with grid reinforcement
or
40 mm HRA wearing course with elastomer additive
140 mm HRA basecourse with elastomer additive

The details of construction for both types of remedial works were investigated and it was concluded that trial sections should be constructed of the different alternatives to assess their ease of construction and effectiveness. It was also appreciated that effectiveness of the remedial works would require medium to long term assessment.

3 Remedial works

Documentation was prepared and the first contract for remedial works was let in June 1990. A total of 7 further contracts then followed to September 1994 details of which are given in Table 1. The individual construction details are shown in figure 1.

The contracts involved removing sections of the existing pavement layers by planing and replacement with combinations of asphalt (sometimes modified) and reinforcement. It was found that the majority of pavement cracking appeared to be restricted to the wearing course layer with probably only 10% to 20% being evident in the upper surface of the lean concrete.

Provision was made for sealing cracks in the lean concrete in those contracts which entailed removal of all existing bituminous materials. In practice very limited cracking was found in some of the locations and sealing was not practicable. Table 1 lists those sections where sealing was carried out.

On the contract carried out in May 1993 a crack sealing technique was carried out on the existing basecourse which entailed enlarging the crack with a 300 mm wide planer to a depth of approximately 50 mm and using a bridge jointing compound to fill the resultant chase.

Treatment number 6 was then applied. Problems were encountered with the stability of the jointing compound when hot asphalt was placed over it and considerable deformation took place during the rolling operations. This method of treatment was abandoned after a short time.

4 Bituminous materials

The initial contracts were carried out using Cariphalte DM binder in the asphalt which is a Styrene Butadiene Styrene (SBS) bitumen blend supplied pre-blended to the coating plant. At the time this was the only polymer modified binder available in this part of England. The unmodified asphalt was manufactured using conventional 50 pen bitumen.

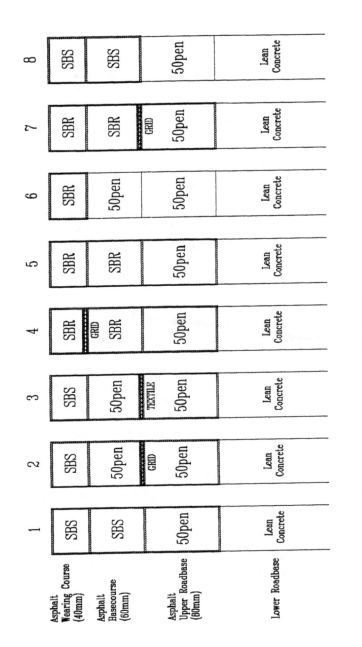

	1	2	3	4	5	6	7	8
Asphalt Wearing Course (40mm)	SBS	SBS	SBS	SBR	SBR	SBR	SBR	SBS
Asphalt Basecourse (60mm)	SBS	50pen / GRID	50pen / TEXTILE	GRID / SBR	SBR	50pen	SBR / GRID	SBS
Asphalt Upper Roadbase (60mm)	50pen	50pen	50pen	50pen	50pen	50pen	50pen	50pen
Lower Roadbase	Lean Concrete	Lean Concrete	Lean Concrete	Lean Concrete	Lean Concrete	Lean Concrete	Lean Concrete	Lean Concrete

Existing materials.

Replacement materials.

Figure 1. Details of remedial treatment types.

During the latter part of 1991 an alternative system of polymer modification for bitumen became available to the coated materials suppliers in the area. This utilised a Styrene Butadiene Rubber (SBR) Copolymer in the form of an aqueous emulsion which could be added to the asphalt during manufacture in the coating plant. This offered the advantages of lower cost and easier handling since the dedicated storage tank for pre-blended binder was not needed. There were however concerns about the adequacy of dispersion of the aqueous emulsion throughout the asphalt during the mixing process and a test method was sought to check polymer dispersion.

After some consideration it was decided to adopt the wheel tracking test in DD184:1990[1] carried out at a test temperature of 60°C on cores cut from the finished road surface. The test temperature of 60°C was chosen because many of the locally available sands are capable of producing material with a high resistance to tracking at 45°C without the use of polymer additives.

Ideally a true "end product" test of the materials ability to resist crack propagation was required but no suitable test method appeared to be available.

The polymer manufacturer and coated materials suppliers carried out laboratory trials and produced asphalt designs which met the wheel tracking requirements. For all the sources tested a mix using 70 pen bitumen and a dry polymer content of 5% by weight of binder content was adopted.

5   Reinforcement

Several materials were available for use as reinforcement in the bituminous layers and two alternatives were used in the first contract. These were a non-woven geotextile and a heat-stabilised polypropylene grid. 300 metre sections of each type were laid and the installation procedure for both was found to be time-consuming and difficult in parts. Despite having followed the manufacturers recommendations for installation it proved impossible to prevent damage to the reinforcement layers being caused by vehicles delivering asphalt to the paving machine. After these experiences further consideration was given both to the necessity for reinforcement and to alternative types of reinforcement. Some of the subsequent contracts omitted the reinforcement whilst the remainder utilised a different type of reinforcement which had recently become available on the market. This was a glass fibre grid with a self-adhesive backing which had both a higher tensile strength and higher modulus of elasticity than the polypropylene grids previously used.

The glass fibre grid proved to be much easier to handle on site and being self-adhesive was not disrupted by delivery vehicles.

When the October 1991 contract was being prepared a question was raised about the necessity of continuing the reinforcement over lengths of apparently sound material between transverse cracks. It was decided to limit the extent of the reinforcement on one of the treated sections to a 1.5 metre length centred on each of the transverse cracks, thus examining the effect of non-continuous reinforcement.

6 Review

In February 1995 a review was carried out of the remedial works contracts undertaken since June 1990 to determine their effectiveness. The review took the form of a visual inspection of all the sections together with a limited amount of coring and rut depth measurement. Reflective cracking was again evident in many of the sections and assessments of crack frequency were made which could be compared with pre-remedial assessments where available. The main points arising from the visual inspection are summarised in Table 1.

Transverse crack frequencies have been classed as follows:

Low	1 or 2 cracks per 100 metres
Medium	3 or 4 cracks per 100 metres
High	5 or 6 cracks per 100 metres
Very High	7 or more cracks per 100 metres

The inspection also revealed two other features. Firstly, two of the sections of wearing course laid in June 1990 had developed ruts. These were located on a south-facing slope towards the top of a long incline. Part of the section which incorporated polypropylene grid had developed ruts up to 10 mm deep and rut depths between 10 mm and 17 mm were found in the section underlain by the non-woven textile. There was some evidence of undercompaction in the basecourse beneath both sections but significantly greater rutting had occurred over the textile. This may be the result of some debonding effect.

Secondly, some small pieces of rubber were found in the wearing course of one section of material laid in May 1993. This phenomenon had been encountered on other sites previously and is thought to be the result of some unmixed Copolymer emulsion getting into the mixed asphalt immediately prior to discharge from the mixer. The pieces of rubber were removed from the surface and the resulting holes filled with a hot poured sealant.

As a consequence of this experience it is likely that in future works a pre-blended polymer modified binder will be specified.

It is hoped to repeat the visual inspection exercise during the early part of 1996.

7 Conclusions

Consideration of the experience gained and observations made to date has enabled a number of conclusions to be drawn, which can be summarised as follows:

- The use of a polymer modified asphalt wearing course without a grid in lane 1 provides only a short term solution to reflective cracking problems.
- The use of polymer modified asphalt wearing course without a grid is so far sufficient to prevent the reappearance of cracking in lanes 2 and 3 (which carry less traffic).

Table 1. Details of remedial works from visual survey - February 1995.

Contract No.	Date	Location Direction/lane	Length (m)	Remedial treatment type	Surfacing material	Grid/textile position	Crack frequency Pre-remedial work	February 1995	Remarks
90/132	Jun-90	N/Bd Lane1	350	1	SBS Wearing/Base Course	None	High to very high	Initial 100m high, thereafter low	Cracks sealed in lean concrete
		N/Bd Lane1	358	2	SBS Wearing Course	PP grid on lower basecourse-full length	Low to medium	Low	Wheeltrack rutting
		N/Bd Lane1	250	3	Conventional Basecourse	Textile on lower basecourse-full length	High	Low	Wheeltrack rutting
90/331	Oct-90	N/Bd Lane1	155	1	SBS Wearing/Base Course	None	High	No cracking evident	
		N/Bd Lane1	87	1	SBS Wearing/Base Course	None	Low	No cracking evident	
91/1092	Oct-91	N/Bd Lane1	1080	4	SBR Wearing/Base Course	GF grid on upper basecourse-full length	Low to very high	Low to medium	Cracks sealed in lean concrete
		N/Bd Lane1	270	4	SBR Wearing/Base Course	GF grid on upper basecourse-on cracks	High to very high	Medium to high	
		N/Bd Lane1	270	5	SBR Wearing/Base Course	None	Very high	High to very high	
		N/Bd Lane1	540	4	SBR Wearing/Base Course	GF grid on upper basecourse-full length	Medium to very high	Low	
92/1009	Sep-92	S/Bd Lane1	2240	4	SBR Wearing/Base Course	GF grid on upper basecourse-full length	Low to high	Low to high	
92/1011	Oct-92	S/Bd Lane1	291	6	SBR Wearing Course	None	High	Very high	
		S/Bd Lane1	400	4	SBR Wearing/Base Course	GF grid on upper basecourse-full length	High	Low	
92/1016	May-93	S/Bd Lane1	200	7	SBR Wearing/Base Course	GF grid on lower basecourse-full length	Low to high	No cracking evident	Cracks sealed in lean concrete
		S/Bd Lane1	570	6	SBR Wearing Course	None	Low to high	Low	
		S/Bd Lanes2,3	200	6	SBR Wearing Course	None	Low to high	No cracking evident	Rubber in wearing course
94/1007	Jun-94	N/Bd Lane1	66	6	SBR Wearing Course	None	No records	No cracking evident	
		N/Bd Lane1	135	5	SBR Wearing/Base Course	None	No records	No cracking evident	
		N/Bd Lane1	447	6	SBR Wearing Course	None	No records	Low	
		N/Bd Lane1	162	5	SBR Wearing/Base Course	None	No records	Low	
		N/Bd Lane1	131	6	SBR Wearing Course	None	No records	Low	
94/1013	Sep-94	S/Bd Lane1	700	8	SBR Wearing/Base Course	None	No records	No cracking evident	

Key PP - Polypropylene . GF - Glass fibre

- Textile reinforcement between asphalt layers appears to reduce the resistance to wheel tracking of overlaying materials.
- Self-adhesive glass fibre reinforcement has found to be the easiest material to handle on site and overlying material has not been subject to rutting.
- Where reconstruction has been carried out with modified asphalts, a grid reinforcement located between the upper roadbase and basecourse appears to be a more effective crack inhibitor than the same materials with a grid between basecourse and wearing course. This latter is in turn better than omitting the grid.
- Better resistance to the re-appearance of cracking was obtained by using a continuous layer of glass fibre reinforcement rather than treating the existing cracked areas in isolation.
- The SBS modified material appears to be more effective than SBR material in crack inhibition. This may be due to the higher proportion of polymer (7% SBS as opposed to 5% SBR) in the preblended SBS binder used in the work and also the different viscosities of base bitumen prior to modification (approximately 160 pen for the SBS binder and 70 pen for the SBR binder). In future work a trial will be carried out using 7% SBR blended with 100 pen bitumen. It is anticipated that this combination will still achieve a satisfactory resistance to wheel track rutting.
- One clear problem with all the work done to date is that although measurements of wheel tracking rate were used to design materials and to assess contract compliance, resistance to rutting, whilst essential, is not the main requirement of surfacing materials when overlying cracked lean concrete roadbase. The main requirement here is resistance to reflective cracking which is believed to increase with ductility and/or fatigue resistance of the mixed material. There is at present no accepted method of measuring these parameters. Clearly a reliable method of measuring resistance to reflective cracking is needed so that this requirement can be specified, different materials can be compared, and contract compliance can be assessed.

8 Acknowledgements

The author thanks the Director of the Northern Network Management Division of the Highways Agency for his permission to publish this paper and to colleagues at Cumbria County Council for their assistance in its compilation.

9 References

1. British Standards Institution (1990) *Method for the determination of the wheel tracking rate of cores of bituminous wearing courses.* BSI, London. DD184.

Reinforcement of bituminous layers with fabrics and geonets

S.S. JOHANSSON
Swedish National Road Administration, Borlänge, Sweden
E.V. ANCKER
Aalborg, Denmark

Abstract

This is the final report from the research & development project: "Reinforcement of flexible pavements". The objective was to find out whether flexible layers can be made more resistant to different types of crack formations if a layer of geotextile is laid on an old cracked asphalt surface with a new flexible wearing course on top. The purpose was besides to examine if this method is a cost effective aid in the struggle against cracking in asphalt pavements.

Study tours to USA showed they have not managed to sort out the cost effectivity over all in spite of more than 25 years experience, and we have not managed to do a fair judgement in Sweden either after 10 years use. More research work has to be done to explore the cost effectivity completely.

The results from the crack surveys though will create a cautious optimism about the method when it is used against load generated reflexion cracks. Properly made the method will delay the crack formation but will normally not be able to stop the crack formation. It has to be emphasized that this is a small test and therefore the results should be handled with care.

The method should be used selectively, and the use should be preceeded by a careful investigation of the damage type and the cause to the damage.

Falling weight deflectometer is an internationally accepted tool to measure bearing capacity of roads. Measurements on our test road has shown that it is not possible to predict pavement performance only with results from falling weight deflectometer.

Keywords: Asphalt, bearing capacity, fabric, geonet, geotextiles, reflective cracking, reinforcement, wearing course.

1. Project description

A big problem with Swedish bituminous pavements of today is the accelerated detoriation process including crack formation, rutting and low bearing capacity. The

Reflective Cracking in Pavements. Edited by L. Francken, E. Beuving and A.A.A. Molenaar. © 1996 RILEM. Published by E & FN Spon, 2–6 Boundary Row, London, SE1 8HN. ISBN 0 419 22260 X.

main reasons for this is an unexpected traffic increase, increased axle loads, super single tires and higher tire pressures. The public resources are limited and do not allow reinforcement- and maintenance works to the extent needed with conventional methods. Therefore it is urgent to develop and test new maintenance methods adapted to Swedish conditions.

As a part of that work a research and development project "Reinforcement of bituminous layers with geotextiles", was started by the contractor Nordic Construction Company (NCC) in Söderhamn 1985. The aim was to find out whether bituminous layers can be made more resistant to different types of cracks by inserting fabrics or geonets between an old pavement and a new wearing course. The project had economic support from Swedish Contractors' Development Fund (SBUF) and four geotextile suppliers.

2. Test road

2.1 Existing conditions
The test road is situated on Road 588 at Mokorset in Söderhamn, which is situated in the middle of Sweden. The test road is 850 m long and 7 m wide. The area has normally cold winters and warm summers. On average the negative daygrades Celcius are about 500, and the frost depth is 1,7-1,8 m in wintertime. Total average day traffic is about 2 700 vehicles, of which roughly 20 % are heavy. The speed limit is 70 km/h, and most of the cars have studded tyres in the winter.

The underground consists of fine natural sand. The unbound base layers are very thin and the total pavement, shown in fig 1, is only 350 mm thick.

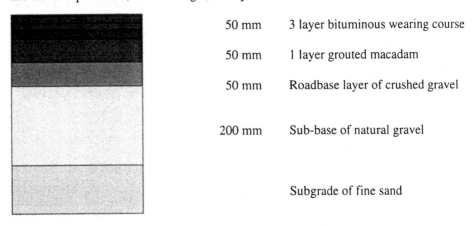

50 mm	3 layer bituminous wearing course
50 mm	1 layer grouted macadam
50 mm	Roadbase layer of crushed gravel
200 mm	Sub-base of natural gravel
	Subgrade of fine sand

Fig 1. The road design.

At the time for the reconstruction in 1987, the test road had extensive longitudinal cracks and alligator cracking. Besides there were a smaller number of transversal cracks and a few potholes. There were also deep ruts partly depending on wear from studded tires, but probably most caused by high traffic loads in regard to the bearing capacity of the road. A contributing reason to the widespread alligator cracking was possibly also binder ageing in the bituminous layers.

2.2 Preparatory work

The bearing capacity of the road was tested with falling weight deflectometer (FWD). The test showed that the road had bad bearing capacity. The first half of the road had a worse condition than the second half. The test road was divided into 17 test areas, each 50 meters long. Every second test area was a reference area without reinforcement. The four different geotextiles were distributed by random sampling so that one of each sort was represented on the first half of the test road and one of each on the second half (fig 2).

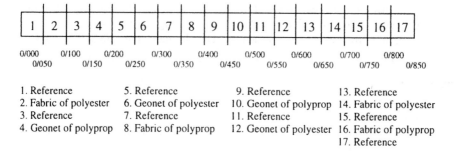

1. Reference
2. Fabric of polyester
3. Reference
4. Geonet of polyprop

5. Reference
6. Geonet of polyester
7. Reference
8. Fabric of polyprop

9. Reference
10. Geonet of polyprop
11. Reference
12. Geonet of polyester

13. Reference
14. Fabric of polyester
15. Reference
16. Fabric of polyprop
17. Reference

Fig 2. Test road divided into test areas

2.3 Maintenance methods

The surface of the test road was tack coated with bitumen emulsion and levelled with dense asphalt concrete, about 35 kg/m^2. After the geotextiles were installed about 15 kg/m^2 dense asphalt concrete was spread and compacted on the whole test road. The spread was done with a conventional chip spreader, and the compaction was done with a 6 tons vibrating roller. Then the wearing course consisting of 80 kg/m^2 dense asphalt concrete was laid and compacted with the same roller as was used for the spread. The four tested geotextile products were:

1. Fabric of polyester, needlepunched, 210 g/m^2. The levelling course was tack coated with three layers of bitumen emulsion, together 1,1 kg/m^2, the fabric was rolled out and another tack coating was sprayed on the fabric.

2. Fabric of polypropylene, needle punched, 140 g/m^2. The levelling course was tack coated with one layer of bitumen emulsion, 0,3 kg/m^2, the fabric was rolled out and sprayed with two layers of bitumen emulsion, together 0,6 kg/m^2.

3. Geonet of polyester, mesh size 30x30 mm, 333 g/m^2. After tack coating with 0,3 kg/m^2 bitumen emulsion the geonet was rolled out and was attached to the pavement with steel nails.

4. Geonet of polypropylene, mesh size 60x75 mm, 210 kg/ m^2. After tack coating the geonet was rolled out and nailed to the pavement with steel nails. The net was very stiff which caused problems when the wearing course was laid. The heat from the wearing course made the net to expand and to form waves which were folded together in the mix. In places the net came up to the road surface and a lot of cracks arose.

2.4 Test slabs

On a levelled gravel road close to the test road small test areas were made on plywood plats 1,20x2,40 m. Three test slabs of each maintenance method was made and three slabs without reinforcement as references. The slabs were made with the same process and the same equipment as was used on the test road. As a bottom layer a 50 mm thick layer of dense asphalt concrete was laid and on the top the same type of wearing course as on the test road. After the small test areas were finished they were sawed into slabs, loaded on a truck with a loader and transported 500 km to the Swedish Road and Transportation Institute for testing.

3. Laboratory tests

Indirect tensile and Marshall stability were tested on cores from the test areas. The results showed no effects from reinforcement materials compared to the references.

3.1 Tensile test bench

The test bench is constructed to be able to take big forces and big samples (fig 3). The equipment creates horizontal tensile stresses and strains in the tested slabs similar to those arising by thermal forces in a road body, and similar to those forces caused by heavy traffic and frost heave in a bituminous layer.

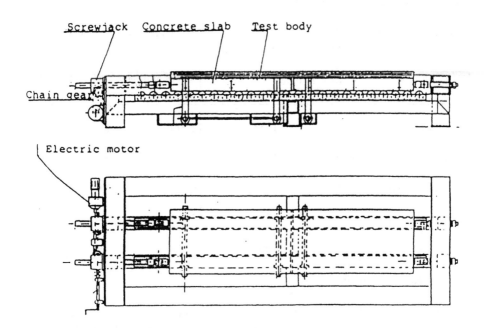

Fig 3. Tensile bench in plan and profile

3.2 Test slabs

In the laboratory the slabs were sawed to 1,0x2,0 m to avoid edge effects. Besides a waist was sawed in the slabs to assure a break in the narrow area (fig 4). It was done to avoid breaks in the ends of the slabs, where they were fixed to the bench with clamps.

3.3 Test results

The tests were performed at -10 °C, which was supposed to correspond the wors
possible situation in the road, late autumn or early spring which means low nigh
temperatures and water enriched unbound layers. For practical reasons the tensile tes
speed was set to 0,5 mm per hour even though we know that the tensile strain grow much
slower in the reality.

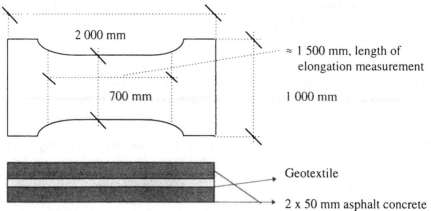

Fig 4. Test slab in plan and profile

Tensile stresses and strains were registered during the whole process and the results
are shown in table 1. The results are the average of two test bodies per method.

Test slab type	Stress at break		Strain at break	
	N/cm^2	in % av ref	μStrain	in % av ref
Fabric of polyester	106	115	5090	83
Fabric of polyprop	107	116	5316	87
Geonet of polyester	112	122	6174	101
Geonet of polyprop	119	129	5759	94
References	92	100	6129	100

Table 1. Results from tensile test

The results indicate that fabric reinforced test bodies have about 15 % higher stress at
break than the references, and that the geonet reinforced test bodies have about
20-30 % higher stress at break than the references. Besides all the breaks in the
reinforced test bodies have occured at equal or smaller strain than in the references. It
seems a bit strange that geotextiles can reinforce an asphalt layer as they normally have
much more elongation at break than normal asphalt layers. Possible explanations of the
reinforcement effect are:

• The geotextile shrinks by the heat at installation and will become prestressed when
  cooling off.

‚ The geotextile is by compaction pressed down in all small unevennesses between the two bituminous layers and becomes prestressed.

‚ The geotextiles have good adhesion to the two pavement layers and thereby has the ability to bridge over the most weak points in the wearing course.

¡. Field tests

¡.1 Measurements with Falling Weight Deflectometer

Bearing capacity has been measured by falling weight deflectometer (FWD) before the maintenance operation and every autumn thereafter 1987-1994, except for the year 1992. The purpose of FWD is to simulate a passage of a loaded lorry on a selected spot and to register the deflection caused by the passage. This is accomplished by a falling weight hitting a steel plate with the diameter of 300 mm on the road surface. The power used is most often 5 tons to simulate the load from a lorry with 10 tons axle load. The falling weight causes a deflection in the road surface which is measured with geophones. In this case the deflections were measured in the deflection centre (D0), and 200 mm (D20), 450 mm (D45) and 900 mm (D90) from the deflection centre.

In different reports it is stated that the deflection between D0 and D30 primarily describes the condition of the bound layers and the unbound base layer, whilst the deflection at D45 and D90 will show the condition in the subgrade. As can be seen in fig 5 our measurements confirm that statement.

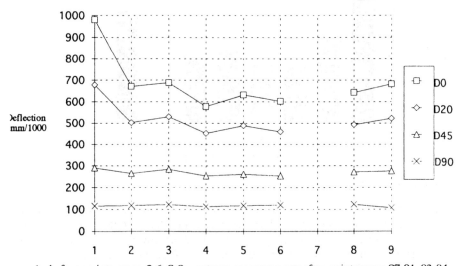

Deflection measurements in D0, D20, D45 och D90

1= before maintenance, 2-6, 8-9 = autumn measurements after maintenance 87-91, 93-94

Fig 5. Deflections on test area 1, reference.

D0 and D20 are affected by the maintenance step whilst D45 and D90 are not. We can presume that D0 - D20 on our road will represent the wearing course and the geotextile reinforcement, and the difference between D20 and D45 is corresponding to the levelling course which was rather thick in the wheel paths. If we look at the difference

D0-D20 on test area 8 and 9 for all autumn measurements (fig 6), we can notice that we have maximum bearing capacities after 1-2 years which is explained by traffic compaction and binder stiffening. After the maximum is reached the bearing capacity is rather constant but going downwords again in the last measurement in October 1993.

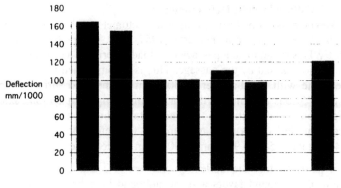

D0-D20, right side, test area 8, fabric of polypropylene

Deflection mm/1000

Measurements with FWD 8707-9310

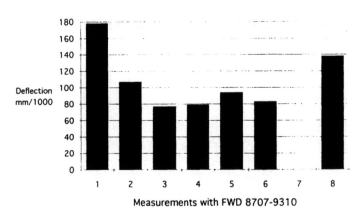

D0-D20, right side, test area 9, reference

Deflection mm/1000

Measurements with FWD 8707-9310

Fig 6. Test areas 8 and 9, D0-D20.

We can assume that when the difference has reached the same level as at the fir measurement in July 1987 the road should have about the same level of distress as it ha at that time. We do not know though how the breaking down process will progress, so is difficult to forecast the construction life time. Probably it will be an acceleratir process in the end. It is a bit surprising that the deflection on test area 8 is just about th same before and after maintenance in 1987. There is a possibility that this fact is due to compression in the reinforcement layer consisting of fabric and bitumen.

Another way to analyze the effect of a reinforcement operation is described in a Swedi: research report (1) as the "radius of curvature" on the road surface. The "radius" calculated by means of the deflection in the centre D0 and the deflection in a spot on suitable distance from D0. The mathematical expression has the formula:

$$R = \frac{r^2}{2 \cdot D0 \cdot \left(\frac{D0}{Dr} - 1 \right)} \qquad \text{where:}$$

R   = radius of curvature (ROC)
r   = distance from D0 to deflection nr 2 (in our case D20 = 200 mm)
D0   = deflection in the load center
Dr   = deflection at the distance r from D0.

The ROC for test areas 8 and 9 is shown in fig 7. By many measurements it is found that a road in bad condition normally has a ROC of less than 100 m. Both the test areas confirm this at the first measurement before maintenance in 1987. At the latest measurement in 1993 both the test areas has passed their maximal ROC and are going down towards the 100 m limit again.

"ROC" D0-D20, test area 8, right side.

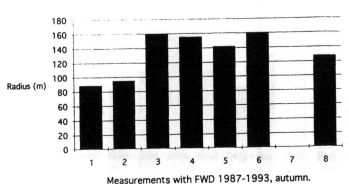

Measurements with FWD 1987-1993, autumn.

"ROC" D0-D20, test area 9, right side.

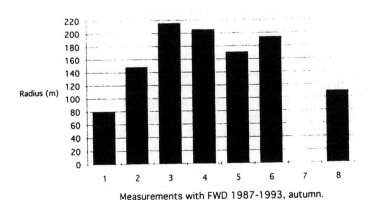

Measurements with FWD 1987-1993, autumn.

Fig 7. "ROC" for test area 8 and 9.

In fig 8 we can see the results from measurement before maintenance in 1987. Test area 1, which is the worst one, has a "ROC" about 50 m and all the test areas have a "ROC" on less than 100 m. We can also see the results of the last measurement in October 1993 on test area 1-9. Test areas 1, 5 and 9 are all close to 100 m but the other two references 3 and 7, and the reinforced test areas are still high up.

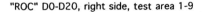

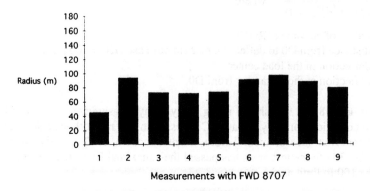

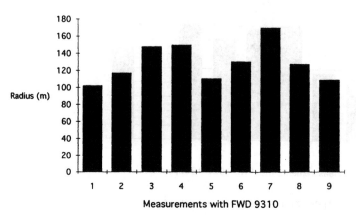

Fig 8. "ROC" for test areas 1-9, 1987 before maintenance and 1993.

4.2 Crack mapping

Crack mapping has been made every year. The crack mapping will say nothing about the severity of the cracks. The state in 1993 expressed in percent of the condition before maintenance operation in 1987 is shown in table 2.

From the table it can be seen that the fabric of polyester and the geonet of polypropylene has done better than the references. The fabric of polypropylene has just about the same crack development as the references. As that fabric has just about the same properties as the fabric of polyester they should have a similar crack development.

Test area	Longituinal cracks 93/87	Transverse cracks 93/87
References	ca 52 %	ca 81 %
Fabric of polyester	ca 13 %	ca 46 %
Fabric of polypropylene	ca 47 %	ca 100 %
Geonet of polyester	ca 16 %	ca 54 %
Geonet of polypropylene	ca 35 %	ca 848 %

Table 2. Results from crack mapping 1993 compared to 1987.

The difference is that the fabric of polypropylene, used a much smaller amount if bituminous binder than the fabric of polyester, which probably is the reason for the different performances. The geonet of polypropylene had because of installation problems extensive cracking from the start and is therefore excluded from the comparison.

5. Experiences from study tours
One study tour has been made to USA, and another one to USA and Canada. The purpose with the study tours was to take part of other countries' experiences concerning different materials, work methods, efficiency on different types of damages and the cost effectivity of the method.

Some findings from the study tours are (2 o 3):

- Special machinery is used to lay the geotextile and to impregnate it with binder

- They use most often pure bitumen for impregnating the fabric

- A reinforcement by impregnated fabric corresponds to about an asphalt layer of 30 mm looking at the ability to delay reflective cracking

- Geotextile reinforced asphalt pavement will resist the formation of reflective cracking 2-3 years longer than unreinforced asphalt pavement.

When using the method it should be pointed out that:

- The geotextile should never be laid out on a wet or moisty road surface

- The geotextile should always be laid full road width to be effective

- The rate of bituminous binder should be adapted to the geotextile used

- Pure bitumen should be preferred.

A bituminous layer thick enough should be placed over the geotextile.

6. Cost effectivity
Cost effectivity is difficult to judge. American experience indicate that the use of a geotextile reinforcement between a new and an old asphalt layer is not effective:

- At large horizontal movements
- At very small and very big vertical movements
- At transverse temperature cracks
- At very cold climates.

The method might be cost effective:

- At small horizontal movements
- At medium big vertical movements
- When a waterproofing membrane is wanted in the road construction
- At alligator crackings principally depending on load related fatigue.

Geonets are 2-3 times more expensive than fabrics and should therefore show better performance to be equally cost effective. A fabric installation will cost about one third of the cost for a new wearing course. On our test road the best geonet and the best fabric show just about the same performance. Therefore in this case fabric looks to be a better alternative.

In USA they have not managed to sort out the cost effectivity over all in spite of more than 25 years experiance, and we have not managed to do a fair judgement in Sweden either after 10 years use. More research work has to be done to explore the cost effectivity completely.

7. Discussion

Many people question if an asphalt reinforcement of the type described in this report on one hand is placed on the right level in the construction and on the other hand if such a construction has any effect at all. To get maximal effect of a reinforcement in a road construction it should probably be placed where the strain is maximal. That should normally take place in the lower part of the bituminous layer. But what is the situation in a cracked asphalt layer?

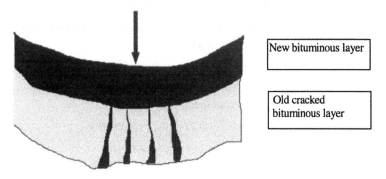

Fig 9. Fatigue cracks caused by traffic

When a cracked asphalt layer is loaded by traffic the stress will fall at the existing cracks and the cracks will expand and increase. If a new asphalt layer is laid on the cracked layer the stresses and strains are moved up in the construction to the lower part of the new asphalt layer. To delay/prevent the cracking in a new bound layer it should be most effective to place a reinforcement layer at the bottom of the new asphalt layer.

If the cracks are caused by frost heave there is another situation. As the cracks are caused by heaves the cracks are probably formed from the road surface and merge down in the construction.

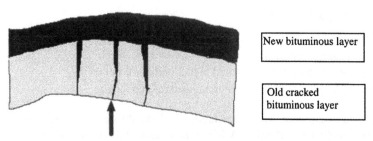

Fig 10. Cracks caused by frost heave in an asphalt layer

Therefore, also in this case a reinforcement placed on top of the old surface and under a new wearing course seems to be proper.

Results from measurements with FWD show that a reinforcement with geotextiles will give no benefit. Nevertheless geotextile reinforced areas seem to be more resistent to reflective cracking than the references. Possibly the reinforcement will affect the fatigue properties in the bituminous layers and in that way might mdelay the reflective cracking.

8. Conclusions

Analyzes from the obtained results show that the bearing capacity measured by falling weight deflectometer is at least as good on the references as on the geotextile reinforced test areas. It is obvious that reinforcement with geotextiles will give no increased bearing capacity measured with FWD compared with references without reinforcement. The test areas with bitumen saturated fabrics even have a lower bearing capacity than the references! The test area with best performance regarding reflective cracking is test area 2 which also has had very low bearing capacity measured with FWD all the time. This is probably explained by an increase in fatigue resistance caused by the reinforcing saturated geotextile.

The results from the crack surveys though will create a cautious optimism about the method when it is used against load generated reflexion cracks. Properly made the method will delay the crack formation but will normally not be able to stop the crack formation. It has to be emphasized that this is a small test and therefore the results should be handled with care.

The method should be used selectively, and the use should be preceeded by a careful investigation of the damage type and the cause to the damage.

Falling weight deflectometer is an internationally accepted tool to measure bearing capacity of roads. Measurements on our test road has shown that it is not possible to predict pavement performance only with results from falling weight deflectometer.

References

1. Djärf, L et consortes (1993). Project *"Model development"*. Work report in March 1992. Swedish Road and Transportation Research Institute.
2. Predoehl, NH:"Evaluation of Paving Fabric Installations in California. California DOT, Sacramento, Californien February 1990.
3. Button, JW:"Engineering Fabric and Asphalt Overlay Performance". Research report 187-17. Texas Transportation Institute, Texas November 1989.

Behaviour of crack preventing systems on roadways with hydraulic layer and cement concrete on the French road network

G. LAURENT
CETE de l'Ouest, Nantes, France
D. GILOPPE
CETE Normandie Centre, Rouen, France
P. MILLAT
CETE Lyon, L.R. d'Autun, France

Abstract.

This document assesses the results of the behaviour of crack preventing systems observed on the French road netwotk.

It briefly shows the part played by the Observatory of Roadway Techniques in the study of roadways and describes the methodology used by the College to monitor roadways.

Every year in winter, about 100 road sections are monitored across the country.
. The crack-preventing systems used are divided into 6 categories :
• sand asphalt interlayer,
• aggregate covered membrane,
• membrane covered by cold asphalt concrete,
• modified binder and geotextile interlayer,
• geotextile manufactured on site,
• fibre modified asphalt concrete.

After a brief description of the different techniques, their behaviour is assessed according to an index of crack development in relation to their frequency before maintenance work. A correction is made in order to take into account the condition of the cracks visually monitored.

In certain sections, the findings are based on observations made over a period of 8 years.

Results are given per category of different systems. They show how the activity of the cracks relates to the behaviour of the techniques; this activity is linked to the volume of heavy goods traffic, amongst other things.

This analysis helps us to understand how these products function and opens the way for product research and improvement. Some solutions are promising, others need to be abandoned or improved.

Finally, using an effective system is all the more necessary depending on how active the cracks are. Experience shows that we must reduce the vertical beating. In addition, a technical/economical compromise solution needs to be found between

Reflective Cracking in Pavements. Edited by L. Francken, E. Beuving and A.A.A. Molenaar. © 1996 RILEM. Published by E & FN Spon, 2–6 Boundary Row, London, SE1 8HN. ISBN 0 419 22260 X.

using thick layers of asphalt concrete and application of a crack preventing complex.

Keywords : Crack-retarding interlayer, Sand- asphalt, Aggregate covered membrane, Membrane covered by geotextile interlayer, Modified binder and geotextile interlayer, Geotextile manufactured in site, Fibre modified asphalt concret.

1 The Observatory of Roadway Techniques

In 1992, the directors of the LCPC (Central Laboratory of Road and Bridge Engineering) and SETRA (Department of Technical Studies for Roads and Motorways) established the Observatory of Roadway Techniques for the French road network.

Eight colleges were created, notably including the college responsible for observing the behaviour of "Systems limiting the development of transversal cracks of hydraulic shrinkage". Amongst other things their missions have been:

• to gather all technical and economic data necessary for defining the technique;

• to objectively analyse the data gathered in order to draft an annual summary.

This summary of the observations made on the French road system with the different techniques used is presented below.

2 The context for using crack preventing systems

Cracking of hydraulic layers is partly caused by thermal contraction when the materials set (rapid single movement), and partly by the effect of external thermal variations (slow and repeated movements). These two phenomena lead to horizontal displacements of the sides of the crack.

Due to the effect of heavy goods traffic these cracks are also subjected to vertical movements which accelerate their development across the wearing courses. On roads with heavy traffic the transversal cracking can cause serious structural disorders, on account of the repeated stresses on the road.

In order to retard the development of cracks, systems are placed at the interface beneath the wearing course.

In order to be considered as effective by the College, the intervention techniques must meet the following objectives over 7 to 10 years:

• **limit the development of cracks to fine cracks that are not eroded;**

• **preserve the watertightness of the road;**

• **limit deterioration of the interface of surface layer - hydraulic layer, by reducing the stresses on cracks;**

In order to achieve this, the principle generally accepted is to introduce a system between the cracked roadway and the wearing course which allows partial uncoupling of the two layers. However, this uncoupling must not be total due to the risk of causing premature fatigue of the wearing course.

The layer placed in between must therefore be able to absorb the cracks due to shrinkage or shear mentioned previously without transmitting them to the wearing course.

3 Classification of the products used in France

They are divided into 6 categories by the College:

-3.1 Sand asphalt interlayers

These have a grain size of 0/2 to 0/6 mm and are generally made from hard aggregates produced from crushed massive rock. They must conform to standard

P. 98 101 and, concerning their mechanical characteristics, the SETRA/LCPC recommendations for asphalt concrete layers [1].

The formulation study comprises the following tests: PCG test, rut test at 30 000 cycles at 60 °C, Duriez test, shrinkage deflection test (specific machine developed by the technical network of the Ministry of Public Works for the whole complex, i.e. system plus wearing course [2]).

The binder content, pure bitumen or more generally modified asphalt, is in the region of 9 to 12.5 p.p.h., the content of fines from 10 to 15% for a composition modulus of 5.5 to 6.

Some formulations contain fibres, either mineral or organic, which allow pure bitumen to be used whilst preserving good stability of the mixture with regard to ruts. These can be fabricated using all types of asphalt plant.

The nominal thickness is 2 cm. Work is carried out using a road finisher with levelling screw with a binding layer previously dosed with 300/350 g/m² of residual asphalt. Compaction is achieved using smooth rim tandem rollers.

In 1995 the observed costs of the sand asphalt interlayer techniques varied between 20 to 30 FF/m² exc. tax depending on their composition (pure bitumen or modified asphalt) and the surfaces to be treated.

3.2- Aggregate covered membranes (or SAMI Stress Absorbing Membrane Interlayers)

- These comprise a layer of binder rich in elastomer, dosed between 2 and 2.5 kg/m² and are coated with a grit, generally with grain size 6/10, in order to ensure working of the asphalt.

The formulation study is limited to rut tests and shrinkage deflection. The observed costs of the stress absorbing membranes vary between 25 and 40 F/ m2 exc. tax essentially depending on the size of the area to be treated.

3.3- membranes covered by cold asphalt concrete

In this case the membrane dosed between 2 and 2.5 kg is made using a binder heavily dosed with elastomer and covered with a cold mix generally with grain size 0/4 then with an asphalt concrete wearing course.

This is achieved using specific machines: high pressure road spreader, cold mixing machine. The overall thickness is approximately one centimetre.

In 1995, the observed costs of the techniques using cold mix asphalt membranes varied between 25 and 35 F/ m2 exc. tax depending on the size of the area to be treated.

3.4- Modified binder and geotextile interlayer

The binding layer is made using a modified binder either in emulsion or more recently in the form of an anhydrous binder spread whilst hot. The proportion is between 0.8 and 1 kg/m² of residual binder.

The geotextile is generally a non-woven fabric material and/or heat sealed polyethylene or polypropylene.

Without precise specifications, geotextiles meet the requirements of ASQUAL (Association for the Promotion of Quality Assurance in the Textile Industry).

The geotextile used for this technique is intended to serve amongst other things as a reservoir for the binder; its mass per unit area is between 120 and 250 g/m². It is packaged in rolls of 100 to 150 m; the width varying from 1.9 to 3.8 m. The glass grid technique associated or not with geotextiles is not (or no longer) employed to treat shrinkage cracks of hydraulic layers. It is more commonly reserved for the treatment of fatigue problems on soft roads

In 1995 the observed costs of the modified binder and geotextile layers varied between 10 and 20 FF/m² exc. tax, depending on the type of technique employed and the area treated.

3.5 Geotextile manufactured on site

This method is similar to the previous one; the geotextile being as it were constructed on site. Two techniques exist at present:

• with continuous fabric threads; in this case the proportion is 0.8 to 1 kg of residual bitumen per square metre, that of the projected threads is 100 grammes per square metre.   In order to facilitate working the asphalt covering the fibres are generally gritted with 6/10 aggregates in the proportion 7 to 8 l/m².

• with short threads of approximately 10 centimetres; in this case the proportion of residual binder is of the order of 600 g/m2 and the proportion of fibres is in the region of 70 g/m2, this is all gritted with 4/6 aggregates then generally coated with a bi-layer dressing as wearing course.

In 1995, the observed costs of the modified geotextile techniques varied from F F/m2 exc. tax, depending on the size of the areas to be treated.

3.6- Fibre modified asphalt concrete

These asphalt concretes with grain size 0/6 or 0/10 are made using pure bitumen. The addition of fibres, generally organic (cellulose) fibres, allows a larger quantity of binder to be set, thus ensuring good rut resistance. This technique is similar to that of the sand asphalt interlayers with better rut resistance due to the granular framework.

This technique, with 3 to 4 cm thickness, also improves the structure of the road. The intended proportion of binder is of the order 6,8 à 7 p.p.h. with the content of fines greater than 10 %.

This asphalt concrete is fabricated using traditional asphalt plants. Particular attention must be given to the proportion of fines, as well as the mixing temperature and the period of mixing (homogeneous mixing).

This asphalt concrete is generally coated with SMA with modified binder, allowing the necessary skid characteristics to be ensured for the surface course under heavy traffic.

In 1995, the observed costs of the fibre modified asphalt concrete techniques varied between 20 and 35 FF/m² exc. tax, depending on the size of the area to be treated.

3.7- Application of the systems

For all of these techniques, it is not advised to apply them in late autumn or unfavourable weather conditions; dry support layer and outside temperature of 10° C.

In addition, as generally concerns thin layer techniques, the support layer, notably for the sand asphalt interlayers, must not present any permanent deformation greater than 1 cm.

3.8- Wearing course

Whichever intervention technique is used, the surface course plays an important part in the behaviour of the complex created. In particular, the parameters of thickness, content and nature of binder, content of mastic etc. have an influence [3].

Different observations and studies have shown a relationship between asphalt concrete thickness and effectiveness with regard to development of cracks. Notably this plays an important role in the reduction of beating effects of the edges of the crack [4] and [5].

However, thick solutions are expensive and the aim of an intervention system combined with a thin wearing course, is to be economical as well as effective.

4 Observation Methodology

Observation of the behaviour of intervention systems is all the more difficult due to the products not being directly observable because they are situated beneath the wearing course.

The variety of the sites and the diversity of parameters (traffic, thickness and nature of wearing course and above all the level of seriousness of the cracking) have an influence and due to this fact monitoring the intervention techniques is only significant on sections where the characteristics are well known.

Monitoring is carried out by means of visual surveys carried out in winter or at the end of winter. It is during this period that the transversal cracks are more open and more visible.

These surveys are carried out on foot or by car at slow speed when weather does not permit, so that all cracks, even those that are just forming, can be noted.

About one hundred sections are observed in this way each year. Only sites where sufficient information exists before and during works are acceptable. The majority of the sections observed are located on the national road system and bear T1 traffic (300 to 750 HGV/day/direction) or TO (more than 750 HGV/day/direction).

In order to take into account the level of cracking, a correction is applied in relation to the state of the crack, according to the following principle:
• fine crack beginning, counted as 0.5;
• definite crack of length greater than 1.5 m, counted as 1 crack;
• branched and splitted crack, regardless of length, counted as 1.5 cracks
• crack very eroded with loss of material, uneven surface, and development of fines and/or crackling, counted as 2 cracks.

The results are used in the form of an index ‚I, which translates the level of corrected cracks that have reappeared in relation to those which existed before the works were carried out:

$$I = \frac{\text{number of corrected cracks observed}}{\text{number of cracks before works}} \times 100$$

The system is considered to be more effective when, for a given number of winters, the index of reappearing is low. Conversely, a system that is not very effective could have an index of cracking greater than 100 after a few years.

The groups of techniques described previously are assessed and then judged as representative of their category. Sometimes there is only one single technique per group.

5 Results per system group

-5.1 Control test sections without system
These are situated on the same sections of road as those which comprise a crack prevention system; the control asphalt concrete serving as wearing course in both cases.

They are generally composed of pure bitumen layers 4 cm thick (type 1 of the NFP 98 132 standard).

On average (fig 1) an index of cracking of 50 between the 4th and 5th winter is observed for 4 cm of pure bitumen covering.

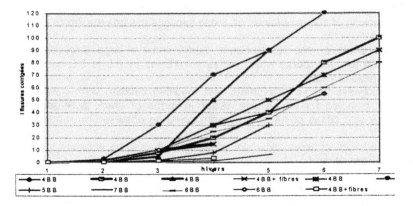

Fig 1   Crack development of control asphalt concrete

The rate of development is more rapid depending on how active the transversal cracks are before works are carried out : beating of 30 to 40/100 mm as is the case for two sections in graph 1 having the fastest index of reappearance.

An index of 100 is obtained on average in the sixth year, for a surface layer of pure bitumen 4 cm thick.

-5.2 Sand asphalt interlayers

For the sand asphalts made with pure bitumen 2 cm thick, covered with 4 cm of asphalt surface layer, the transversal cracks develop in approximately the same way as the control sections above with one or two years difference (graph 2). If one compares their behaviour with the control section (when it exists), their growth is parallel.

On average an index of 50 is observed between the 5th and the 6th winter with a surface layer of 4 cm of pure bitumen.

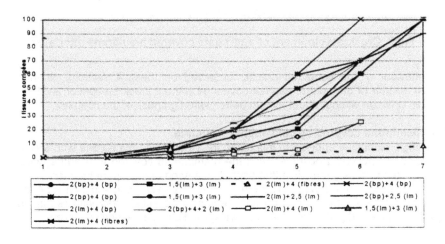

Fig 2   Crack development of sand asphalt interlayers

In addition, rut development is often observed on pure bitumen sections, on roads with traffic T0, regardless of the thickness of the surface layer (from 2.5 to 8 cm). The index of cracking of 100 is obtained on these sections towards the 7th or 8th winter.

Core samples taken on the maintenance works show a large dispersion of thickness of sand asphalt course which varies sometimes by 0.5 to 3.5 cm for a theoretical thickness of 2 cm;

With equal thickness the sand asphalts with modified bitumen with a wearing course also of modified binder behave better in terms of crack development. However, a few rare cases of rut development could be observed on roads with very heavy traffic.

Likewise, the addition of fibres allows improved behaviour to be obtained, due to the increased binder content.

-5.3 Aggregate covered membrane interlayers (or SAMI)

This technique is not used very much in France today (no works recorded in 1994 and 1995). The results are from previously recorded observations and monitoring of 4 previous sections, with an asphalt layer generally of 4 cm pure bitumen.

An index of 50 is obtained at 5 years and 100 at the 7th winter (fig 3) with a surface layer of 4 cm pure bitumen.

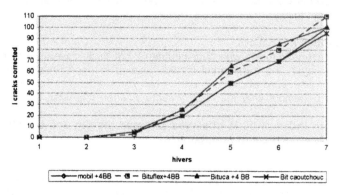

Fig 3 Crack development of aggregats covered membrane interlayer

The behaviour of the corrected cracks, observed on these 4 sections, is thus identical to that of the control asphalt coverings without crack prevention system. However, one can without doubt expect better watertightness of the complex.

-5.4 membrane covered by cold asphalt concrete

One single technique appears in this category. Coating the membrane with a cold mix avoids the punching observed with the membranes covered with aggregates. The wearing course is generally a thin asphalt covering of 4 cm pure bitumen.

This technique is more recent and therefore has only been observed over a period of winters. At the moment, its behaviour seems to be satisfactory with regard to crack development (fig 4), when the supporting medium is an asphalt concrete.

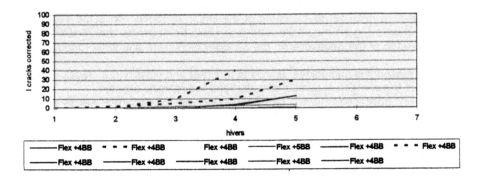

Fig 4 Crack development of membrane covered by cold asphalt concrete

In comparison, the results are less favourable when the supporting medium is a dressing. We can assume that there is a punching effect by the support dressing, as in the case of the aggregate covered membranes. This difference was observed in particular on a section comprising part asphalt concrete and a dressing as support course.

It is not possible to prejudge the behaviour over a longer period, notably the age reached for an index of cracking of 50, but on account of the appearance of the curves one can assume that at 7 winters the index could be of the order of 10 to 30 on asphalt concrete support.

-5.5 Modified binder and geotextile layer

Geotextiles are adhered to the supporting medium either to anhydrous modified binder or modified emulsion. They are generally covered with 4 cm of pure bitumen.

The reappearance of cracks is relatively slow for the first 3 years (index 25) and then develops more rapidly.

It can be assumed that on average the index of 50 is reached at the 6th winter (fig.5) with a surface course of 4cm of pure bitumen.

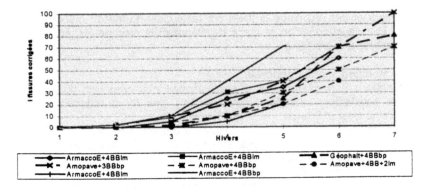

Fig 5 Crack development of geotextile

The development of cracks is sometimes closer to fine crackling, which can be interpreted as a diffused development of the crack. Moreover, on some sections, generalised crackling on the whole road lane could be observed due to difficulties with adherence of the geotextile on the supporting medium and local formation of folds.

For this reason it is sometimes difficult to interpret and classify the cracks. When the crack is fine one can assume that watertightness remains assured.

In addition, good performance has been observed for geotextile sections plus dressing on roads with average traffic (T3 - T2).

-5.6 Geotextile manufactured on site

The only technique observed at present is that with continuous fabric threads. The method of using short projected fabric threads is at present the subject of a so-called innovative process and the results are not yet available.

On a reduced sample (5 recent sections observed) we can estimate that the index of 50 is obtained between the 5th and 6th winter, with a pure bitumen surface layer.

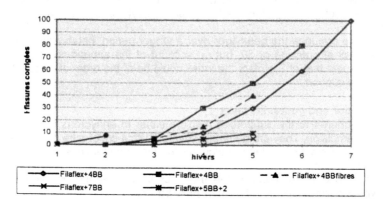

Fig 6 Crack development of geotextile manufactured on site

The behaviour of this technique is therefore fairly close to that of manufactured geotextiles. The visual condition of crack development is also similar; fine cracks with a tendency to fine crackling

-5.7 Fibre modified asphalt concrete

This technique is derived from that of sand asphalt interlayers with fibres which have shown satisfactory behaviour, however, this was for a reduced sample, essentially in the west of France.

The asphalt concrete course of 3 cm ensures a greater average thickness than with the sand asphalts and the granular structure with 0/10 aggregates allows rut development to be avoided with increased binder contents. The wearing course is composed of SMA with modified binder.

The recent use of this technique does not allow its long term behaviour to be known. However, the results are promising for the oldest (4th winter), and to date the index of cracking varies from 0 to 1. In addition, no ruts have been observed on any of the sections treated bearing a traffic load T0 (more than 750 HGV/day/direction)

6 Conclusion

In maintenance of roads with hydraulic layer the systems tested and observed, sometimes on a small sample, present the following behaviour:

• **sand asphalt layers** of pure bitumen are sensitive to rut development and only slightly slow down the development of cracks. It is necessary to prohibit asphalts that are too soft of the type 180/220 and grain sizes 0/2. Using modified binder generally avoids rut development. They have improved performance when they are combined with a thick wearing course along with modified binder.

The use of fibres allows increased binder content and hence improved performance with regard to crack development whilst avoiding the problems of ruts. For sand asphalt interlayers, this technique has the best performance.

• **aggregate covered membranes** have not been satisfactory on the sections observed. Today, their use has practically been abandoned in France.

• **membrane covered by cold asphalt concrete** (one single technique of this type at present) shows promising performance on asphalt concrete supporting medium but findings based on a period of only 5 years do not allow a definitive judgement to be made.

• when they are well adhered to the supporting medium (without folds and on a dry supporting medium) the modified binder and **geotextile interlayers** have comparable performance to the pure bitumen sand asphalt interlayers for a lower price. The quality of the installation work is essential to the performance of the system. The development of cracks sometimes in the form of fine crackling or branched cracks is penalising when it comes to judging them.

• **geotextiles manufactured on site** (one single technique observed of this type at present) have similar behaviour to manufactured geotextiles, with an easier method of working.

• the 0/10 **asphalt concretes with addition of mineral or organic fibres** allow good performance with regard to ruts whilst ensuring a structural contribution. Not enough time has elapsed to make a definitive judgement from the findings, but their behaviour is promising as in the case of the sand asphalt interlayers.

The observations show that the rate of development of transversal cracks of hydraulic shrinkage of the treated asphalt layers is influenced by the cumulative effect of heavy goods traffic born by the road (frequent vertical movement).

Using a system aimed at limiting the effects of this cracking is therefore all the more important and necessary depending on the level of erosion of the cracks - condition assessed by beating greater than 30 hundredths of a millimetre - **and the volume of traffic. Experience shows that in this case a structural contribution is necessary to reduce the beating.** A technical and economic compromise must be found between thick layer asphalt concrete materials and the implementation of a crack prevention interlayer system combined with a wearing course, generally 4 cm thick.

Conversely, for roads that display cracking that is not very active, binding is suitable and generally sufficient to ensure watertightness of the wearing course, particularly when traffic is less than T1 (300 HGV/day/direction). It can then be completed using a surface dressing.

It is therefore up to the chief engineer to judge when it is appropriate to use a crack preventing interlayer system in relation to the state of the road to be treated, the seriousness of the cracks and the volume of heavy goods traffic. A technical/economical study is necessary to contrast the different products mentioned previously and traditional solutions of thick layer asphalt concrete.

Bibliography

[1] - Highways Department - Implementation of hot mix materials - 1985.

[2] - G. Colombier LR AUTUN - Test machine for crack preventing complexes for roads treated with hydraulic binders. RGRA n°680 June 1990.

[3] - LPC Road and Bridge Engineering Laboratories report n°156 et 157 September 1988.

[4] - M.DAUZATS - Mechanism of surface crack development of wearing courses, LCPC report n°154, March 1988.

[5] - H. GOUACOLOU - Theoretical study of development time of cracks across wearing courses. Internal report.

On application of reinforcing and anticrack interlayers of road pavement structures in the Republic of Belarus

V. YAROMKO, I.L. ZHAYLOVICH and P.A. LYUDCHIK
Belavtodorprogress Association, Dorstroytechnika Institute, Minsk, Belarus

Abstract
The article deals with experience of application of reinforcing and anticrack interlayers in flexible road pavements to improve crack resistance. Comparison with standard sections is available and also analysis of new techniques.
Key words: anticrack, geotextile, glassgrid, interlayer, pavement, polyethylene, reflective cracking, reinforcement, resistance.

1 Introduction

In recent years there has been a tendency in the world practice to implement reinforcing and anticrack interlayers which provide elimination of reflective cracking and reduction of depth of asphalt pavement courses. Normally various types of geotextile materials (continuous or in the form of grids) are used for the above interlayers. It is considered that the availability of a thin interlayer between old cracked pavement and new one prevents the crack movement upwards or, in any case, delays the period of its penetration into a new structure. Behaviour of the above interlayers has not yet been completely estimated, but the examination of experimental sections conclusively indicates a positive influence of interlayers on limiting of cracking or its substantial delaying in time. On Minsk-Mogilev-Cherikov road experimental section cracks have not been revealed for 5 years. At the same time through transverse cracks appeared after 1 year and 9 months on another experimental section where no interlayer was applicated.

In the Republic of Belarus the application of reinforcing and anticrack interlayers relates to 1987. Until now 9 experimental sections have been

Reflective Cracking in Pavements. Edited by L. Francken, E. Beuving and A.A.A. Molenaar. © 1996 RILEM. Published by E & FN Spon, 2–6 Boundary Row, London, SE1 8HN. ISBN 0 419 22260 X.

constructed. The interlayers are made both of continuous geotextile stripes and of grids (polyethylene and glassgrids).

Analysis of experimental data indicates the effectiveness of interlayers of interlayers made of glassgrids, providing higher strength parameters, reinforcing effect, proper adhesion of new and old pavements and lower cost in comparison with interlayers made of synthetis materials.

2 Description of experimental road sections

In Belarus issues of application of synthetic geotextile materials have been studied since 1987. Initially the studies were related to construction of experimental sections. Totally 9 experimental   sections were constructed (Table 1) from 1987 to 1994. Reinforcing and anticrack interlayers were applicated using unweaved synthetic materials (geotextile of Syktyvkar manufacture, Russia), SO-L grade glassgrids and polyethylene grid (Novopolotsk, Belarus).

Experimental sections were constructed on Minsk - Naroch road (km12 - km13), Minsk - Mogilev - Cherikov road (km34 - km35, km40 - km41, km180 - km181), St.Petersburg - Kiev - Odessa (km588 - km589), Minsk - Gomel (km273 - km274).

3 Results of field inspections on sections with continuous reinforcement

Regular field inspections (1 - 2 times per year) are being conducted with the objective to evaluate pavement roughness and cracking.

Good results have been obtained on experimental section No.3 (Minsk - Mogilev road, km180 - km181) constructed in June 1989. Heavy repair project assumed pavement reinforcement by double 12 cm thick asphalt layer. On test segment of 65 m length (one traffic lane) a layer of geotextile was laid on the existing 8 cm thick asphalt pavement.

Analysis of inspection data indicated arising of fisrt three cracks on the segment without geotextile interlayer after 702 days. The distance betwen cracks is 57 and 56 m. On the segment with geotextile interlayer no cracks were revealed after 1800 days (5 years) of inspection. It is interesting to note, that on the reference segment without geotextile interlayer, where the thickness of the upper asphalt layer is 4 cm, transverse cracks appeared a little bit later (on 854 day), but the distance between them is less: 18 m. The smaller thickness of of aspahlt on this segment (4 cm) was specified on structural grounds due to sufficient strength of the existing pavement.

Experimental section No.5 (St.Petersburg - Kiev - Odessa, km588 - km589) was constructed in August 1987. The main objective was to reduce the thickness of asphalt layers and to provide crack resistance of road pavement with base course made of natural sand and gravel mixture stabilized with fly-ash. The section includes 10 segments, the first being a reference segment and the rest of experimental character. The thickness of asphalt on experimental segments was reduced in comparison with design from 16 cm to 13 and 12 cm, three layer pavement was substituted for double layer pavement. Geotextile was applicated both on fly-ash stabilized base (segment 2) and between asphalt layers (segments 3, 5, 8).

First three cracks along the entire width of pavement emerged 1040 days after completion of the surface layer on segment without geotextile interlayer. Only one transverse crack of 0.3 m length emerged on each segment with geotextile (Nos.2 to 8) after 1040 days. Further on the crack on segment 2 extended up to 1 m and on segment 8 (with pavement thickness reduced up to 12 cm) two transverse cracks emerged along the entire width of pavement after 2124 days, i.e. 5.8 years.

Similar data were obtained on section No.9 (Moscow - Minsk - Brest road, km408 - km409) constructed in 1989. Anticrack interlayer of Syktyvkar geotextile was laid on the existing pavement with thickly and widely spread cracks, overlaid by one 5-6 cm layer of asphalt. Analysis of inspection data indicates, that during first four years there were no reflective cracks on the section. After five years three transverse reflective cracks emerged along the entire width of pavement and in six years the number of cracks increased up to 13. On reference segment transverse cracks started to emerge on the second year and after five years their number accounted for 167 per 1 km. The distance between cracks was 3-11 m.

4 Results of inspection on sections with local reinforcement of cracks

On experimental sections Nos. 2, 6, 7 local reinforcement of individual transverse cracks was provided using glassgrids and polyethylene grids.

On experimental section No.2 (Minsk - Naroch road) three schemes of local reinforcement of individual transverse cracks were implemented using glassgrids.

The first scheme included reinforcement of cracks along the entire width of pavement with application of asphalt underneath the grid (fig. 1a).

The second scheme assumed reinforcement of cracks along the entire width of pavement without application of asphalt (fig. 1b).

The third scheme included reinforcement of cracks only on one traffic lane without application of asphalt (fig. 1c).

During constructon period cracks on one traffic lane were overlaid by 5 cm asphalt layer and on the second lane by 1.5-2 cm asphalt layer for protection against traffic with further application of 5 cm asphalt layer.

Three cracks were reinforced following scheme 1 and two cracks on scheme 2. Works were completed in 1990. Constant inspection has been arranged since.

Out of cracks, reinforced on scheme 1, in a year one crack reflected on the surface and two cracks developed on the half width of pavement (fig. 1a). On the traffic lane, where grids were not overlaid by asphalt but just sprinkled with a very thin layer of asphalt, cracks were reflected on the third and forth year after construction.

Out of cracks, reinforced on scheme 2, one crack reflected in a year and only on half of pavement width. In five years the crack reflected completely. The second crack during 5 years did not reflect completely but only on 1/4 of pavement width (fig. 1c).

Thus the most effective is the second scheme of local reinforcement prodiving good anticrack interlayer (fig. 1b): crack No.4 completely developed after five years and crack No.6 did not completely reflected during this period. Cracks Nos. 1, 8 completely reflected on the surface after three years and crack No. 2 after four years.

It should be noted that effect of reinforcement for the above schemes was revealed on the right traffic lane, where the grid was not immediately overlaid by asphalt but just manually sprinkled to provide anti-traffic protection.

Reinforcement on scheme 3 (fig. 1b) is of slight effect expressed only during first two years thus permitting to recommend full width reinforcement of cracks by glassgrids.

Inspections were conducted on experimental sections Nos. 6 and 7 (Minsk - Mogilev - Cherikov road), where individual transverse cracks were reinforced by polyethylene grids with no effect revealed. After first winter period five cracks reflected on the surface out of seven reinforced.

5 Conclusion

Implemented inspections prove the effect of reinforcement both of geotextile and glassgrids. It should be noted that on section No.3 (Minsk - Mogilev - Cherikov road) there have been no cracks on the test segment during more than 6 years. On the reference segment cracks reflected after 1-2 years.

Through transverse cracks appeared on section No.5 with reduced asphalt layer (13 cm instead of 16 cm) after 2.8 years and on segment with geotextile 2.5 m long transverse crack reflected after 3.9 years and during further 4 years it did not extended thus proving the effect of reinforcement.

Effect of reinforcement has not been revealed in case of reduction of asphalt thickness from 16 to 12 cm because of considerable reduction of pavement depth.

Effect of reinforcement has been revealed on Moscow - Minsk - Brest, km408 - km409, where there were no reflective cracks during first four years and first three transverse reflective cracks along the entire width of carriageway appeared after five years. Upon completion of 6 years period the number of reflective cracks increased up to 13 per one kilometer.

Results of inspections of previuously constructed experimental sections of road pavements, incorporating reinforcing and anticrack interlayers, allow to conclude that depending on schemes of reinforcement there may 3-5 years reduction in rate of reflective cracking.

Table 1 Description of experimental road pavement sections with reinforcing interlayers

Section No	Road title	Traffic volumes (vpd)	Condition of pavement before interlayer application	Description of experimental sections			Thickness of asphalt layer (cm)	Year of construction	Notes
				Length (m)	Type of interlayer material	Scheme of interlayer application			
1.	Radoshkovichi - Sloboda	1,930 - 2,270	Thickly spead cracks	51.5	Glassgrid	Continuous reinforcement	5	1990	
2.	Minsk - Naroch	10,250	Widely spread cracks	200	Glassgrid	Local reinforcement		1990	
3.	Minsk - Mogilev - Cherikov	4,700	-	65	Geotextile	Continuous reinforcement (one lane)	8	1989	
4.	Approach road to Khatezhino	-	Thickly and widely spread cracks	80	Glassgrid	Inclined reinforcement (5m pitch)	5	1994	
5.	St.Petersburg - Kiev - Odessa	2,000	-	500	Geotextile	Continuous reinforcement of various structural layers	12-16	1987	Interlayer was applicated during new construction
6.	Minsk - Mogilev - Cherikov	3,500	Widely spread and single cracks	1000	Polyethylene grid	Local reinforcement	5	1991	
7.	Minsk - Mogilev - Cherikov	3,900	Widely spread and single cracks	1000	Polyethylene grid	Local reinforcement	5	1991	
8.	Minsk - Gomel	3,700	Thickly spread cracks	255	Geotextile	Continuous reinforcement	2	1988	No more inspections as the section does not exist (thin asphalt layer)
9.	Moscow - Minsk - Brest	7,000	Thickly spread cracks	1000	Geotextile	Continuous reinforcement along 2 lanes	5 - 6	1989	

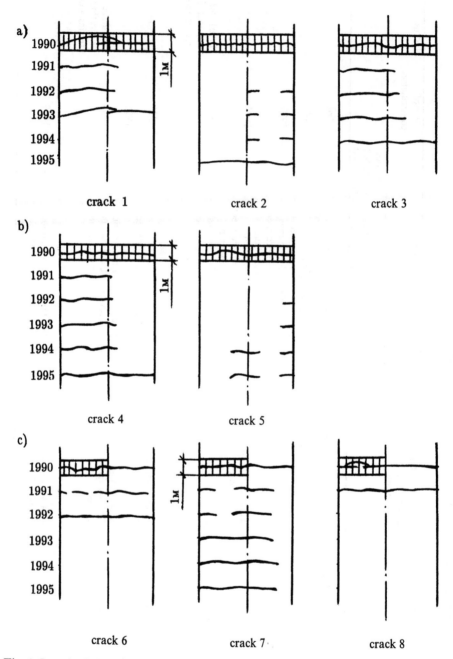

Fig. 1 Local reinforcement of transverse cracks

a - with sprinkling of asphalt underneath the grid along the entire width of pavement
b - without sprinkling of asphalt underneath the grid
c - on half width of pavement and without sprinkling of asphalt

Key word index